International Federation of Automatic Control

# PRODUCTION CONTROL IN THE PROCESS INDUSTRY

IFAC Workshop Series, 1990. Number 8

# IFAC WORKSHOP SERIES

*Editor-in-Chief*
Pieter Eykhoff, University of Technology, NL-5600 MB Eindhoven,
The Netherlands

Other IFAC Publications

*AUTOMATICA*

the journal of IFAC, the International Federation of Automatic Control

*Editor-in-Chief:* G. S. Axelby, 211 Coronet Drive, North Linthicum,
Maryland 21090, USA

*IFAC SYMPOSIA SERIES*

*Editor-in-Chief:* Janos Gertler, Department of Electrical Engineering,
George Mason University, Fairfax, Virginia 22030, USA

*Full list of IFAC Publications appears at the end of this volume*

# PRODUCTION CONTROL IN THE PROCESS INDUSTRY

*Proceedings of the IFAC Workshop,*
*Osaka, October 29–31, 1989 and*
*Kariya, November 1–2, 1989, Japan*

Edited by

## E. O'SHIMA

*Tokyo Institute of Technology,*
*Yokohama, Japan*

and

## C. F. H. VAN RIJN

*Shell Laboratorium*
*Amsterdam, The Netherlands*

Published for the

## INTERNATIONAL FEDERATION OF AUTOMATIC CONTROL

by

## PERGAMON PRESS

Member of Maxwell Macmillan Pergamon Publishing Corporation

OXFORD · NEW YORK · BEIJING · FRANKFURT
SÃO PAULO · SYDNEY · TOKYO · TORONTO

| U.K. | Pergamon Press plc, Headington Hill Hall, Oxford OX3 0BW, England |
| U.S.A. | Pergamon Press, Inc., Maxwell House, Fairview Park, Elmsford, New York 10523, U.S.A. |
| PEOPLE'S REPUBLIC OF CHINA | Pergamon Press, Room 4037, Qianmen Hotel, Beijing, People's Republic of China |
| FEDERAL REPUBLIC OF GERMANY | Pergamon Press GmbH, Hammerweg 6, D-6242 Kronberg, Federal Republic of Germany |
| BRAZIL | Pergamon Editora Ltda, Rua Eça de Queiros, 346, CEP 04011, Paraiso, São Paulo, Brazil |
| AUSTRALIA | Pergamon Press Australia Pty Ltd., P.O. Box 544, Potts Point, N.S.W. 2011, Australia |
| JAPAN | Pergamon Press, 5th Floor, Matsuoka Central Building, 1-7-1 Nishishinjuku, Shinjuku-ku, Tokyo 160, Japan |
| CANADA | Pergamon Press Canada Ltd., Suite No. 271, 253 College Street, Toronto, Ontario, Canada M5T 1R5 |

Copyright © 1991 IFAC

First edition 1991

**Library of Congress Cataloging in Publication Data**
IFAC Workshop on Production Control in the Process Industry
(1989: Osaka, Japan, and Kariya-shi, Japan)
Production control in the process industry: proceedings of the
IFAC Workshop, Osaka, October 29–31, 1989 and Kariya, November 1–2, 1989, Japan/
edited by E. O'Shima and C. F. H. van Rijn.—1st ed.
p. cm.—(IFAC workshop series: 1990, no. 8)
Includes indexes
1. Production control—Congresses. 2. Manufacturing processes—Congresses.
I. O'Shima, E. (Eiji), 1931– . II. Rijn, C. F. H. van.
III. International Federation of Automatic Control. IV. Title. V. Series.
TS157.A1I33 1989 685.5—dc20 90–23206

**British Library Cataloguing in Publication Data**
Production control in the process industry.
1. Production. Automatic control
I. O'Shima, E. II. Rijn, C. F. H. van III. International
Federation of Automatic Control IV. Series
670.427

ISBN: 9780080369297

*Transferred to digital print 2009*
*Printed and bound in Great Britain by CPI Antony Rowe, Chippenham and Eastbourne*

# IFAC WORKSHOP ON PRODUCTION CONTROL IN THE PROCESS INDUSTRY

*Sponsored by*
International Federation of Automatic Control (IFAC)
Technical Committees on
   Applications (APCOM)
   Economics and Management Systems (EMSCOM)
   System Engineering (SECOM)
International Federation of Operational Research Societies (IFORS)

*Co-sponsored by*
The Japan Institute of Systems Research (JISR)
The Society of Chemical Engineers, Japan (SCEJ)

*Organized by*
National Organizing Committee on behalf of
   National Committee of Automatic Control, Science Council of Japan

*International Programme Committee*

E. O'Shima, Japan (Chairman)
C. F. H. van Rijn, The Netherlands
   (Co-chairman)
O. Asbjoernsen, USA
J. E. Doss, USA
R. Genser, Austria
I. Hashimoto, Japan
J. König, FRG
M. Kümmel, Denmark
E. Kunugita, Japan
L. Lasdon, USA
H. Madono, Japan

H. Makabe, Japan
M. Matsushita, Japan
H. Matsuyama, Japan
Y. Naka, Japan
E. Nakanishi, Japan
N. Nishida, Japan
D. Seborg, USA
G. Stephanopoulos, USA
T. Takamatsu, Japan
P. Uronen, Finland
W. F. Wanger, USA
Y. Yoshitani, Japan

*National Organizing Committee*

T. Takamatsu (Chairman)
E. Kunugita (Co-chairman)
H. Nishitani (Secretary)
O. Kageyama
Y. Kojima
K. Konoki

S. Matsuhisa
H. Morikawa
E. O'Shima
K. Ueda
T. Umeda

*Advisory Committee*
Y. Sakurai (NMO)
Y. Sawaragi (JISR)
M. Yoshida (SCEJ)

# PREFACE

Production Control nowadays receives strong attention, especially in the parts manufacturing industry. It addresses the essential problem of a production manager; how to ensure that the right amount of products of the right quality at the right moment in time is available to the customer such that in a societally responsable way maximum profit is generated for the company. Solving (or assisting in solving) such a problem requires a multidisciplinary approach. Apart from human resource management, production control addresses a wide range of technical fields, such as ligistics, production planning and scheduling, operational control, quality control and assurance, plant and/or site optimisation, (maintenance) engineering and safety engineering. Various scientific disciplines potentially have a rewarding contribution, such as Chemical Engineering, Operational Research, Control Theory, Statistics and Reliability Engineering.

The IFAC APCOM Working Group on Chemical Process Control realised that, although there exists a large number of all kinds of meetings on specific disciplinary areas, virtually no effort is made to focus on integration aspects. A real danger then exists that disciplines evolve in isolation, focussing more and more on detail and loosing sight on the overall context. The APCOM "contemplative stance" recognises that the future lies in, for instance, synergising the process control skills with those of the computer scientists and statisticians. The Computer Integrated Manufacturing approach is only feasible if the conceptual problems of controlling and optimisation of production processes can be solved.

It was this mood that led the APCOM Working Group to the initiative of organising an IFAC Workshop focussing on such integration aspects. The IFAC approval and the co-sponsoring of IFORS was readily achieved. The group deliberately selected Japan as the host country, recognising the difference in management style compared with Europe and the USA, now attracting such widespread attention. Whereas the Western world relies significantly on quantitative, computational techniques, Japanese management puts strong attention on human involvement at all levels of staff, integrating their specific skills.

The workshop was centred around invited papers; for each theme involving industrial speakers from both the Western World and Japan. These papers were followed by invited comments from academia. The combined input led to ample public discussion on the four session themes: production control, quality control and quality assurance, operational control and maintenance strategy. Good results were also obtained by having the shorter papers being presented in two poster sessions. The availability of sufficient time and the freedom of choice led for each presentation to intensive discussions between presenters and part of the public with a specific interest. A successful novelty was to have part of the workshop taking place on an industrial site with organised visits to an automotive parts producing industry and a modern refinery.

The workshop to some extent beared the stamp of

being the first of a kind. It appeared quite difficult to properly define production control in a sense that identifies the interfaces between vested disciplines and invites for combined research effort. As such, it could not be prevented that the presentations, each of high quality, were not always in the same tune. The Japanese attendance was quite large and should be acknowledged for their extensive input in discussions; a larger number of foreign attendants would have been appreciated.

On this occasion we wish to express our appreciation to all of the session chairmen for their contribution toward the success of the workshop:
(Production Planning)
  C. McGreavy, Leeds Univ.
  T. Ibaraki, Kyoto Univ.
(Quality Control & Quality Assurance)
  O. A. Asbjornsen, Univ. of Maryland
  I. Hashimoto, Kyoto Univ.
(Operational Control)
  C. H. White, É.I. Du Pont
  E. Nakanishi, Kobe Univ.
(Maintenance Strategy)
  J. König, Bayer
  H. Matsuyama, Kyushu Univ.
(Final Discussion)
  C. F. H. van Rijn, Shell Laboratorium
  E. O'Shima, Tokyo Inst. of Tech.
(Plant Tour 1)
  J. U. Clemmons, ChemShare
(Plant Tour 2)
  N. Jensen, Tech. Univ. of Denmark

All together, the workshop turned out to be a rewarding initiative, both for the attendants and the organising committee. The APCOM Working Group on Chemical Process Control is grateful to the International Program Committee for their effort to make the workshop a success. A special word of thanks is in its place for Prof. E. O'Shima of the Tokyo Institute of Technology and Prof. T. Takamatsu of Kansai University. It was a pleasure for me to cooperate with such academics with insight, ·influence and perseverance. The National Organising Committee should be complimented for the flawless organisation of the meeting, Prof. Kunugita and Nishitani acting as active co-chairman and secretary. The daily organisation benefitted greatly from the friendly and cooperative support of the NOC members and a great number of Japanese students. The APCOM Working Group realises that these persons, after the delegates satisfied went home, still had a large job working out the transcripts of the discussions and preparing the proceedings of the meeting.

We also wish to thank the Japanese company participants for their financial support allowing a.o. for impeccable simultaneous translation during the four days of the workshop. We feel indebted to the management of Aisin Seiki and Idemitsu Kosan for providing hospitality and organising plant tours, thereby bringing the subject of the meeting in a realistic context.

Cyp F. H. van Rijn
Shell Research BV
Amsterdam

# CONTENTS

## SESSION 2 - QUALITY CONTROL AND QUALITY ASSURANCE

## SESSION 3 - OPERATIONAL CONTROL

## POSTER SESSION 2

## SESSION 4 - MAINTENANCE STRATEGY

## PLANT TOUR 1

## PLANT TOUR 2

## SESSION 5 - FINAL DISCUSSION

# LATE PAPER

# PRODUCTION CONTROL:
# A MULTIDISCIPLINARY PROBLEM

## C. F. H. van Rijn

*Koninklijke/Shell-Laboratorium, Amsterdam (Shell Research B.V.), P.O. Box 3003,
1003 AA, Amsterdam, The Netherlands*

Abstract. In this opening paper we will show that in order to assist
managers in taking decisions on production control an integrated
approach of hard-core disciplines is required. Starting with a general
problem formulation, we will address a number of specific subjects which
would benefit from such an approach.

Keywords. Production control; planning and scheduling; process control;
optimisation; safety; product quality; reliability; availability;
maintenance engineering; flexibility; business needs.

## 1. INTRODUCTION

In 1987 the IFAC APCOM Working Group on
Chemical Process Control published a
"contemplative stance" (Kümmel, 1987) in
reaction to an internal IFAC document.
The tone of the latter document was
rather defensive, expressing the need
"to stay at the forefront" and "not let
other organisations make the running on
the border area" between control and
other disciplines. In contrast, the
APCOM document recognises that the
future lies in synergising the process
control skills with e.g. the skills of
the computer scientists and
statisticians. It was this sense of mood
that led to PCPI '89.

Production control, the main task of a
plant or site manager, aims at creating
and sustaining conditions such that, in
a socially and environmentally accepted
way, the right amount of products of the
right quality at the right moment is in
time is produced to make maximum profit
to the company (Fig. 1). A number of
scientific techniques and tools based on
these techniques are available to
support this decision making process:

1. Operations Research (OR) techniques
   have become widely accepted for
   solving planning and scheduling
   problems. It is nowadays quite
   common to note that small and large
   production facilities are planned,
   scheduled and optimised with
   (possibly, large scale) LP's. On
   the product distribution side, OR
   techniques support decisions on
   depot location, inventory levels
   and optimal vehicle routing.

2. A similar maturity can be observed
   for Process Control (PC). The
   number of applications of
   multivariable, constraint,
   self-tuning and/or adaptive control
   is steadily growing. A significant
   number of control and
   identification packages are
   available and used in industry.
   With the advent of powerful process
   computers application of plant-wide
   control and optimisation is
   becoming feasible.

3. Statisticians successfully
   introduced control charts for
   Statistical Process Control (SPC),
   favoured by the current drive for
   product quality assurance.

4. With the dwindling societal
   interest for nuclear power and the
   increasing concern for major
   technological hazards, safety
   engineers are increasingly trying
   to apply their tools (hazard and
   operability studies (HAZOP),
   quantitative risk assessment (QRA),
   etc.) in the CPI.

5. Reliability engineering, until
   recently applied mainly in
   areospace and military defence, is
   receiving growing attention in the
   process industry because of the
   potentially significant cost
   savings attainable.

These skill areas clearly have
distinctive roots and showed a different
evolution in time. Consequently, they
now employ different approaches,
techniques and jargon. On the other
hand, they also show commonalities. One
common aspect is that they focus on
similar operational and engineering
aspects of the manufacturing process,
such that overlap of problem areas
occurs. Another common factor is that
they focus on that part of the
managerial and/or engineering problem
which is quantifiable, measurable and
lends itself to mathematical formulation
and problem solving. Based on this
commonality, these disciplines lend
heavily on the strong developments in
computing and information technology.
Experts in that field now form part of
the site workforce, in many cases
solving problems that before were
handled by hard-core disciplines such as
OR or PC.

The above may support the view that
"process design and operation requires
insight into an extremely wide variety
of professions and complicated
interactions between the process
engineering fundamentals" (Asbjörnsen,
1988). For the non-expert manager,
however, such skill variety creates

confusion. One should also recognise that developments in these skill areas take place rather in isolation and do not, in many cases, benefit from cross-fertilisation. There exists a real danger that, on basis of such a sectarian evolution, research may focus on problems whose priorities have shifted in time or on problems that can better be solved with other skills. The fact that Industry in general is passive in its contact with academia (possibly with the exception of the USA, (Prett and Morari, 1987)) and mainly reports on successful projects rather than on failures of novel approaches at scientific topical meetings, aggravates this issue.

It is the intention of PCPI'89 to focus on the integral aspects of production control. The Organising Committee invited a number of responsible industrial managers and international consultants to sketch the business problems, and their experience with the success and failure of approaches to solve them. Academic commentators were asked to provide information to what extent the various scientific disciplines can provide effective techniques.

The APCOM working group deliberately selected Japan as the location for this meeting. Whereas production control in the US/European part of the world is supported mainly by quantitative, computational techniques, Japanese management relies more heavily on human involvement of all levels of staff. Their success with approaches like Kanban, Just-in-time, Total Quality Movement and Total Productive Maintenance, has attracted considerable attention and found follow-up in the Western World. At this meeting, we hope to compare and discuss these two approaches.

This opening paper aims to give an overview of the actual production control problems. Not trying to be exhaustive, it focuses on some specific subjects where integration of disciplines may bear fruit. It should be noted that the opinions given are those of the author. Although based on fruitful discussions with colleagues they do not intend to represent the views of the Royal Dutch/Shell Group.

## 2. THE PROBLEMS OF A PRODUCTION MANAGER

The proceedings of the recent 12th World Petroleum Congress provide a valuable source of information on the needs perceived for existing and future refineries to remain profitable. Among the presentations of staff of major companies such as Exxon, BP, Texaco and Total, one can easily identify a number of common elements:

- a strong focus on the refinery margin, hence a drive to reduce refining costs (utility, operations, maintenance)

- in view of the volatility of the market, more flexibility with respect to planning and processing, and growing attention for logistics

- reduced crude-oil "residence time" (less inventory)

- more intelligent control and optimisation, with process models used by engineers with less specific skill

- consequently more attention towards intelligent measurements and reliability of models and data

- more automated operation

- a strong focus on product quality control

- an open eye for managing the maintenance process and costs

- a need for performance monitoring

- emphasis on reliability, safety and environmental aspects.

In general, "only the most efficient and technologically able refineries will continue to prosper" (Kohlhase et al., 1988). "The challenge today and in the future lies in how best to incorporate the coming generation of computer techniques in refineries to improve process and energy efficiency and further improve response time" (Bentley, 1988).

This vision is not greatly different from that in the CPI or the paper and pulp industry. Schlenker (1987) mentions the same flexibility, efficiency and reliability aspects, the focus on product quality control and advanced control and optimisation. He, however, also emphasises shorter product life cycles, decreasing from 10-15 years in the seventies to 2-3 years in the nineties. A similar observation can be made with respect to major process changes, such as those caused by the use of new generations of catalysts. Such variations in lifetime lead to a different notion of process and manufacturing optimality. No longer will the CPI in these cases primarily rely on long-term optimisation, using advanced techniques: the main problem then will be to fastly generate new process understanding and have new plants on-stream in a short period. These plants may be simpler in construction and "should be inherently self-regulating, requiring little or no process control" (Benson, 1988). For the petro-chemical industry another flexibility aspect arises with the tight interfacing with refineries (Davies, 1988). Additional profit is intrinsically available in the oil-chemical interface, but calls for process integration and proper scheduling.

Developments in industry cannot be regarded in isolation from those in society in general. An all-encompassing managerial problem is to keep a workforce motivated and acting as an effective group. This requires that the workforce members regard their job as being interesting, see where their activity fits in the overall effort and be able to recognise their contribution. Staff at all levels increasingly want to exercise _responsibility_ and be _accountable_ for _performance_. These aspects have strong consequences for the design of decision support systems (DSS). Human decision making in many cases is a team effort, with members choosing from a variety of possible decisions, and not necessarily going for the one that mathematically may be optimal. Any support to that process should be in such a form that the decision strategy proposed is transparent and understood by the user(s).

Another general aspect concerns the growing availability of information (read: data). Modern information technology and data communication strongly affect our business decision making. They foster competition between industries, will change the industry structure and boundaries and can dramatically reduce costs. However, the quality and speed of such decision making can well be negatively affected by misdirected and information-poor data. It appears that neither the OR, nor the computing scientists pay attention to this problem. As information is neither free, nor by definition reliable, information analysis should form an inherent part of decision support techniques.

## 3. PRODUCTION PLANNING AND SCHEDULING (PPS)

In the petroleum industry, PPS is an inherent part of the downstream logistics problem (Langeveld, 1989) (Fig. 2). Based on forecasts of products demands, decisions have to be made on the volume and type of crude to be available at the refinery fence at a specific moment in the production schedule. The need for and economic value of the crude are based on the product slate required, the refinery flexibility and processing costs and existing feedstock commitments. Various uncertainties play a role here; price volatility of crude and product, unsure product demand, crude arrival and lifting times, availability of refinery units, and an imprecise knowledge of product recovery from a crude.

Refineries usually operate in modes where processing units are trimmed to yield maximum economic benefit. It is the task of the scheduler to define these modes of operation in time such that both the production target is met and process constraints (feedstock availability, tankage, equipment conditions) are not violated.

Current refinery planning/scheduling models are based mainly on linear programming (LP) techniques and are of considerable size (typically 3-10 000 variables and constraints for a 30 period model covering one month of operation). To evaluate changes in market and/or process environment there exists a strong need for fast LP solving. The economic importance of the interfacing between refineries and associated chemical plants increasingly requires further modelling detail, frequently involving non-linear descriptions (NLP) and combinatorial problems (MI(N)LP).

A similar trend to more complexity is caused by networking between manufacturing processes. Consequently, new algorithmic developments like Interior Point Methods (Karmarkar, 1984) and advances in computing speed due to parallelisation (Van de Vorst, 1989) are followed with vivid interest.

In the author's opinion we here clearly face an evolutionary process. Where PPS traditionally relied solely on mathematical programming (MP) techniques, now a number of problems are met:

- MP techniques are being used more and more by non-specialists. Even with LP techniques this requires good interfaces, easy to understand algorithms and solution approaches and robustness against changes in operational environment. Expert systems are being developed for this purpose (Greenberg, 1989)

- users complain about the complexity of such systems; it is difficult to obtain the required data. The spreading availability of databases and communication networks may, however, overcome part of this hurdle

- a major problem lies with the definition of the objective function for the optimisation. It is often far from clear what the "optimal" objective is in this highly dynamic and uncertain environment

- the translation of this global objective to that of a local decision maker, say, the manager of a site unit, is not transparent. Neither does the approach lend itself to fast "what-if" analyses.

- LP assumes a static, deterministic world; the (dynamic) effect of sequential decision making and that of uncertainty are difficult to incorporate. It also misses the concepts of feedback and learning and thus does not address the problem of performance monitoring.

In line with the growing drive for personal responsibility it is therefore no surprise to see an alternative development towards the use of simple spreadsheets (White, 1986) as a pragmatic engineering solution. Such spreadsheets, related almost one-to-one with the refinery flow diagram the user is familiar with, have the transparency required. A drawback, however, is that optimisation completely relies on the insight and creativity of the user.

There is another fundamental problem, viz. that between PPS and on-line unit optimisation. The LP models at the planning level are approximate representations of the production process involving averages of daily operations, whereas control and optimisation is concerned with stabilisation and optimisation on a continuous basis. The actual behaviour of a unit may differ considerably from that in the (coarse) planning model. The marginal values derived from the latter are thus, in general, inadequate for the purpose of on-line optimisation (Baker, 1984).

A fresh look at the PPS problem is required. The effects of successive decisions at various levels in an inherently uncertain, opportunistic environment should be addressed. The role and tasks of the different decision makers, as well as their interaction and consistency with each other, should then form essential elements. Large scale system theory (Titli 1980, Malinowksi 1987) conceptually could provide such local autonomy and responsability but scientific progress in this area is limited. Other hierarchical techniques leaving (restricted) autonomy to local decision makers such as multi-level programming (related to game theory) deserve attention.

Even present techniques may allow for some form of decomposition. An analysis of the site PPS objective function may reveal that parts of the cost-benefit balance are restricted to single units. Such optimisation may be effected by local optimising control, relieving the site optimisation complexity and simultaneously providing tighter optimisation. This line of reasoning may be extended to those parts of the objective function that deal with process parameters in more than one unit. These relationships may change with the mode of operation, actual process conditions and product requirements. Making these relations visible to decision makers will improve understanding and motivation. A basic requirement then is that the PPS model more truthfully represents unit behaviour. In the next section we will revert to the value of having (non-linear) process models based as far as possible on first principles and process kinetics. Here, it suffices to state that with the current trend in computing power and costs such models become feasible for planning and operating processes simultaneously.

PPS in the chemical industry differs from that in a refinery. For a number of chemical processes running on well-defined feedstocks and producing a known and restricted product package simple calculations will do. For larger continuous processes, such as ethylene cracking, scheduling is restricted mainly to choosing between feed alternatives and optimising process conditions. Batch processes require a different approach and considerable literature on job shop scheduling is available (White, 1987). Simulation, math programming, queueing theory and heuristics are being applied. There is a link here with the Flexible Manufacturing Systems (FMS) in discrete part manufacturing.

It is interesting to note that two essential keywords from Section 2 are missing in the above: flexibility and inventory optimisation. The first is "despite appearing to be intuitively obvious and straightforward, an ill-defined concept and closely related to individual perceptions of what is desirable in a rather nebulous fashion" (Mc Greavy, 1989). Grossmann and co-workers developed techniques (Swaney and Grossmann 1985) that allow a well-defined, quantitative trade-off between design and control variables in the light of uncertain process parameters. This approach appears to be powerful, allowing extensions to non-linear systems and stochastic aspects (Pistikopoulos and Grossmann, 1988 a,b) and conceptually towards product availability assessment. Inventory control and optimisation cannot be isolated from such flexibility assessment; buffer inventories have the objective to bridge the gap between (stochastic) demand and process flexibility. LP PPS models, in general, either account for the time-average value of inventory levels (planning) or regard these as constraints (scheduling). In view of the economic importance the subject of (operational) inventory control and the associated (design) issue of tankage warrants further research effort. We gladly noted progress in stochastic OR at this point (Federgruen, 1989).

Flexibility also is an important factor in decision making. In the math programming approach, where time is discretised in periods, it is assumed that subsequent sets of decisions are simultaneously taken at the beginning of a period. Where these decisions apply to stochastic variables (feed quality, plant capacity, product demand), decisions furtheron in the future may become sub-optimal or even infeasible. Rather than to restart the deterministic optimization with the then known realisations of these variables, it would be better to take these uncertainties into account. The size of the decision problem and the lack of information on these uncertainties in many cases prevents the use of stochastic decision models. Lasserre and Roubellat (1985) attempted to deal with this situation by extending the deterministic optimization with an account of the volume of the choice set of decisions as a measure of the first-period decision flexibility.

## 4. CONTROL AND OPTIMISATION

To judge from the available literature and the number of simple and advanced applications, process control engineering appears to approach maturity. One can no longer imagine a modern plant being operated without regulatory control, either single-loop, or adaptive, or multivariable and/or locally optimising control. Most of these control problems can be solved by conventional PID algorithms in modern instrumentation and control systems. Only in those cases where appreciable non-linearities, dynamic steering, strong interactions between outputs and steering variables and/or constraint handling play a role, has the control engineer to revert to more advanced techniques requiring additional computing facilities. For these dedicated applications a good, economic pay-back is usually easily achieved (1-3% on raw material costs).

However, to conclude from the above that the process control community has solved all outstanding problems and received full recognition would be a misconception. Process control in the CPI at present may best be regarded as an add-on discipline, coming into play only at the later stages of process design or in actual operation. One thus may observe that advanced control is used as a cure for problems that better could be removed upfront in the design process (Benson, 1988). Many of the reported advanced control applications replaced existing process control schemes. In many of such cases the relative improvement is marginal (a few percent) but economically rewarding due to the large throughput. Note that, in order to reap such benefits, the advanced control system has to be robust with respect to large disturbances and changes in process and plant characteristics. The economic advantage is easily lost if the control system fails and causes a proces strip!

The notion is growing that such robustness is difficult to achieve (Prett and Garcia, 1988). In conventional control one relies heavily on the supervision of the operator, interacting with the process through simple, but well understood means of regulation. In the case of advanced control this transparency and thus the possibility of relying on operator supervision is easily lost. Guaranteeing operational robustness then involves extensive diagnostics of disturbances, of control behaviour and of changes in process and plant characteristics.

Many software protections have to be built in. For those cases which are not foreseen by the control engineer, an ill-defined responsibility exist by trusting the operator to take effective action (cf. Section 5).

Many control researchers will recall the fierce debates on "the gap between control theory and practice" at topical meetings in the late sixties. Industrial involvement since then has singificantly changed this perception (Cutler, 1979) but again we now observe signs of "ill-health in the control community" (Boardman, 1989). Could it be that the recent interest for robustness theories, robust identification, non-linear multivariable control (Fleming, 1988) are a result of above mentioned observations? Can an effective solution be obtained via proposed methodological approaches (supposedly more complex than hitherto used) or is a more process oriented view (Committee on Chemical Engineering Fronties, 1988) more rewarding?

In the author's opinion, the add-on character of to-day's process control is the kernel of the problem. In present applications, shortcomings in basic process know-how impair effectiveness. With fundamental decisions in design taken without consideration of control problems and opportunities, PC is doomed to play a contributing, remedial role. This is at variance with the insights and experience build up in systems theory in general and in control in particular. Note that this situation clearly differs from, for instance, that in aircraft industry, where the acceptance of "fly-by-wire" concepts provided a quantum leap in development.

The mission should be to emphasise process understanding, starting as early as in the process research phase. Chemists, traditionally empiricists, now get used to employ experimental design techniques to increase the efficiency of experimental work. Rather than with long tables of experimental data, they now come up with a (regression) model. The real goal in the end is to have a "reasonable" kinetic description (Grievink, 1986) via "learning models" (Fig. 3). Together with similar engineering models based on first principles and "cold flow" experiments an optimal process environment based on complex unit models may be obtained. These models in turn can be used for detailed engineering (process configuration, equipment selection, scale-up) with simultaneously accounting for process economics, control and safety aspects and the future requirements for the plant information system (economic performance monitoring, process diagnostics, data reconciliation). Implicitly, one then builds the (basis for) unit models which later can be used for planning, scheduling, control and operation support. Many of the current control problems (nonlinear, adaptive, optimisation) could profit from such a (admittedly nonlinear, complex) model running in parallel with actual operation. Gain scheduling could take place on the basis of calculated sensitivities between model input and output. As process dynamics are determined by the physico-chemical events (fast time scale) and equipment hold-ups, major dynamics, (in many cases adequate for ensuring control stability), may be estimated.

Moreover, such models form an excellent source for providing training and information interpretation to the operator. Deviations between model and plant unit will either increase our understanding of the process (the latter being the perfect model) or pinpoint evolving phenomena, otherwise going by unnoticed, until critical.

Obviously, such an approach is based more on mission than reality. It does, however, shift our emphasis from a single discipline driven to a more integrated approach. In our own research we observed that identification of a linearly parameterised dynamic regression (black-box) model frequently leads to unrealistic results, due to non-ideal, limited experiments. Including even a basic notion of the underlying physics (such as sign of loop gains or the existence of inherent stability) leads to "grey-box" models (Tulleken, 1989) that are far superior with respect to physical relevance and variance of model parameters.

## 5. PRODUCT QUALITY CONTROL

A number of clients of the CPI have introduced (originally, Japanese) philosophies of Just-in-Time and Quality Assurance. In this approach the receiving firm relies on strict quality control of materials/feedstocks purchased, eventually doing away with acceptance sampling. To become a preferential supplier, the company should be able to demonstrate such capability. Based on this business need, Statistical Process Control (SPC) has rapidly become an important management tool. Relying on the involvement of all levels in the production process and using transparent and simple statistical tools (control charts), one can detect deviations from quality in an early phase. The cause of deviation has to be found separately; here, too, one should rely on the insight and experience of the workforce.

The connotation "control" in SPC is rather misleading. SPC does not address the mechanism of deviation, nor does it indicate corrective measures. Whereas the CPI has to fulfil for each product a set of quality specifications, which are in general correlated and measured and controlled in different ways, SPC treats each variable in isolation. Its basic premises of dealing with a single variate being independently and identically distributed are even violated in case of a single product specification, because in continuous processes the dimension of time leads to correlation.

A fundamental analysis of single variate statistical process control was already presented in the early seventies (Van Der Grinten, 1973). Extensions to multivariate problems are just showing up in statistical literature. Moore (1988) and Mac Gregor (1988) have recently given a valuable contribution from a control theoretical view. Their analyses provide good examples of what cross fertilisation between disciplines could achieve:

- statisticians use the basic assumptions of the process being stochastic ; a value being in control means that the observed variance does not exceed expectation
- PC engineers regard the process as deterministic with assignable cause and effect relationships. Stochastic terms are introduced to represent model and measurement errors. A value being in control means that it is at or close to a reference (set point). The absolute value of this reference is not taken into account.
- SPC aims at finding the root cause of persistent quality deviations. Triggered by statistically not acceptable variance a separate, independent analysis is started to eliminate the source. PC acts reactively to all process disturbances (possibly, even to noise) and only in case of severe instability, is further analysis applied.

- SPC disregards process dynamics; at present it mainly considers single variate problems, with development in multivariate statistical quality control progressing slowly. PC employs this structural knowledge and identified process dynamics to effectively keep qualities close to set point and to prevent violating constraints.
- SPC relies on involving staff at all levels and therefore employs simple, easy-to-understand tools. If successfully introduced, users are the technique owners. PC aims at automatic decision making. Their tools try to take up tasks from the operator, eventually to run the process fully automated. The control engineer owns the technique, in general requiring tools of a greater complexity and requiring more education that those in SPC. The responsibility associated with this ownership leaves much to be desired, the PC engineer trusting the operator to intervene if control cannot cope with process disturbances.

It is advantageous to combine these skills. We did so in our in-house research on in-house multivariate product quality control. We focus on a set of product specs, using a process model describing main influential variables taking into account their dynamic interaction, and the effect of sampling and delay in availability of measurement results. The uncertainty of model structure and parameters and that due to inaccuracies in measurements are treated in stochastic terms. The statistical contribution was most valuable in analysing information theoretical aspects in a Bayesian framework. The contribution of PC was essential in model building and in defining the control structure and algorithms. Their combined effort leads to a technique to guarantee (in absolute, acceptance sampling terms) product specs, to <u>decide</u> on sampling frequency, allowable information delay and required measurement accuracy. Finally, it directly leads to <u>control design</u>.

## 6. PROCESS SAFETY

Managing safety is an integral and essential part of management in the process industry, not only because there is a legal obligation, but also because in most cases the company philosophy is grounded on top management business principles. Although the CPI has a good safety record based on long accumulated experience, there exists a continuous drive by both Management and Governments to improve the safety record.

Safety is to a great extent a man-related problem but system engineering aspects also play a significant role. Quite frequently one overlooks the integration between man and machine (process) in so-called human errors. The nuclear industry traditionally puts greater emphasis on the systems aspects. In view of the extreme danger of core melt-down, probabilistic techniques were developed to provide a quantitative risk assessment (QRA) for new designs. Such studies are obligatory and require substantial effort by (mainly contracted) specialists. With the dwindling societal interest for nuclear power these safety and/or risk engineers now look for application of their techniques in the process industry. They are supported by Governments setting out policies how to analyse risk aspects of existing and new hazardous installations. There is a distinct trend in such analyses towards quantification.

Industry at large is rather sceptical on the public use of QRA, pointing at the shortcomings of modelling approaches like fault trees (lack of dynamics, the restrictions of Boolean logic, the difficulty to handle correlated events). From incident records and accident analysis suspicion arises to what extent improbable sequences of actions could be taken into account a priori. There exists a serious lack of probabilistic data on basic events and reported analyses of comparable installations show considerable spread in outcomes (Mandl, 1983). The use of safety

engineering techniques such as HAZOP, FTA, FMEA for in-house evaluation of design alternatives, on the other hand is growing. However, the information obtained in such analyses is not, in general, systematically carried forward to further design stages, nor is it readily available to the process operator.

This lack of generic systematic techniques is felt in actual operation, where safety is controlled by process safeguarding, alarm systems and the human operator. The performance of alarm systems leaves much to be desired. In our ergonomic evaluations of several generations of systems (Swaanenburg et al., 1988) we invariably noticed a high

alarm load, with operator intervention rates varying between 5 and 275 actions per hour. A significant part of that load (between 30 and 75%) is caused by oscillating alarms, another part due to (predictable) follow-up alarms. Techniques reported in the literature on alarm interpretation are vulnerable with respect to dynamic effects, but also, more seriously, to a possible existence of a non-unique relationship between alarm information and the event(s) that caused the alarm.

An interesting conceptual question then is: "can safety be controlled [*] in the sense of control theory paradigms?" In PC a process is required to operate at a number of process states $X(t)$, which may vary in time t. These states manifest themselves through the process structure as a process output vector $Y(t)$. (Fig. 4). The process states $X(t)$ may not readily be measurable and then have to be reconstructed from the output vector $Y(t)$ and the input vector $U(t)$. After comparing the reconstructed state vector $X(t)$ with its reference value the control system (based on a process model normally obtained by identification) will steer the process to the desired values.

How does the above description of process control fit in with safety control? To that end we will compare a number of relevant parameters:

(continued overleaf)

---

[*] We prefer to use here the term "control", rather than "managment" as it focuses our attention on observa - bility, controllability, stability and dynamic aspects.

C. F. H. van Rijn

| Items | Process control | Safety control |
|-------|-----------------|----------------|
| reference | desired process states, continuous | process values exceeding prespecified range (0-1 information) |
| states | process parameters, hold-up, conversion, quality, etc. | safety parameters, distance to explosion limit, max. pressure, etc. |
| information reconstruction | either direct (classical control) or by estimator (model based) | by safeguarding system or operator (alarm interpretation) |
| control signals | continuous, or discrete in small steps, "linear" | operator action, relief valves, trips, etc. "non-linear" |
| process outputs | temperature, pressure continuous variables | limit or tolerance transits, eventually hazard |
| interaction between steering variables | known, compensated or exploited | ? |
| dynamics | known or identified, essential for proper control | static approach dynamics hardly taken into account |
| control and stability | based on feedback using control laws with guaranteed stability | ? human reliability? interpretation effect? timing? |

Let us consider some of the main differences.

First of all there is a striking difference between the type of information and the information content. Whereas process control relies on continuous process information and its dynamics in time, safety control is based mainly on information that certain levels have been (alarm) or are close to being (pre-alarm) passed. This information is discrete and does not contain any dynamics.

A second basic difference lies in the systematic approach. Process control has evolved into a science, where based on information theoretical concepts and models of the process, at least asymptotical stability can be proved. As a consequence, requirements are imposed on the structure of the measuring instruments. It is doubtful whether such a structured approach to accessing information from the process is also used on the alarm/safeguarding side. Judging from the interpretation errors made by operators and the difficulties perceived in computerised alarm analysis, one is inclined to believe that a more structured approach on where to place what type of safety device would be beneficial. In

fact, one would like to see a safety model of the plant similar to the process models used in control and for state reconstruction.

A third major difference lies in the steering part of the problem. Whereas in PC linear, continuous signals are used, in safety control because of the nature of the problem, quite drastic, frequently discrete, in general non-linear, actions are taken by operators or by safeguarding systems. Again, problems may arise because of overlooking the dynamic aspects.

In our opinion there lies a challenge here for the PC community engrained with the concepts of observability and controllability. In our own research on incipient fault detection we achieved promising results by combining statistical data processing (Zhang, 1989) with a model reference approach. Using only existing process measurements (no add-on instrumentation) we obtained an alarm function with better robustness characteristics (less nuisance alarms) towards process dynamics than conventionally achievable. Moreover, analysis of the residuals between model and process data provides valuable information on the characteristics of the fault. The use of model reference techniques implicitly ensures uniqueness of the relation between alarm and cause.

Safety controllability aspects are even more interesting for research. An operator usually can choose from a gamut of possible actions to bring a process back into a safe state. Hence, the safety control problem is highly dimensional and usually non-linear. There is a considerable business incentive because solving such problems is a pre-requisite for further automation of process supervision.

## 7. PLANT MAINTENANCE

Maintenance costs of refineries and chemical plants can currently amount to 30-50% of the operating expenses (excluding energy). This figure refers to direct costs, the (appreciable) economic value of lost production or impaired quality not being taken into account. It is not unusual to note that the maintenance manpower to equal or exceed that in operations. Maintenance costs have a strong tendency to increase (especially with automation); the increase in the US CPI over 1987 was 11% per annum (Basta and Morris, 1988). Maintenance managers only have limited degree of control; Pierce (1987) of Union Carbide mentions an average of 50%. We will certainly receive more information on this issue furtheron in the workshop from Mr. Lee Solomon.

Although a significant interest in maintenance is thus growing on maintenance, we believe it should be regarded in a broader context, that of Reliability, Availability and Maintenance (RAM) (Van Rijn, 1987). Reliability defines the ability of an item to perform a required function under stated conditions for a stated period of time. It may also be denoted as a probability of survival. As such, reliability can directly be associated with process availability, product quality, safety and environmental conservation. Availability is defined as the fraction of time a system is capable of fulfilling production needs and thus adresses reliability and maintenance. System effectiveness, moreover, also takes into account unit production capacity.

Maintenance may be defined as "the combination of all technical and administrative activities to retain an item in or to restore it to a state where it can perform its required function" (British Standards 4778). Maintenance thus is a typical managerial problem and as with all managerial problems, it should focus on the why's (effectiveness) and how to's (efficiency). It is a complex problem for a number of reasons:
- in RAM the maintenance manager is faced with realisations of inherently stochastic processes. Engineering faculties to a great majority, however, present an almost completely deterministic education. Maintainers thus look for a deterministic cause and effect relation. Such a cause may indeed be a design imperfection, wrong use of material or operating without a design window. For a large

part, however, only stochastic trends exists, like wear-out, fouling, loss of strength due to shock loading, etc. Data on these processes are limited.

- the number of maintenance activities is so large, that decisions have to be taken in a short time and frequently at a low hierarchical level
- the maintenance function has a low profile, being a cost centre and not belonging to the firm's key activities. Hence, graduates starting in the maintenance function strive for swift transfer to engineering, regarded to posses a higher status
- part of this low profile can be explained from the fact that maintenance traditionally has been unable to quantify its costs and benefits. Budgetary control then is one of the few tools for higher management
- the maintenance manager thus lacks the insight, the data required for, and the estimate of the consequences for proper decision-making in a stochastic environment
- in his task he is confronted with essential decisions taken at design level. Here too, techniques to assess the future maintenance cost-benefit balance are imperfect. Moreover, the traditionally used economic evaluations (DCF, NPV) favour reductions in capex and neglect later maintenance costs.

Scientific interest in this area is restricted mainly to reliability engineering, a subset of mechanical engineering and stochastic OR. Although a significant number of papers and textbooks discuss a multitude of models, practical application (and thus relevance) is limited. This image is changing, abeit at a slow rate. First coordinated activities are noted to introduce reliability engineering tuition in university and vocational training. Car manufacturers are quite active to improve reliability and maintenance costs. In the US, the DOD has sponsored major research and the Air Force is installing a large scale reliability programme.

The efficiency of the maintenance process is now being promoted by the use of Maintenance Management Information Systems (MMI). Such MMI's provide excellent databases for equipment information, spare parts, planned maintenance activities and overhaul results. They provide the maintenance manager with a good overview of past, actual and outstanding activities and automate the clerical process (job cards, reports, time-based preventive maintenance). What is still missing is decision support on effectiveness: why and when should what activities on which unit be done?

Our research (Van Rijn, 1987) focuses on three levels of effective decision-making:
- pseudo-analytical mathematical techniques have been developed to evaluate system interval effectiveness (the probability of a process meeting required demand in a time interval selected by the user) as a function of process configuration, unit reliability and maintenance characteristics, buffer storage, and operational and maintenance strategies (Van Der Heyden and Schornagel, 1986). These techniques were implemented in a pc package that helps to define minimum equipment and assess future maintenance needs in the <u>design</u> phase (Fig. 5)
- decision support tools have been developed for the <u>audit</u> phase (Fig. 6), to help engineers on deciding on maintenance strategy (break-down, preventive maintenance, either time-based or using opportunities)
- a package PROMPT-II was developed and fieldtested (Fig. 7) to provide <u>daily decision support</u> on opportunity-based preventive maintenance and inspection. This prototype package found an excellent user acceptance, led to better use of opportunities and thus reduced costs and eliminated backlog.

RAM is still to a great extent a scientific terra incognita. Research would benefit from the interest of other disciplines. For example, condition monitoring of equipment can be proven to lead to economic savings <u>if</u> the observed variables (vibration, temperature, debris analysis, etc.) can be related unambiguously to a specific failure mechanism (diagnosis) and have such predictive power that maintenance can be scheduled at an opportune moment. At present, most of the tools offered by instrumentation manufacturers lack such a relation with the failure modes of a specific piece of equipment. Hence, observability and predictability characteristics cannot be guaranteed and have to be evaluated in practice. The danger exists that the effectiveness of condition monitoring is thus restricted to prevention of excessive, follow-up damage (alarm). Condition monitoring would be an interesting research subject for PC with its experience in state estimation.

Maintenance optimisation, apart from structural integration from design to operational level, also needs to be embedded in production planning and scheduling. Using opportunities (moments in time with zero or small downtime costs) for preventive maintenance significantly affects the cost-benefit balance. Proper use of such opportunities can reduce the duration of three-yearly shutdowns. It calls, however, for another element to be addrssed in the PPS objective function.

## 8. EPILOGUE

In this paper we have touched upon a few subjects whose importance is regularly stressed by business managers. This survey was by no means exhaustive.
We have deliberately left out such a principal item as performance monitoring, which is possible only after integral problem formulation and agreement on target or optimality by various decision makers.

We have shown that most of the production control problems require a multi-disciplinary approach, that such a symbiosis can be mutually rewarding and may lead to new solutions. Industry has its own responsibility to foster such an in-house climate. For academic disciplines it "is very, very difficult to make the pieces come together" (Rhein, 1989). Hopefully, the presentations at this meeting and the atmosphere during the discussions will effectively contribute to this process.

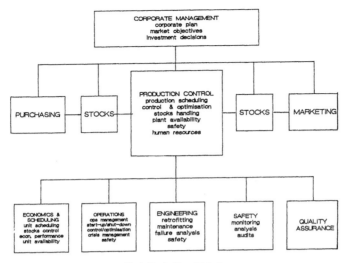

Fig.1  Production Control

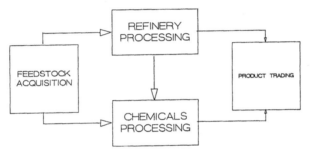

Fig. 2 Logistics of downstream oil and petrochemicals

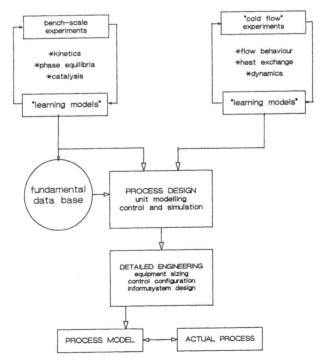

Fig. 3 Models in Process Development, Design and Operation

C. F. H. van Rijn

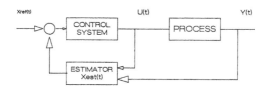

Fig.4 Process with feedback control system

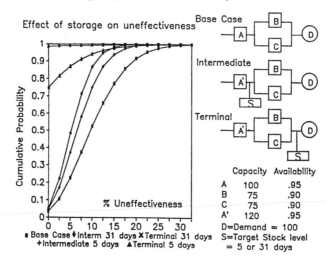

Fig. 5.  System Interval Effectiveness

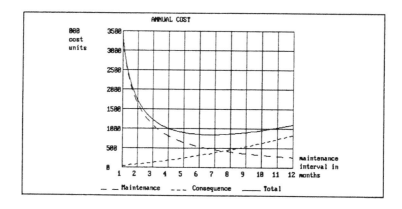

Fig. 6.  Maintenance Decision Support

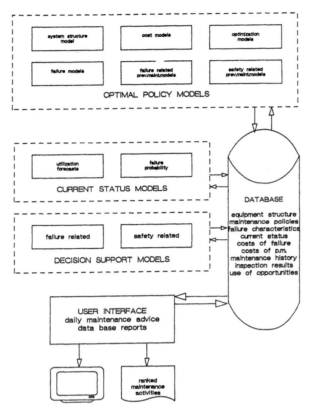

Fig.7 Block diagram of PROMPT II

## REFERENCES

Asbjörnsen, O.A. (1988). A Systems Engineering Approach to Chemical Process, Design and Operation. Paper presented at PSE-88, Sydney, 1988

Baker, T.E. (1984). The Integration of Planning Scheduling and Control. Chesapeake Decision Sciences Inc. PO Box 65, Berkeley Hts, NJ 079022

Basta, N., and Morris, G.D.M. (1988). Managers tackle Maintenance Problems. Chemical Engineering. December 19, 1988. p. 30-33

Benson, R. (1988). Process Systems Engineering: Past, Present and a Personal View of the Future. Paper presented at the Third International Symposium on Process Systems Engineering, Sydney, 28 August - 2 September 1988

Bentley, F.W.P. (1988). Downstream Developments in Established Markets Proceedings of the Twelfth World Petroleum Congress, Vol. 5, p. 183-193, John Wiley & Sons, Chicester.

Boardman, J.T. (Chairman) (1989). An Interim Report from the Working Party on "Strategic Direction" for Control". IEE Newsletter Spring 1989

Committee on Chemical Engineering Frontiers (1988). Frontiers in Chemical Engineering. National Academy Press, Washington DC

Cutler, C.R. and Ramaker, B.L. (1979). Dynamic Matrix Control - A Computer Control Algorithm. AICHE National Meeting, Houston, Texas

Davies, K. (1988). Oil-Chemical Interface. Presentation to the Oil and Gas Seminar in Jakarta, Indonesia, June 1988, available from Shell International Petroleum Company Ltd. (PAC/233), London, SE/1 7NA.

Federgruen, A. (1989). Methodologies for the Evaluation and Control of Large Scale Production/Distribution Systems under Uncertainty. Proceedings of the Shell Conference on Logistics, p. 143, Pergamon Press, Oxford

Fleming, W.H. (Chairman) (1988). Report of the Panel on Future Directions in Control Theory: a Mathematical Perspective. Society for Industrial and Applied Mathematics. Philadelphia, PA.

Greenberg, H. (1989). Intelligent User Interfaces for Mathematical Programming. Proceedings of the Shell Conference on Logistics, p. 198-223, Pergamon Press, Oxford

Grievink, J. (1986). De Rol van Modellen. Het Ontwerpen van Reactoren. I²-Procestechnologie, nr. 2, p. 10-15, (in Dutch)

Karmarkar, N. (1984). A New Polynomial-time Algorithm for Linear Programming. Combinatorica 4, 373-395

Kohlhase, K.R., Lawrence, P.A. and Waller, G.J. (1988). The Challenge to Refining Technology to meet New Requirements, Proceedings of the Twelfth World Petroleum Congress, Vol. 4, p. 247-256, John Wiley & Sons Lts., Chicester

Kümmel, M. and Seborg, D., (1987). A Contemplative Stance for Chemical Process Control, Automatica, 23, 801-802.

Langeveld, A.T. (1989). Downstream Oil and Chemicals Conference: Where Ends have to meet. Proceedings of the Shell Conference on Logistics, p. 1-10, Pergamon Press, Oxford

Lasserre, J.B. and Roubellat, F. (1985). Measuring Decision Flexibility in Production Planning. IEEE Transactions on Automatic Control, 30, p. 447-452

Mac Gregor, J.F. (1988). Statistical Process Control and Interfaces with Process Control. In: Prett, D., Garcia, D., and Ramaker, B. The Second Shell Process Control Workshop. Solutions to the Shell Standard Control Problem, Butterworth, Stoneham MA (to be published in 1990.

Moore, C.F., (1988). What and Who is in Control? A Process Control, Ibid

Malinowski, K. (1987). Steady-State Systems: On-Line Coordination. Systems & Control Encyclopedia, Pergamon Press, Oxford

Mandl, C. and Lathrop, J., (1983). LNG Risk Assessments; Experts disagree, Risk Analysis and Decision Processes. The siting of Liquefied Energy Gas Facilities in four Countries, Ed. H. Kunreuther, Springer, Berlin

Mc Greavy, C. (1989). An Interactive Graphical Environment for an Evolutionary Approach to Process Design. Dechema Monographs Vol 116 - VCH Verlagsgesellschaft (1985)

Pierce, F.R., (1987), Summary of Invited Comments in Foundations of Computer Aided Process Operations. Proceedings of the First International Conference. Park City, Utah, July 5-10, 1987, Elsevier, Amsterdam

Pistikopoulos, E.N. and Grossmann, I.E. (1988a). Optimal Retrofit Design for Improving Process Flexibility in Linear Systems, Comp. Chem. Eng. 12, p. 719-731

Pistikopoulos, E.N. and Grossmann, I.E. (1988b). Stochastic Optimization of Flexibility in Retrofit Design of Linear Systems. Comp. Chem. Eng. 12, p. 1215-1227

Prett, D.M. and Morari, M. (1987). Shell Process Control Workshop. Butterworths, Stoneham MA

Prett, D.M. and Garcia, C.E. (1988). Fundamental Process Control. Butterworths, Series in Chemical Engineering, Stoneham MA

Rhein, R., (1989). Engineering Research Centers: The Jury is still out, Chemical Engineering. May 1989, p. 41-46.

Schlenker, R.P. (1987). Process Industry Automation. The Challenge of Change. Proceedings of the First International Conference on Foundations of the Computer Aided Process Operations, Elsevier, Amsterdam, p. 1-27

Swaanenburg, H.A.C., Zwaga, H.J. and Duynhouwer, F., (1988). The Evaluation of VDU-based Man-Machine Interfaces in Process Industry, Third IFAC Conference on Man-Machine Systems, Oulu, Finland

Swaney, R.E. and Grossmann, I.E. (1985). An Index for Operational Flexibility in Chemical Process Deisgn. AIChE Journal, 31, 621

Titli, A. and Singh, M.A. (Editors) (1980) Large Scale Systems Theory and Applications, Pergamon Press, Oxford

Tulleken, H.J.A.F. (1989). Grey-box Modelling and Identification, using Physical Knowledge and Bayesian Techniques, IFAC Symposium on Adaptive Systems in Control and Signal Processing, Glasgow UK, 19-21 April

Van der Heijden, M.C. and Schornagel, A., (1986). Interval Uneffectiveness Distribution for a k-out-of-n Multi-state Reliability System with Repair, European Journal of Operational Research, 36, 66-77

Van der Grinten, P.M.E.M., and Lenor, J.M.H., (1973). Statistische Procesbeheersing (in Dutch), Prisma Technica 50.

Van Rijn, C.F.H., (1987). A Systems Engineering Approach to Reliability, Availability and Maintenance, in Foundations of Computer Aided Process Operations, Proceedings of the First International Conference, Park City, Utah, July 5-10, 1987, Elsevier, Amsterdam

Van de Vorst, J.G.G. and Bisseling, R.H. (1989). Parallel LU-Decomposition with Partial Pivoting on a Mesh Network of Transputers Journal of Parallel and Distributed Computing

White, J.R. (1986). Use of Spreadsheets for Better Refinery Operation. Hydrocarbon Processing, October 1986, p. 49-52

White, C.H. (1987). Applications of Operations Research Methodology to Process Operations, Computer Aided Process Operations, CACHE Elsevier, 1987

Zhang, X.J., (1989 Auxiliary Signal Desgin in Fault Detection and Diagnosis. Lecture Notes in Control and Information Sciences, Vol. 134, Springer Verlag

# PRODUCTION MANAGEMENT IN
# DOWNSTREAM OIL INDUSTRY

## A. G. Hop,* A. T. Langeveld** and J. Sijbrand*

*Shell Internationale Petroleum Maatschappij B.V., The Hague, The Netherlands
**Koninklijke/Shell Laboratorium (Shell Research B.V.), Amsterdam, The Netherlands

Abstract. The term "downstream oil industry" here refers to the acquisition of
crude oil and other feedstock, its conversion in a refinery complex and the
subsequent marketing/distribution of oil products.

Long before the term "Computer Integrated Manufacturing" was coined, computers
were applied in the oil industry to plan, monitor and control their refining and
marketing/distribution activities. Many computer systems are now being used,
mostly in a distributed, stand-alone fashion.

Meanwhile, forced by the changing business climate most Oil Companies realize
that in order to get acceptable business results it will be crucial to apply
managerial control to oil trading, -supply, -manufacturing and marketing
activities as an integrated business. And the necessary computing systems will
have to work together in some logical fashion.

In order to get proper insight many companies now apply structured Information
Planning which, starting from an analysis of the business, tries to establish
the right computing system framework and an action plan for system development
and implementation.

The paper follows the same logic and will first draw attention to the
"downstream oil" management control process thereby focussing on the necessary
horizontal and vertical integration of activities.
Most of the attention will be given to the activities that are carried out
within the Refinery fence.

The paper will consider the different types of computing systems used: process
-control,-supervision and -optimization systems, decision support systems for
planning and scheduling, and a variety of information systems for management
support.

The paper then considers in some more detail logistic systems, here meant to be
the decision support systems for use in downstream oil operational planning to
make trading/supply/manufacturing decisions, for refinery scheduling and for
process optimization.

Generally each such logistic system will contain a model, a solution algorithm,
a data base and a user interface.
It will be illustrated that the description of the activities in an oil refinery
in mathematical terms and the acquisition of reliable input data are extremely
complicated:
The feedstock is a complicated mixture of hydrocarbons, many process
relationships are non-linear or not fully known, many of the processes are
linked and influence each other. And uncertainties e.g. in the future market
requirements, price structure, etc., do not facilitate decision making either.

Finally the paper will indicate unfilfilled wishes and the type of developments
that still have to be done to arrive at a mature framework of Refinery
Management/Control systems, such as:
More accurate and faster mathematical programming procedures by the use of
better algorithms and faster computing, better user interfaces and
interpretation aids, better technological models, feedstock- and product
characterization and quality measuring instruments.

Also, proper solutions for data management in this multi-level optimization process are not trivial; and the amount of detail required at the various levels, in the light of the prevailing uncertainties needs further study too. There is the challenge to avoid unnecessary complexity and make the computing tools as simple as possible in order that the user can understand why the answers are as they are.

Keywords. Computer applications, Computer Integrated Manufacturing, Decision Support, Information Planning, Logistics, Oil Refinery, Optimization, Production Control.

## INTRODUCTION

The term "Downstream Oil" in the title of this title of this paper refers to the acquisition of crude oil and other refinery feedstocks, its conversion in a refinery complex and the subsequent marketing/distribution of oil products.

Long before the term "Computer Integrated Manufacturing" was coined, computers were applied in the oil industry to make optimal plans for their supply and manufacturing activities. Linear Programming was one of the first widely used algorithms on industrial mainframe computers. With the advent of the mini-computer in the early 70s, many Oil Companies started to use these to monitor and control refinery processes and marketing distribution activities.

In the late 70s computing systems were also developed to support the day to day scheduling activities of refineries, traditionally a process that had been very much a "back of the envelope" affair, based largely on the intuition of the experienced refinery scheduler.

Also in the 80's systems running on Personal Computers have made inroads in the above mentioned application fields.

Indeed, computing systems and tools of many different types are now being used, mostly in a stand-alone, dedicated fashion.

Meanwhile, forced by the changing business environment, many Oil Companies started to realize that in order to get acceptable business results it would be crucial to apply managerial control to trading, supply, manufacturing and marketing activities in some integrated fashion.

Fortunately Information Technology (Computing and Communication) has now reached a level of maturity that one may start thinking in terms of "Computer Integrated Manufacturing" in the Downstream Oil Industry.

However, it has already become clear that setting this up is a very challenging, multi-disciplinary task. It requires the joint exploitation of all we know about oil business management, oil processes and products, engineering, mathematics, operational research and, last but not least, information technology.

This paper is no scientific essay. It is a snapshot of the way we look at integrated downstream oil activities and how these could be supported by computing systems.

## INFORMATION PLANNING

In the past, many computing systems have been developed in an atmosphere of "technology push" and appeared to fail because of insufficient clarity about functional specifications, insufficient follow-up and failure to see the organizational consequences. Often computing systems were put in operation without proper documentation being available. And training was often given only to the first-time users, not to their successors in the job.

Indeed, the management of Information Technology has become quite an issue because of the ever increasing costs involved, the difficulty to establish convincingly the benefits and the speed of development in the Technology itself. Many companies now apply structured Information Planning which, starting from an analysis of the business, tries to establish the right computing system framework and an action plan for system development and implementation.

One should not only look into system and facilities requirements but also consider what obstacles have to be worked on to really get the benefits that are potentially there (Uronen, 1988).

## TRENDS IN DOWNSTREAM OIL ACTIVITIES

Traditionally in most Oil Companies the downstream activities are carried out in three business segments: Refining, Supply/Trading and Marketing/Distribution. In each of these business segments there is a hierarchy of mainstream (oil) management control activities: Long range planning, short term planning, operational scheduling, optimization and supervision of actual operations. It has already been realized for quite some time that within each business segment there should be a lot of cooperation between these different management control activities in order to be able to overcome sudden difficulties and to arrive at consistent management decisions (Vertical Integration).

In many Oil Companies, only the planning activities (at the top of the hierarchy) are carried out for the three business segments on an integrated basis. And each business segment is accountable for and appraised on its economic performance against the planned targets (Fig. 1).

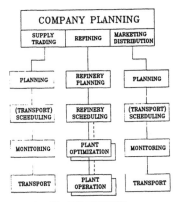

FIG.1: DOWNSTREAM SYSTEMS INTEGRATION

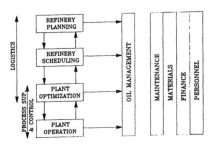

FIG.3: REFINERY COMPUTER SYSTEMS

By the same token the refinery scheduler
indicates how the processing plants and blenders
should be run, and each plant manager can be held
accountable for plant performance against
targets.

The economic reality of the recent past , i.e.
price volatility, economic opportunities with
ever shorter time windows, "make or buy" choices,
etc., has forced the Downstream Oil Business  to
work in an increasingly integrated fashion, also
at the "lower" hierarchical levels, in order to
maximize the overall business results.
(Horizontal Integration, see Fig. 2)

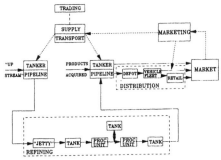

FIG.2: HORIZONTAL INTEGRATION

Such integration can have uneasy consequences: it
influences segmental schedules and interferes
with organizational accountabilities.

In addition to the above two types of integration
of the "oil production management" activities,
there is also Lateral Integration with the
maintenance, technological, personnel,
contractors, materials, finance and other
activities.

This trend to integrate Downstream Oil activities
has a significant impact on the requirements for
the supporting computing systems in the sense
that these systems not only have to comply with
their inherent tasks but have to communicate with
one another. Therefore they should fit in a well
designed computer system framework which can only
be established after a thorough information
analysis.

## REFINERY COMPUTER SYSTEMS

There are three main classes of computer systems
involved with refinery production management,
i.e. decision support systems, process
supervision and control systems, and management
information systems. (Fig. 3)

The decision support systems support the refinery
programming and scheduling activities. These
systems are most directly affected by the
increasing integration of the Downstream Oil
business and they are therefore most sensitive to
changes in the "rules of the game" in Downstream
management. These systems are based on (a
combination of) mainframes, minis or even PC's.
High powered personal workstations may be
expected to make inroads in this field. The
decision support ("logistics") systems will be
dealt with in more detail below.

The process supervision and control computers
support the operator in his task of carrying out
the production programme advised by the refinery
scheduler. In this area we find mini-computers
which carry out the more advance tasks such as
optimization with the help of online process
models, trending and reporting. In this area we
also find the distributed control systems
consisting of groups of microprocessors each
serving one or more automatic control loops in a
plant via built in but adjustable control
algorithms. Sometimes more sophisticated
algorithms, such as those used for multi-variable
control, may be found in the distributed control
environment but they may also appear in the
supervisory (mini) computer.

The third category of systems involved in
refinery production management are the
"management information systems". They take care
of such tasks as oil accounting, mass balancing,
customs and stock reporting; calculation of
technical statistics on efficiency, upgrading,
product quality, waste water cleanliness;
reporting on refinery economics and on the
contribution made to the overall Downstream
profit; generating performance indicators of all
sorts. These systems are typically implemented on
minis, though sometimes on mainframe computers.

In addition to the three classes of refinery
production "oil" management systems mentioned
above there are a number of significant computer
applications in the engineering and
administrative fields, e.g. for maintenance
management, materials- and contractor management,
shutdown- and project planning, cost control and
financial accounting, and for personnel
management. These systems are becoming
increasingly more relevant for production
management proper, especially the maintenance
management system.

LOGISTIC SYSTEMS

The definition of "logistic systems" is not always clear and unambiguous. For some the term refers to systems to support the control of the movement of goods through a manufacturing complex. Others use it to indicate total production management systems.
We usually use the term to indicate the decision support systems for use in downstream oil planning to make trading/supply/manufacturing decisions, for refinery scheduling and process optimization (The higher levels in the hierarchy of control picture).

Our Operating Company and Refinery planning systems are still largely based upon LP techniques for which we use one of the commercially available packages. We have developed our own, model generating, pre-processor system, which is very popular with our Operating Companies.

For daily scheduling we use a variety of techniques, including multi-period LP and simulation. As described in (Langeveld, 1981) the results of Refinery Planning need quite some adaptation in order to arrive at a workable realistic daily schedule.

For process optimization we use our own proprietary system, based on non-linear optimization techniques and stemming from our own in-house Research and Development work.

Generally each logistic system will contain a model, a solution algorithm, a data base and a user interface. In the past we have given much attention to the development of process models, including feedstock characterization models, and to solution algorithms (simulation, LP, multi-period LP, NLP,etc.). Less attention has been given to data base design and the user interface. Yet, especially at the higher levels in the hierarchy of control the user has to interact intimately with the system and the success of the system will depend heavily on the quality of the user interface. Also, the timely availability of reliable data appears often to be a problem.

In the hierarchy of logistic systems there is also a hierarchy of objective functions, models and data. At the higher levels they will have a more aggregated character, but in order that decisions taken at the different levels are consistent, the objective functions, models and data should be consistent too.

The rigorous modelling of the activities in an oil refinery is an extremely complicated affair. The feedstock is a complicated mixture of hydrocarbons of which the exact properties are only known upon arrival, many process relationships are non-linear or not fully known, many of the processes are linked and influence each other, many products (bitumen, lubricating-oil) have performance specifications which are not readily measurable in the refinery.

It has to be investigated carefully what level of detail is required in the models for each type of logistic system. Generally, the lower the level in the hierarchy of control the more detail is required. However, if the input data are inaccurate it does not seem to make much sense to use very accurate models. In this respect we have seen interesting examples of "garbage-in is garbage-out".

EXPERIENCES AND NEEDS FOR FURHTER DEVELOPMENTS

In this section we summarise our experiences with the different production.management systems in oil refineries and indicate areas for further development.

a. Measurement systems

The effectiveness of a production management system depends critically on the quality of the data used in the systems. The data capture in the oil processes aims at the timely provision of accurate data about the state of the processes and the streams they produce.
There are still quite a few unfulfilled wishes in this respect. There is first of all the limited accuracy of measured data. Calibration of plant instrumentation is often not stable; therefore historical data stored in process computers are often not useful for estimating trends in plant conditions like catalyst deactivation. There are insufficiently reliable techniques for the on-line measuring of stream compositions and measuring or estimating quality characteristics. Therefore at least part of the measurements related to stream composition or qualities, are obtained in a laboratory from samples taken from the streams. This means that there is a time lag before the measurements become available, which makes it difficult to use this information for corrective action.

Fortunately, the current trends in measurement technology (Amundson, 1988), give hope for the future. New sensor principles are being investigated, based on solid-state , biological and optical techniques.
Also the "smart sensor" concept is being developed, which is capable of error checking, automatic compensation for interferences, and periodic self-calibration. Also, sensor configurations exist that provide redundancy, fault detection and data reconciliation. All of this sounds interesting, but we should recognise that it will take considerable time before our plants will be equipped with these "ideal" advanced measurement systems!

b. Control systems

Most of the control techniques in our plants are based on PID controller algorithms. Process control computers and ICS systems are installed in many of our refineries, and some of them are equipped with multi- variable control systems. Our experiences with the latter are limited, but some concern exists about their complexity, the black-box nature of the underlying linear models, and their robustness. For this reason, further development and practical experience are needed related to robustness and automatic adaptation of multivariable control techniques. This is indeed a recognised area for attention (Amundson, 1988).

In the longer run we should perhaps aim at control techniques based on physico-chemical process models that include non-linear relations. This is an area of active research, but perhaps too immature to count on in the short term. However, it is fully in line with developments in chemical engineering, where science-based process models are increasingly used and where computers invade the trade (Amundson, 1988). Also from a mathematical viewpoint this so-called "non-linear control" is receiving attention (Fleming, 1988).

## c. Process Supervision systems

The last twenty years have seen a trend towards centralised control rooms, an increasing span of control for plant operators, and increased automation. With the present emphasis on product quality, and environment/health/safety, operator action is more than just keeping the plant at the required operating conditions and reacting to alarms. Modern supervision systems should provide the operator with sufficient information for him to monitor the plant's activity, should be capable of providing quickly more precise and detailed information at the request of the operator, and should be able of assisting the operator in diagnosing problems. With respect to the last requirement some research projects are going on related to alarm interpretation and diagnosis with the aid of logic programming (expert systems, knowledge-based systems). Similarly, statistical techniques are being used for data reconciliation and error detection, which are useful for early warnings of instrument malfunctioning or plant problems (e.g. leaks). A useful area for research could be the treatment of safety as a control problem: rather than discriminating between two levels of safety only (safe and unsafe) the present state of technology allows the definition and detection of areas of safe, less safe, marginal and unsafe operation, and by anticipating on a potentially unsafe course, one can avoid a situation which otherwise could lead to emergency shutdown.

## d. Plant optimisation systems

As mentioned earlier, plant optimisation systems provide guidance to the operator and/or control systems as to the best way to run the units in order to satisfy the plants production target. Most systems are at present still in off-line mode, and therefore are run infrequently. Since on-line optimisation requires a considerable effort to develop a robust model and to tune the optimisation algorithm, this is only done for selected processes. It is clear, that further development is needed to simplify the effort and also to reduce the run-times required for non-linear optimisation on process computers. A positive trend with respect to the first requirement is the increasing knowledge about oil processes as embodied in process models, while the increasing power of computers will address the second element.

## e. Refinery scheduling systems

Current refinery scheduling systems are based on simulation or linear programming. Their aim is to make and maintain an operations scheme for each refinery plant in accordance with the month's production plan.
The scheme should specify for each plant and for any moment in a given time interval (say 4 weeks) which feedstock it takes, which way it should be processed, which tanks are involved in the processing, and which qualities of the products are to be expected (Langeveld, 1981; Ballintijn, 1986). A simulation model, often in spreadsheet form, may be a handy tool. In the past we have experimented with LP-based systems, in which a minicomputer was used as an interface to the user, with a dial-up connection to a mainframe computer for running a multi-period LP model. The advantage of this system was the user friendliness of the mini-computer part, which could be used as a simulation tool, but the mainframe connection appeared a bottleneck: turn-around times were too long and the LP output had limited value as a starting point for the scheduler.

At present the scheduling job is more complex and dynamic than 15 years ago: the attention to product quality means that the scheduling models have become non-linear, and the changes in the market have resulted in much more frequent updating of plans (e.g. unexpected crude oil acquisitions, unplanned product deliveries). Therefore much better tools are needed to handle the increased complexity and the requirement for fast results. Part of this will be remedied by faster computing equipment. However, even if the computational element is negligible, there are the problems of collecting the right input data and of interpreting the system output, which already accounts for more than 50% of the human effort involved in scheduling. It needs no explanation that the quality of the user interface of scheduling systems is essential for their acceptance.

## f. Planning systems

Almost all our planning systems are LP based. These planning models provide indications about acquisition of feedstocks on the basis of product demand and product price forecasts. Two decades ago it was sufficient to run the planning models once a month, leaving the refinery scheduler with the day-to-day aspects of operations. Due to the greater variety in crude oil sources, the local spot markets, and the abundance of crude oil, there are many opportunities during the month to make an extra profit compared to the plan; but there are also deviations from the assumed price levels leading to lower than expected margins. Therefore there is a tendency to run the planning model more often. This leads to the same needs for development as for scheduling systems: faster turn-around, better input and analysis tools, user friendly interfaces.

For LP-based systems the importance of training should be mentioned too: the techniques are a lot more complicated than simulation based tools: a simple database and a spreadsheet are easier to master than a matrix generator and the properties of an LP solution! This training is an on-going cost, because, as stated earlier, not only the first-time users need to be trained but also those that take over their jobs.

## g. System integration

Many of the production management systems for "downstream operations" have been developed on the basis of automation of existing activities. The challenge we are now facing is their vertical, horizontal and lateral integration. For instance, in the context of vertical integration: Can we be more effective and efficient by streamlining the information flows between the systems supporting the different levels? Certainly we can simplify and shorten the data streams using information technology (computer networks), for it will soon become clear that the interfaces between systems leave gaps; or that there are overlaps in their formal aims. A structured information analysis is required to reveal this.

A specific aspect of vertical integration needs attention. We have mentioned the trend to use process models for almost all levels in the production management hierarchy. At this moment there is little commonality between the models used at the control level (black box model obtained from responses of the plant to disturbances), the plant optimisation level (steady-state model based on physico-chemical and engineering knowledge), and the planning, scheduling level (empirical relationships from production records). It would be very useful if all these models could be derived from a single reference model containing all scientific and engineering knowledge and practical experience.

Now let us turn to horizontal integration. The gaps and overlaps mentioned above could also hold for horizontal connections between systems. At present, the horizontal integration within a refinery is done at the scheduling level, and within a plant at the optimisation level. However, if a plant, for whatever reason, deviates from the scheduled production, this disturbance will affect the next plant, which will have to deviate from the schedule, etcetera. This may upset the whole refinery. Ways should be devised to avoid this, e.g. by adjusting the schedule (if this can be done fast enough!) or by coordination links and procedures between plant systems.
Similarly, the integration of systems that support crude oil acquisition activities ,refinery operations and product trading and market operations is still an area for further attention. Here,uncertainties in the future market requirements, price structure, etc., should be taken into consideration.

Finally, a word about lateral integration. Of course the whole production management system is not a closed system: there are links with other systems from which information is taken, or to which information is given. A production management system is assumed to have up-to-date information about the production process, and this is of interest to general management. Good automatic links to management information systems need to be available; and in a similar way links to finance systems need to be installed.

In this context the relationship with maintenance systems also has to be mentioned.
Maintenance is important in two ways: its part in a refinery operating budget is considerable (30-50% is not uncommon), and it contributes directly to the production capacity. Recently, studies appeared suggesting that maintenance activities could be made more effective and save 25% or more, by using planning techniques based on probability models for failure (Rijn van, 1987; Basta, 1988). This is definitely worth further investigation.

It may be clear from the above that we have only made a first step towards an integrated production management system. Many further steps are needed: a process of evolution, that in total may turn out to be a revolution: CIM.

## CONCLUSIONS

We believe that there is much benefit in considering the Production Management of Downstream Oil activities as one integrated set of business issues that therefore has to be supported by a framework of computer systems that are logically connected in one way or another.

A lot of work still has to be done to improve each individual class of system in this framework, which boils down to faster solution algorithms, faster computing, more reliable input data, better technological models, better user interfaces.

Systems integration means the coupling of human activities that traditionally have been in different "worlds". Therefore, in order to make it work one has to start with rather mundane things like agreeing on a common glossary of terms and data definitions.

Also, one should agree on a proper computing/communication architecture.
Indeed many issues in connection with the system integration process do not require very sophisticated academic research but rather the willingness of people from different disciplines to work together towards the common goal : A higher overall profit margin.

Systems integration can only succeed if it has been preceeded by a proper business and information analysis and if it has been made very clear throughout the organization that this work has top management commitment.
After all, they are the real owner of the problem.

## REFERENCES

Uronen, P. (1988). Introduction to plant-wide automation, Journal A: Vol. 29, No 3.
Amundson, N.R. (1988). Frontiers in Chemical Engineering: Research Needs and Opportunities, National Research Council, National Academcy Press, Washington D.C.
Fleming, W.H. (1988). Future Directions in Control Theory: A Mathematical Perspective, Society for Industrial and Applied Mathematics, Philadelphia.
Langeveld, A.T. (1981). Operational use of multiperiod LP models for planning and scheduling, in Large-Scale Linear Programming, G.B. Dantzig, M.A.H. Dempster, M.J. Kallio (eds.), IIASA, Laxenburg, Austria.
Ballintijn, J.F., A.T. Langeveld, R.P. van der Vet (1986). Planning and scheduling: a multi-disciplinary effort, in Quantitative Methods in Management, C.B. Tilanus, O.B. de Gaus, J.K. Lenstra (eds.), Wiley, New York.
Rijn, C.F.H. van (1987). A systems engineering approach to reliability, availability, and maintenance, in Computer Aided Process Operations, G.V. Reklaitis, H.O. Spriggs (eds.), Elsevier, Amsterdam.
Basta, N., G.D.L. Morris (1988). Managers tackle maintenance problems, Chem.Eng., December.

# PRODUCTION PLANNING AND CONTROL — A HIERARCHICAL FRAMEWORK AND TECHNOLOGY STRUCTURE

### C. H. White

*Engineering Department, E. I. du Pont de Nemours & Co. (Inc.), P.O. Box 6090, Newark, DE 19714-6090, USA*

Abstract. Production Planning and Control is becoming increasingly important as production companies strive for world-class status. This paper explores a planning and control decision hierarchy, describes an integrated decision framework via examples, and calls for proper focus of efforts versus an over-reliance on technology and automation.

Keywords. Inventory control; linear programming; operations research; philosophical aspects; production control.

## INTRODUCTION

Production planning and control decisions are key components to the successful operation of any enterprise. Early in the life of a production enterprise, these decisions can be handled by a small group of people working closely together. In a small developing manufacturing enterprise, the founder/owner often makes these decisions. As the enterprise grows, a group of trusted assistants help with these decisions; and the group shares both goals and information. As the organization continues to grow in size and complexity, this centralized planning and control becomes too large; the enterprise then tends to move toward a more hierarchical organization with decentralization of the planning and control decision making. The enterprise may split down to a group of smaller individual enterprises much like the original one; but much more likely, it will adopt a more functional decentralization form which requires a structured hierarchy of decision making. It is the latter type of enterprise organization that is the topic of this paper.

## MANUFACTURING FUNCTIONS

Among the critical functions for a production enterprise are: product and process development, plant design, production and process control, materials procurement, materials inventory management, materials requirements planning, production planning and master scheduling, and customer order management. For a developing business, these functions are often handled by one or a few people. Many businesses began to separate these functions as they grow and evolve toward complex organizations with multi-level hierarchies for planning and control. One representation of a planning and control hierarchy is shown in Fig. 1.

| Level<br>(Time Frame) | Hierarchical Functions |
|---|---|
| Level 4:<br>(years-months) | Corporate Systems<br>+ Corporate Planning<br>+ Coordinate Information<br>+ Allocate Tasks to Plants<br>+ Inventory and<br>  Distribution |
| Level 3:<br>(weeks-days) | Plant-wide Systems<br>+ Plant Resource<br>  Management<br>+ Technical Support<br>+ Coordinate Areas |
| Level 2:<br>(days-minutes) | Supervisory Control<br>+ Area Control<br>+ Coordinate Level 1<br>  Systems |
| Level 1:<br>(Distributed Control)<br>(minutes-seconds) | Regulatory Control<br>+ Unit/Machine Control<br>+ Monitor and Control<br>  Process Instruments<br>+ Interact with<br>  Supervisory Systems |
| Level 0: | Instruments and Sensors<br>+ Valves<br>+ Flow Meters |

Fig. 1. An information-planning-controlling hierarchy.

## PLANNING AND CONTROL HIERARCHY

As companies strive for manufacturing excellence, they must better plan and control the cost, quality, and timing of their products and processes. This will require understanding, improving, and compressing the hierarchy of functions. The key is not how many levels or functions should be uniquely labeled; but, rather, (1) recognition and understanding of these interacting decision levels, and (2) that for each business care should be taken to identify its critical needs and then to focus attention on the appropriate decisions. Recognition of how these functions interact will aid in encouraging the appropriate focus in planning so as to avoid problems at lower levels.

Level 4 decisions are the mission of corporate or division management. These are strategic decisions such as entering a market, building a new plant, and expanding

an existing plant. They deal with setting
goals and providing resources (people,
equipment, capital, ...) so the lower level
functions can carry out the plans and meet
the specified goals.

Level 3 is the mission of plant management,
and efforts at "plant-wide control" should
focus on coordinating the actvities of all
production and support resources to meet the
goals passed down from Level 4 utilizing the
provided resources. Examples would include
scheduling the release of work orders for
production campaigns to achieve smooth
production flows. This may be accomplished
via sophisticated computer-based systems or
with simple lot release (kanban) systems. In
either case, the objective is to have
production flow smoothly through the
available resources to meet the production
requirements.

Level 2 deals with supervisory control of
areas within the plant. Needs may seem to
differ for complex chemical processes and
discrete manufacturing processes; but, in all
cases, the objective is to manage the major
plant resources (equipment, material, tools,
people, energy, ...) to meet the allocated
tasks. In the case of a stamping press, this
may focus on tool life and product quality;
whereas for a distillation column, it may
focus on energy and material costs and
product yields as well as quality. For many
processes, this may result in dealing with
selecting the best alternative (set of
process conditions) simultaneously
considering a number of constraints. A
number of tools (optimization, simulation,
decision trees, expert systems, ...) have
been used to handle such problems [1].
However, it is often useful (and profitable)
to also focus on the constraints and see if
they can be relaxed. It is often better to
invest capital to reduce or remove a
constraint than to continuously pay a penalty
for its existence. Examples range from
overly long set-up/changeover times and costs
to the lack of sufficient resources such as
high pressure steam, or production operators,
or automatic-guided vehicles for a materials
handling system.

One of the major points of this paper deals
with proper and early analysis of
alternatives looking at the whole set of
production system decisions with the desired
feedback of information. Specifically, it is
often possible to modify the design of
processes and/or plants to make it both
easier and cheaper to operate them. Examples
will be used to demonstrate this.

Level 1 deals with the direct regulatory
control of the processes and equipment. It
is clearly critical as the higher levels
cannot be automated without consistent Level
1 control. This is an area of established
technology and will not be addressed further
in this paper.

### PRODUCTION PLANNING AND CONTROL

When dealing with the broad task of
Production Planning and Control, it is
appropriate to look at the full range of
decisions, determine all of the objectives
and constraints for the specific situation,
and then consider what is best for the whole
enterprise.

A useful pair of planning paradigms are shown
in Fig. 2. The one on the left is the
traditional practice of strategic, tactical,
and operational planning levels. At the
strategic level; objectives are set,
resources are provided, and "marching orders"
are given. At the tactical level; these
resources are allocated, specific broad tasks
are assigned, and action is initiated. At
the operational level, these tasks are
executed using the resources allocated.

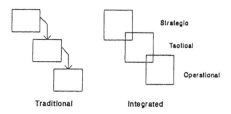

Fig. 2. Decision making levels.

The traditional practice (shown on the left)
can be labeled a "waterfall" scheme where
goals, resources, and constraints flow
downward. With this practice, the lower
levels are simply directed to carry out the
plans established at the higher levels using
the resources provided. The integrated
planning (shown on the right) can be labeled
a "feedback" scheme in that goals and
resources are passed downward while
information is fed back up allowing for
adjustment of plans to better utilize this
shared information.

An example would be the one of entering a
market (strategic), designing a plant
(tactical), and operating the resulting plant
(operational). In a feedback planning
system, information is shared, and both goals
and resource allocations can be reconsidered
using better information. For instance, if
it was found that a plant to make 80
different products would cost $60MM dollars;
but that a plant to make 70 of these would
cost just $32MM, then the question of
modifying the number of products could be
considered, market forecasts could be
reexamined, and it might prove better to
build the $32MM plant and then consider the
10 products separately. Another case of
beneficial feedback would be if it was found
that by adding additional equipment and
increasing the plant investment, it would be
possible to greatly reduce operating costs.
Then the investment decision can be
reevaluated with solid information about
operating costs, and the investment could be
increased in order to achieve the provable
reduced operating costs.

I submit that the feedback system of decision
making will both achieve better overall
results, and it can be faster and more
realistically responsive. If information is
encouraged to flow from the front lines to
field headquarters and then quickly on to
central headquarters; better decisions will
be reached, realistic goals will be

established, adequate resources will be
provided, and the total organization will be
better able to function as an aligned entity.
In a production enterprise; marketing and the
corporate leaders will better target market
opportunities, engineering will design
appropriate processes and plants, and
manufacturing will have adequate facilities
and will thus be able to effectively produce
at the desired levels.

### BATCH PRODUCTION EXAMPLE:
### PROBLEM DESCRIPTION

A plant produces over 100 different products
in a 5-stage process. Each product passes
through one of the units in each of the 5
stages. The equipment areas for each of the
5 stages have many units of differing sizes
and capabilities. As shown in Fig. 3, the
physical product flow is through 1 of the 59
mixing tanks; 1 of the 9 filling machines
(after being released by the Quality
Laboratory); 1 of the 7 packaging machines;
and then, 1 of the 5 palletizers. In
addition to the equipment not being
identical, there were also limitations
because not all equipment was interconnected.

## BATCH FLOW LAYOUT

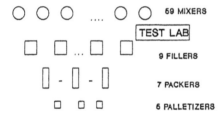

Fig. 3. Batch Processing Facility.

The plant equipment and its configuration of
piping interconnections evolved over time as
the product mix grew and volume expanded.
Mixing tanks had different sizes, production
capabilities, and piping interconnections to
the filling machines. The filling machines
had different capabilities in both the types
of containers that could be filled and in
which of the packaging machines were
accessible via the installed conveyor lines.
Batch sizes, product specifications, and
container sizes varied widely; consequently
the mixing and fill-out times varied widely.
Fill-out required the simultaneous use of a
filler, a packager, and a palletizer. Once
batches were released by the laboratory (the
specifications are tight) and a
filler-packager-palletizer "train" was
assigned, a production operator was required
to connect the equipment and then monitor the
fill-out operation.

Since batch orders arrived periodically with
specified product type, batch quantity,

container size, and due date; the task of
scheduling when and where to make the batch
was complex. Further, the batch mix times
were uncertain as variations in either the
ingredients or ambient conditions could
result in extended times.

### PLANNING AND CONTROL DECISION MODELING

Modeling is becoming a widely valued tool for
aiding planning and control decision making.
When developing any model, it is important to
first determine the purpose of the model, the
questions it is going to be used to help
answer, and the level and time frame for the
decisions. Strategic decisions (Level 4)
generally do not require highly detailed
models; but, rather, deal with the task of
objectively evaluating and comparing many
alternatives. Tactical decisions (Level 3)
deal with how best to achieve the specified
goals and allocate provided resources; they
generally need to get into more detail and
need to both generate and evaluate
alternatives. Tactical decisions also deal
with the coordination of plant-wide product
flows and strive to balance loads so there
are not bottlenecking problems. Operation
decisions (Level 2) deal with supervisory
control, scheduling issues aimed at
optimizing area operations and rebalancing
loads so as to smooth out the flows of
product and information. Regulatory control
(Level 1) deals with interfacing with the
process instruments and sensors and
controlling the process conditions and
quality.

For the multi-product batch processing
example described earlier, we were initially
asked "How can we get more product?", "How
much more can we get out of this plant?", and
"What is the impact of adding more equipment
versus changing how I operate and schedule
the plant?". Initially, we focused on the
higher level issue. The plant had attempted
to determine if there was a consistent
"bottleneck" area that could be
"de-bottlenecked" and, thus, attain a
capacity increase. The "strategic plan" was
to add another filler based on the perception
that mixing was frequently blocked and,
therefore, batches could not be filled out
immediately upon being released by the
laboratory. After more careful analysis of
the past plant performance, a number of us
felt that the filling machines were not a
consistent bottleneck, and, further, that
some procedural and scheduling changes could
be effective in increasing the throughput
without having to add capital equipment. At
this point, a more detailed modeling effort
was initiated as described in [2]. The point
is that feedback from the lower levels
pointed out that the early decision to add
capital investment to increase throughput was
not the most effective way. When the
question changed from "determine the capacity
increase from adding another filler" to
"determine how to most effectively increase
overall capacity", the decision then required
a more detailed model and the resulting
feedback showed that modified allocation and
operations scheduling decisions could better
accomplish the overall objective.

Shortly after this detailed simulation model
of the operations was completed, the
marketing people decided to take on a set of
new product orders. The plant schedulers

(Level 3) felt these orders could be produced to meet the specified due dates; but the filling-packaging area people (Level 2) felt it would not be possible without causing significant delays in the routine production. The model was used to show that the container sizes and batch sizes of the new orders would indeed result in large backlogs. However, we also determined that by just changing the container sizes of a few of the new orders, smooth production flows could be insured. Marketing was able to work this out with the customers; the result was that some new orders were modified, the plant was able to take on all of the new orders while meeting all previous orders with the result that both the customers and the plant people were happy. This is indeed "feedback planning" as discussed earlier and shown on the right in Fig. 2.

## CAMPAIGNED PRODUCTION EXAMPLE: PROBLEM DESCRIPTION

This example deals with design of a large continuous chemical complex where several parallel subsystems feed a packaging area. Each subsystem consists of a sequence of operations. This equipment is cleaned out between campaigns of different products; and there are both major cleanouts between product families and then minor cleanouts for products within the same family. Major times are in the 12-30 hour range while minor times are in the 2-8 hour range. The packaging area consists of blending silos, packaging silos, and packaging units which fill bags, drums, and large tote bins.

A traditional planning process had resulted in making specific investment and plant design decisions. However, one of the senior plant people offered the opinion that this plant would be difficult to operate and would not be as flexible as might be needed. It is commendable that his opinion was heeded and, ultimately, a highly detailed simulation model of the operations was developed with a number of embedded production planning algorithms. Demand forecasts were used to develop detailed production demand patterns, and production schedules were generated that campaigned the products over the subsystems. Product flowed from these subsystems into the blending silos; and when filled and blended, these were rapidly transferred into the packaging silos and then packed out. The packers also needed to be cleaned out between campaigns with major and minor transition times and costs.

Without getting into details, the model was developed to generate optimum production and inventory campaigning for the major units, while considering the cycling of the blending silos, and focusing on the use of both the packaging silos and the packers. Operational interferences, the effect of operator duties, and equipment outages were all found to be important considerations. This forced us to develop a detailed operations-oriented model with embedded decision making modules to handle both plant campaigning and unit operations scheduling.

After significant effort our final results indicated that the blending silos were too large, and there were too few packaging silos. Considering the operational details led to changing the plant design as the original one did not meet the projected production needs, and we were able to lower the required plant investment by over $1.3MM. It also highlighted the production units and areas where process changes would be most helpful. These included reducing changeover times, better scheduling of packing area operators, improving equipment availability, and modifying production campaigns to better meet needs with available equipment and manpower. A key point is that the detailed model allowed objective economic evaluation of these changes. There is often a strong reluctance to make such changes without having solid evidence of their economic and operational impact. The message again is the importance of using an integrated decision making framework and the value of considering operational characteristics early in the design process. Design should be aimed at producing plants that both achieve processing excellence and are flexible and easy to operate.

## PLANNING AND CONTROL HIERARCHY: AN ADDED DIMENSION

Earlier, we described a four-level planning and control hierarchy where the higher levels dealt with corporate decisions and the lower levels range down through plant, area, and unit regulatory decisions. Another dimension to decision making is the current condition or stage of the business enterprise. These deal with the maturation of the enterprise as it moves toward manufacturing excellence. In the early stages, there is a struggle for effective control. Large in-process inventories are tolerated, line supervisors reprioritize and control production on the shop floor and they use "hot lists" and overtime to meet production needs.

As the business matures toward manufacturing excellence, the management recognizes the need to monitor and control all of the resources, not just a few. The Production and Inventory Management tends to become dominated by Material Requirements Planning (MRP) and Manufacturing Resource Planning (MRP II) [6]. These systems are aimed at managing the resource utilization based around launching periodic runs of production plans. This means work centers are scheduled independently with orders moved through each center as a capacity opens up. Such systems are "push" oriented with reprioritization, expediting, and large buffer inventories.

The need for expediting, reprioritizing, and maintaining large inventories goes down as the maturing enterprise moves more toward a "pull" system. Production is pulled through the system to meet sales demands and in-process inventories decrease as the product moves more quickly and smoothly through the system. Attention is also expanded to include areas in both directions along the supply chain. The intent focuses more on how to effectively and quickly move materials from suppliers through manufacturing/processing and distribution and out to the customers. Emphasis is on synchronizing the flow of production, having available the resources that are required to insure quick efficient flow, and on eliminating the nonvalue adding steps such as in-process inventory handling and storage.

Further, maturing brings efforts to have the employees and the technical innovation programs focused on Continuous Improvements and Responsible Automation. Continuous Improvement can be accomplished in many ways. These can be initiated by employees seeing ways to improve operations and also by using models to explore the overall system behavior and, thus, generate alternatives. As production systems grow more efficient and tightly integrated, predicting the effect of a suggested change on the total system becomes increasingly difficult. The intent is to move toward a production system that uses simple rules to smoothly move the material through the total production chain. This is often not easy to achieve and clearly requires an integrated approach. If material is to move smoothly; resources must be available as needed, equipment must be well maintained, operators must understand and follow the rules, and the materials handling systems must be responsive.

We have just explored a dimension of a production enterprise that we have labeled the stage of maturation along the path toward Manufacturing Excellence. Assessment of where a business is along this dimension can be helpful in developing an improvement program. If a business is at an early stage, efforts can be directed at moving the plant away from "hot lists" by focusing attention on all of the resources. If the plant is at an advanced stage, technical innovations and automation are more appropriate.

### VALUES OF MODELS

Models have proven to be a valuable tool for developing a good solid understanding of how the total system behaves. Once thoroughly validated, the model can be used to conduct sensitivity studies and explore the effects both of small changes and of larger automation projects. Models can be used to objectively evaluate and compare proposed changes; but they can also be an effective tool for exploring the system and developing new concepts for improving the total system.

### PLANNING, CONTROL, AND INFORMATION NEEDS

Regulatory Control (Level 1) operates with real-time monitoring of process equipment and very short-term control. This means both the information and the control philosophy needs to essentially be real-time where the decisions involve set-points and driving the units in the desired direction. Goals are passed down in the way of targets.

Supervisory control (Level 2) involves area-wide control such as spreading area loads across parallel equipment. Goals are passed down from the higher levels; but, there can be some area-wide optimization. The time span is longer, and more constraints and economic information need to be incorporated.

Plant-wide control (Level 3) deals with the smooth coordinated flow of production through the area. Additional factors and resources are considered and changes in one area can affect the others. Time span needs are quite problem-specific; some plants require reoptimizing every shift while others require

it only weekly with area level decisions handling fine-tuning adjustments.

Corporate or division planning (Level 4) has higher level information needs; it neither needs the detailed information nor does it need the high frequency refreshing of the data. Control is less of an issue whereas planning is critical. If the appropriate high level decisions are made poorly, then the lower level controls will closely adhere to poorly set goals.

### CONCLUSION

The area of Production Planning and Control is both broad and deep. It ranges from strategic planning aimed at selecting the right markets and plant designs down to operations control of unit processes. The mid-range tasks of Production and Inventory Planning and Production Scheduling are often the major ones considered; but, this is an overly restrictive practice. Production and Inventory Planning models generally attempt to plan production out over several months while looking at demand forecasts, production capabilities, current inventory levels and the appropriate costs. Mathematical Programming models often are appropriate for such decisions ([1, 3, 4]. Production Scheduling models generally deal with the shorter term synchronized sequencing of production flows. They deal with moving production through the plant while considering both due dates and resource/equipment availabilities [5]. Here the information requirements are tighter, the focus shifts toward control with more frequent rescheduling as plant conditions change. Costs are important but insuring a smooth and orderly flow of materials is paramount. At the higher levels of decisions, the focus is on economic issues such as market revenues and material and energy costs; at the lower levels, it is on technical feasibility issues such as equipment constraints, unit optimization, synchronized flows, and customer due dates. These should not be viewed as separate problems, but rather, as ranges in a broad spectrum of manufacturing decisions of different scope, time frame, and information requirements.

The need for better planning and control is growing. Competitive forces require that firms better understand, plan, and control their cost, quality, and time to respond. Those firms who excel will be those that move to align their organizations toward common goals, simplify their processes, share relevant information about the enterprise and the manufacturing functions, integrate their decision making processes, and then introduce technical innovations and automation where it is cost beneficial. The decision making needs to be effectively focused based on the needs of the organization with consideration of its stage of maturity, the availability of key information, and the appropriate balance across strategic tactical and operational decisions.

Computers are becoming faster and cheaper, data monitoring and retrieval systems are becoming more powerful and available, decision-support modeling systems are becoming more accepted, and organizations are

and others will have an accelerating impact
on Production Planning and Control systems.

What will the world look like in 20 years?
Developing efforts at integrated systems will
flourish. Materials will be ordered as
needed from responsive suppliers and move
quickly through the plants. Plants will be
flexible and able to respond quickly to
customer demands. Facilities will be
designed for ease of operation and in-process
inventories will be low. Production Planning
and Control systems will be a critical
resource for survival.

The roots for the Integrated Production
Planning and Control systems of the 2010's
have been planted. Such systems will
certainly range over the full spectrum of
manufacturing decisions from facilities
investment to unit optimization. Integration
and data sharing will become routine,
investment and operating costs will be
considered together, materials will flow
smoothly and inventories will be small,
plants will become flexible and responsive
and all employees will work together toward
common goals. How fast will the world move
in the next 20 years? If history is to be
heeded, it will be faster than in the past 20
years. Planning and control systems are
indeed just at the beginning of a period of
accelerating and exhilarating change.

## REFERENCES

1. White, C. H. (1987), Application of
   operations research methodology to
   process operations. Process
   International Conference Foundations
   Computer Aided Process Operations
   (FOCAPO), Park City, Utah, July 1987,
   Elsevier Publishing.

2. White, C. H. (1989). Productivity
   analysis of a large multiproduct batch
   processing facility. Computers in
   Chemical Engineering, Vol. 13, No. 1/2,
   pp. 239-245.

3. Bradley, S. P., A. C. Hax, and T. L.
   Magnanti (1977). Applied Mathematical
   Programming, Addison-Wesley Publishing.

4. Aronafsky, J. S., J. M. Dutton, and M. T.
   Tayyabkhan (1978), Managerial Planning
   with Linear Programming: in Process
   Industry Operations, John Wiley & Sons,
   New York.

5. Faccenda, J. F., and H. Singh, Scheduling
   Multi-Stage Batch Facilities; ORSA
   National Meeting, October 1989.

6. Vollman, T. E., W. L. Berry, and D. C.
   Whybark (1988). Manufacturing Planning
   and Control Systems, Irwin, Homewood,
   Illinois.

# APPLICATION OF OR METHOD TO PRODUCTION PLANNING

## K. Murata

*Control & Systems Engineering Section, Technology Engineering Department,*
*Tonen Wakayama Refinery, Arida City, Wakayama Prefecture, 649–03, Japan*

**Abstract.** Our production planning is consisted of long range, middle range, short
range production planning and daily scheduling. Generally, it is suitable to apply
linear programming method to the operation planning of petroleum refinery, as it is
not so difficult to construct linear refining models. Thus, linear programming method
is widely applied in production planning. For short range production planning and
daily scheduling, case study function such as balance calculation with friendly man-
machine interface is more suitable than optimization function like OR method. And,
there is batch production procedure in a part of petroleum refinery. As it is hard
to apply linear programming in the area, expert system combined with balance calculator is used.

**Keywords.** Oil refining, production control, operation research, optimization,
linear programming, expert system

## INTRODUCTION

Before the explanation of OR application, let us
describe petroleum refining procedure and planning procedure.

Petroleum refining is consisted of mainly distillation and blending. Crude Oil is distilled into
semiproducts and products. Some semiproducts are
hydrotreated and/or cracked into semiproducts and
products. Finally, semiproducts are blended into
products. As main refining procedures are separation and mixing and some cracking, it is not
difficult to linearize refining procedure.

Next, let us introduce about planning procedure.
Our production planning is consisted of long range,
middle range, short range production planning and
daily scheduling. The purpose of each plan and
scheduling is shown in Fig. 1. The intervals of
long, middle and short range plan are year, quarter
and month respectively. Daily schedule is made
every week. Long and middle range production planning are made by the head office, and short range
production planning and daily scheduling are made
by each refinery based on the head office production planning.

Fig. 1. MANUFACTURING MANAGEMENT & CONTROL SYSTEM

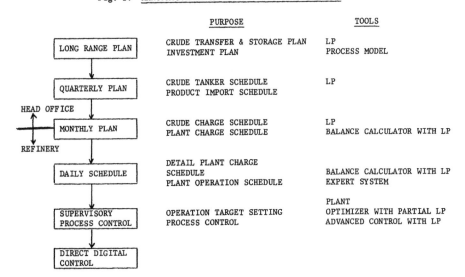

| | PURPOSE | TOOLS |
|---|---|---|
| LONG RANGE PLAN | CRUDE TRANSFER & STORAGE PLAN<br>INVESTMENT PLAN | LP<br>PROCESS MODEL |
| QUARTERLY PLAN | CRUDE TANKER SCHEDULE<br>PRODUCT IMPORT SCHEDULE | LP |
| MONTHLY PLAN | CRUDE CHARGE SCHEDULE<br>PLANT CHARGE SCHEDULE | LP<br>BALANCE CALCULATOR WITH LP |
| DAILY SCHEDULE | DETAIL PLANT CHARGE<br>SCHEDULE<br>PLANT OPERATION SCHEDULE | BALANCE CALCULATOR WITH LP<br>EXPERT SYSTEM |
| SUPERVISORY<br>PROCESS CONTROL | OPERATION TARGET SETTING<br>PROCESS CONTROL | PLANT<br>OPTIMIZER WITH PARTIAL LP<br>ADVANCED CONTROL WITH LP |
| DIRECT DIGITAL<br>CONTROL | | |

HEAD OFFICE

REFINERY

PRODUCTION PLANNING

As described before, it is easy to linearize refining procedure. So, it is suitable to apply linear programming method to long, middle, and short range planning for the head office. However, as cracking process is not linear, we must linearize cracking models.

There are three LP models for each refinery. These models are base models of production planning. In production plannings made by the head office, combined models with three basic models are used. Coefficients of LP model are revised from actual data of plant operation every year to keep model accuracy.

Short range plan made by the head office is informed to each refinery. This plan is monthly production and inventory balance. Each refinery breaks up into three ten days detailed plans to check feasibility. Also, the tool for refineries short range plan requires the function of case study to calculate the material and inventory balance whenever constraints for plants and tanks

change. Therefore, it is not suitable to use LP, because of its long run time and batch type characteristics. In other words, it is suitable to be balance calculator with friendly man-machine interface rather than optimization purpose as OR method. It goes without saying that it is necessary to optimize partly. Then, balance calculator is combined with LP model.

The details of each production plan are shown in Fig. 2, 3, 4.

And, there is batch production procedure in a part of petroleum refining. This area is lube plant operation. There are many kinds of feed and it is required not to contaminate each products from each feed. In other words, it is required to be short time feed change. Also, in this kind of operation, there are constraints which are very difficult to present quantitative equation. Then, as it is hard to apply LP in this area, expert system combined with balance calculator is used.

Fig. 2.  LONG RANGE FACILITY PLAN

            o PURPOSE         :  TO MAKE FACILITY INVESTMENT PLAN
                                 TO MAKE CRUDE OIL CARGOS CAPACITY PLAN

            o CYCLE           :  5 YEARS AHEAD, YEARLY ROLLING

            o INPUT           :  LONG RANGE PRODUCTS DEMAND FORECAST

            o TOOLS           :  CORPORATE LP MODEL
                                 CALCULATIONS (ex. SUPPLY/DEMAND FORECAST BY INDUSTRY)

            o RESPONSIBILITY:  HQ REFINERY PLANNING DEPT.


Fig. 3.  MIDDLE RANGE PRODUCTION PLAN

            o PURPOSE         :  TO MAKE SALES PLAN (YEARLY)
                                 TO SET YEARLY BUDGET (YEARLY)
                                 TO SELECT & BUY CRUDE OILS (QUARTERLY)

            o CYCLE           :  1 YEAR AHEAD, YEARLY; 3 MONTHS AHEAD, QUARTERLY

            o INPUT           :  PRODUCTS DEMAND FORECAST
                                 FACILITIES AVAILABILITY
                                 INVENTORY TARGET

            o TOOLS           :  CORPORATE LP MODEL
                                 CALCULATORS

            o RESPONSIBILITY:  HQ REFINERY PLANNING DEPT.


Fig. 4.  SHORT RANGE PRODUCTION PLAN

            o PURPOSE         :  TO MAKE CRUDE CHARGE PLAN
                                 TO MAKE PLANTS OPERATION PLAN
                                 TO MAKE PRODUCTS DELIVERY PLAN

            o CYCLE           :  1 MONTH AHEAD, MONTHLY

            o INPUT           :  PRODUCTS DEMAND (MARKETTER'S REQUIREMENTS)
                                 CRUDE ARRIVAL SCHEDULE
                                 OPENING INVENTORY & CLOSING INVENTORY TARGET

            o TOOLS           :  COMBINED REF LP MODEL & REFINERY LP MODEL (HQ)
                                 CALCULATORS (REFINERY)

            o RESPONSIBILITY:  HQ REFINERY COORDINATION DEPT.
                                 REF. COORDINATION & CONTROL SECT.

DAILY SCHEDULING

Short range production plan of refineries is broken up into daily scheduling. The tool for daily scheduling is being developed now.

The functions required for daily scheduling tool are almost same as short range planning tool. Then, we choose balance calculator as daily scheduling tool. Inventory balance as each tank or tank group must be checked. Therefore, man-machine interface is very important because it must display more information in a screen. And, the actual data of the day before scheduling date has to reflect some forecast data based on previous scheduling to keep accuracy of data for starting point of scheduling. Then, the functions added to short range planning tool are daily automatic calculation, constraint check/forecast and capability to display more information in a screen.

APPLICATION OF OR METHOD TO OTHER AREA

As other applications of OR method, there are plant optimizer with process model and partial LP and advanced control with LP.

Plant Optimizer is consisted of process model, identification of model coefficients and LP. At first, model coefficients are identified by actual data. And, this completed process model is linearized at initial value partially. This partial LP model is solved. Partial linearization and LP run are repeated one by one.

In advanced control with LP, LP is applied to control interactive multi variable process.

CONCLUSION

There are many OR methods such as LP, DP, simulation, expert system and so on. It is very important to choose suitable tool. Since several decades age, LP method has been our main OR method, but we began to use expert system recently. We are going to try to apply suitable OR methods to subjects we have now.

Certainly it seems that suitable OR methods are shifting from "hard" OR methods such as LP, simulation and optimization to "soft" OR methods such as expert system and human interface technologies.

# PRACTICAL USE OF OPERATIONS
# RESEARCH FOR PRODUCTION PLANNING

## N. Nishida

*Department of Management Science, Science University of Tokyo, Shinjuku-ku,
Tokyo 162, Japan*

**Abstract.** This paper firstly presents fundamental modelling procedures which
will be employed in production planning in process industries. Secondly, various
optimization problems are discussed, which arise in the hierarchical levels of
production planning. Then possible operations research methods to solve them are
presented. Lastly, since there exist many combinatorial optimization problems in
production planning activities, practical considerations of integer programming
methods are discussed. Negative aspects of practical application of integer
programming are also described.

Keywords. Operations research, Integer programming, Production control,
Production planning.

## PROBLEM-SOLVING IN PRODUCTION CONTROL

In every level of a hierarchy of production
control, various problems to be solved exist. It is
widely recognized that modern management requires
problem-solving talent that is more than a simple
combination of intuition and experience.

To accomplish various activities of problem-solving
successfully, the following major steps have to be
undertaken (Shapiro, 1984).

1. Understanding
2. Formulation
3. Solution
4. Interpretation

1. Understanding. A necessary step before attempt-
ing the formulation of a mathematical model is
understanding. Any study begins by determining the
objectives of the project for the "management" or
the "user". Objectives should be defined in terms
of the decision which is the goal of the study.

2. Formulation. Formulation consists of extracting
from complex, real-world circumstances those
elements and relationships essential to the
operation of the system of interest, and of
decreasing the great number of elements and issues
to those essential to solving the problem. The
first issue of problem formulation is often to
decide whether or not the total problem should be
decomposed into smaller subproblems which can be
solved individually.

The simplification process is important. If the
simplification were not made enough, the solution
effort would be time-consuming and costly. The
worst situation would be that no solution technique
cannot be found because of the complexity or large
size of the model. Too much simplification may
cause the decision makers to be unwilling to use
the results, due to the lack of realism and
accuracy in the model specification.

Fig.1 illustrates the conceptual formulation of
decision process in process industries. Five major
models are indicated:

1. Forecasting model
2. Warehouse or tank inventory model
3. Annual planning model
4. Production planning model
5. Allocation and distribution model

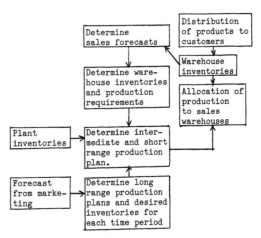

Fig.1 Formulation of decision process in process
industries.

3. Solution. Solution activity follows formulation.
The selection of the appropriate solution technique
is made first based on the characteristics of the
formulated system models. However, without the
knowledge of the solution technique, the task of
formulation cannot be made because the analyst who
does not have enough knowledge about operations
research usually tends to formulate a complex
mathematical model, so that no solution methodology
can handle the model. Therefore, it is necessary to
consider how the model is to be solved numerically
during the model development. This involves the
alternatives, for example, of a simulation approach
or optimization approach, a linear approximation or
a nonlinear model, use of commercially available

solution method or development of a new method.

A special concern with this step is :

1) whether an optimization approach to model
   solution is feasible, or
2) whether a heuristic approach is necessary.

Simulation or                    Optimization
heuristic approach               approach

    Increasing optimality of the solution
    ————————————————————————————————————————→

    Increasing computational time required
    ————————————————————————————————————————→

    Increasing difficulty in interpreting
              the solution
    ————————————————————————————————————————→

Fig.2 Heuristic vs optimization.

As illustrated in Fig.2, if an appropriate optimization approach can be used or developed, the
optimality of the model solution would be better
than that obtained by a heuristic approach. This
statement is true provided that both approaches are
applied to the models having almost the same
complexity. Conversely, the computational time
required to solve the problem with the optimization
approach increases.

Another aspect concerns the choice between:(Agin,
1978)

1. An optimal solution to a simplified model of the
   problem.
2. An approximate solution to an exact formulation
   of the problem.

From a practical point of view, the second choice
is desirable. A complete incorporation of important
elements, even if the resulting solution with a
heuristic approach is nonoptimal, will usually
involve more value than the optimal solution to a
simplified formulation of the problem.

The collection of data follows solution, and the
actual running of the appropriate algorithm comes
next to obtain a numerical solution. The
appropriate collection of data is important, and
the quality or preciseness of data must be
consistent with those of the formulated system
model. Actually, the collection of data is the most
time-consuming effort.

4. Interpretation. The solution results require
interpretation. Whereas formulation requires the
simplification of the real-world system,
interpretation suggests expanding the abstract
meaning of the solution results to apply them to
the actual situation. Interpretation implies
recognizing the simplifications made and testing
the solution results for them.

In addition to the above validation in light of the
formulation, recognizing the key elements which
constitute the solution results is important:

    Why is this solution optimal or approximately
    optimal?
    What are the key constraints in this solution?
    What are the key rules leading this solution?

Only after these questions have been answered can
implementation begin. Although interpretation task

is important in practicing the operations research,
most textbooks do not share many pages with the
interpretation. This is partly because most
optimization techniques except linear programming
and nonlinear programming with restricted
structures do not or cannot furnish the information
for the parameter change (e.g., sensitivity
analysis). In contrast to the optimization
approach, heuristic approach would draw the rules
constructing the solution. Thus, the decision maker
recognizes the answers.

USE OF OPERATIONS RESEARCH FOR PRODUCTION PLANNING

At the higher hierarchical levels, activities of
planning are undertaken. Since the performance to
be monitored is economy, managerial models would be
of help. The manager or decision maker might
describe the interrelation of the system variables
via a set of equations. The equations described,
for example, are supply of feedstocks to the site,
the production process, the distribution of
finished products to the warehouses, and to the
distribution center. Using the managerial model,
planning activities are undertaken by the human
problem solver or decision maker.

Most oil companies as well as the petrochemical
companies use linear programming techniques for the
production planning (Hop, 1989; Murata, 1989).
Since the model size becomes very large, they have
developed utilities such as model generation and
pre-processor systems. The solution is usually
obtained using commercially available packages.
Interpretation of the solution results is made
using information for sensitivity analysis,
typically available from most commercial LP
packages.

Although interpretation of the solution will be
gained, the use of that interpretation might not
directly follow the real-world situation. Because,
the number of parameters associated with the system
matrix, objective function and right-hand side is
too large to look at what happens as they change. A
post-processor will be useful, which summarizes the
results through which the decision maker can gain
insight into underlying factors. Thus he will find
the next task such as which parameter change is to
be tested, what alternative assumptions are to be
tested, and so on.

Furthermore, such post-processor is desired to be
furnished that generates models which follow the
longer range production planning; for example,
those of middle and short range production
planning. These models with more narrow range take
into account of inventories such as warehouses and
tanks as illustrated in Fig.1. In this case multi-
period linear programming is usually applied.

Frequently, the decision maker necessitates
investigating the feasibility of the higher level
production plan, which was given by solving LP of
its model. Such task will be accomplished by
solving the succeeding models of production
planning, which are the multi-period aggregate
model. In this situation, a post-processor which
automatically generates the succeeding models will
be highly helpful.

At the top level of a hierarchy of production
planning, activities are conducted such as the
feasibility studies of construction of new
facilities, or expansion or closing of the current
facilities. The objective of these activities is to
determine the optimal location of supply facilities
(e.g., plants) or demand points (e.g., warehouses).
This is facility (e.g., plant, warehouse) location

model, which may be the most successful application of integer programming.

At the lower hierarchical level, short range production planning are made. In this level the more detail model is required compared to higher levels. Frequently nonlinear models have to be treated for process optimization.

In the chemical industries, a variety-small quantity manufacturing has been enforced in accordance with consumer's demand. The activities of production scheduling has become difficult because, in a given time horizon, a variety of products have to be manufactured with limited resources, such as units, workers, materials, and so on.

The production scheduling is a class of combinatorial optimization problems. The characteristics of the production scheduling are that they have many variables including integer variables and constraints. Thus, it seems that integer programming methods are difficult to apply on the daily scheduling. Actually, industries usually use simulation methods and/or expert systems. Apparently it is ambiguous that whether solutions obtained by these methods are optimal or not. Therefore, it will be necessary to check how far away the obtained solutions are from optimal ones. The integer programming methods may be applied for this purpose. Although a considerable computational time will be consumed, it is worth doing this. Because it is only necessary to do at long intervals; for example, once after the prototype simulators or expert systems have been developed, and then once in several months after they are online.

COMMENTS ON THE PRACTICAL CONSIDERATION OF INTEGER PROGRAMMING METHOD

As described in the preceding section, various combinatorial optimization problems arise in the production planning. Frequently, decision maker faces situations that require models having integer-valued variables. Two obvious situations arise: One requires decisions about indivisible products/resources--for example, workers, machines, vehicles, and so on. Another situation calls for integer variables to be essential in the formulation of any problem with "yes-no" variables or "either-or" constraints. The former case is, in practice, handled by relaxing the integrality restriction, and then rounding the results to the appropriate integer. However, the latter case, where the integer variables are restricted to either 1 or 0, cannot be handled by the same manner as the former case.

There are three general approaches to combinatorial optimization problems:

1. Branch and bound technique.
2. Cutting plane technique.
3. Group theoretic technique.

The branch and bound technique is used today by most commercially available mathematical programming packages (e.g., LINDO). The success of the branch and bound procedure relies highly on

1) giving the initial good feasible solution, and
2) exploiting the good procedure to estimate the optimal objective function value as the solution proceeds.

In addition to these, branching rule on the integer variables is also important. If the algorithm which

successfully could exploit these subjects can be developed, the great reduction of the computational time would be expected. Most commercially available codes allow an initial solution to be input.

As for the second subject, the procedures implemented in most commercial codes simply neglect the integrality restrictions, solve the resulting linear programming problem (relaxed problem) by a simplex method, and let the resulting objective function value be the estimate. When the estimates obtained by the above procedure are very poor, the convergence will become unacceptablly very slow. Thus, such situation sometimes causes the practical use of commercial code to be abandoned.

It is wellknown that good estimates to the optimal objective function values can be frequently attained by applying duality theory in mathematical programming, so that implementing its estimate scheme in the algorithm could shorten the overall computational time.

The relaxed problems, which are usually linear programming problem, are solved repeatedly, and those have the same algebraic structure but different parameters. Thus, reoptimization procedure, if installed, would shorten the computational time.

In addition, the relaxed problems involve a special structure frequently, for example, assignment problem, transportation problem, some network optimization problem, and so on. Obviously, the ordinary simplex method can solve these problems. However, if the specially characterized algorithms for these problems were applied, the great decrease in the computational time could be expected.

As above, there are several methodologies to accelerate the branch and bound algorithm for the solution of integer programming. Unfortunately, most commercial codes do not allow the installation of such optional scheme. In summary, the development of the sophisticated algorithm is a key to the success of the practical application of the branch and bound method to combinatorial optimization problems.

The computer implementation of the algorithm is also very important. Especially, the data structure has to be carefully determined, for example, the computer handling techniques for the sparse matrices. The author's experience shows that considerable difference in the computational time is caused by the computer implementation technique. This trend accelerates as the algorithm becomes complex.

There are several negative aspects in the practical use of integer programming approaches although some are common in the optimization approaches. The followings are crucial among them:

1) Only the optimal solution is obtained.
2) Prohibitive computer time is required for the complex and large size problems.

Suppose that several constraints are not formulated because of, for example, difficulty in the quantitative expression in the model. Thus, after obtaining an optimal solution, it has to be tested for these constraints. If some constraint is not satisfied by this solution and the satisfaction of the constraint is crucial to the real application, the neighboring or suboptimal solutions will be required. However this is difficult because of the inherent nature of the branch and bound method as well as some other optimization techniques.

Inherently, some integer programming problems are difficult to solve, even if some approach (e.g., branch and bound method) can be applied in theory. Furthermore, as the problem size becomes large, even though the model is simple enough to solve, such as traveling salesman problem, an optimal solution cannot be obtained in a reasonable length of computer time.

CONCLUSION

In this paper, solution methods for various optimization problems which arise in production planning were discussed from various aspects; modelling procedure, optimization approach vs heuristic approach, practical consideration of operations research methods, especially, integer programming method.

## REFERENCES

Shapiro, R.D.(1984). Optimization models for planning and allocation. John Wiley & Sons.

Agin, N.I.(1978). The conduct of operations research studies, in Handbook of operations research, J.J.Moder, S.E.Elmaghraby (eds.), Van Nostrand Reinhold Company.

Hop, A.G., A.T.Langeveld, J.Sijbrand (1989). Production management in downstream oil industry, Proceedings of IFAC Workshop on Production Control in Process Industry, 30 October-2 November, JAPAN.

Murata, K. (1989). Application of OR method to production planning, Proceedings of IFAC Workshop on Production Control in Process Industry, 30 October-2 November, JAPAN.

# OPTIMIZATION OF PROCESS PLANT PRODUCTION THROUGH COMPUTER-AIDED-MANUFACTURING

## Using Real Time Data with Rigorous Models to Improve Performance

### J. Clemmons

*ChemShare Corporation, P.O. Box 1885, Houston, TX 77251–1885, USA*

## ABSTRACT

The development of a fast, robust equation based rigorous simulator coupled with the advent of the low cost, high speed micro-computer signals the beginning of a new era in on-line process analysis and optimization. It is now within our reach to integrate the same level of precision employed for design into real-time analysis and optimization of individual processes. The implications for an on-line modeling system applied to day-to-day operating decisions point to a significant change in the way process plants are managed.

Previous process optimization efforts have focused on optimizing the "design". Many engineering manhours and computer processing units have been expended evaluating the trade off of capital dollars versus product recoveries and operating cost, only to see unforeseen economic factors such as product demand and energy cost change the optimum point, in some cases, before plant start-up.

Our current topic focuses on the actual operating plant with fixed arrangement and sizes of equipment to determine on-line what are the best operating conditions to give the most valuable products at the lowest costs.

A research program was begun at ChemShare in the early 1970's to develop an overall process model which was not dependent upon the traditional sequential modular approach of solving one module at a time. It was realized that the sequential modular method was too slow and lacked the robustness necessary for on-line work. A more subtle problem was that due to multiple internal convergence loops, solutions sometimes were not accurate enough to provide the extremely smooth response surfaces needed for multi-variable optimization.

In 1974, we introduced the first simulation program designed to solve the interlinked distillation columns found in refinery crude units as a single device rather than several linked through recycle algorithms. In 1976, our heat transfer optimization program become the first commercial program to solve and optimize the design of a network of heat exchangers simultaneously.

Research along these lines continued and by 1980 we could simulate entire processes with rigorous models solved simultaneously through an equation based technique. We have continued to enhance this technique, improve robustness and incorporate an optimization routine based upon a user defined economic function.

Concurrent with our development activities has been the evolution of the micro-computer and distributed control systems. Micro-computers have decreased in cost while increasing in memory and calculation speed, eliminating previous limitations to performing on-line optimization with rigorous models.

Distributed control systems continuously monitor and electronically log mounds of process data here-to-fore unattainable. The process unit data base is now directly accessible in a "real-time" mode.

Process optimization involves six principal elements: 1) data reconciliation of process measurements, 2) rigorous mathematical model of the process, 3) determination of mechanical constraints, 4) economic objective function, 5) multivariable optimization analysis and 6) implementation by the control system of the new manipulated variables.

Data reconciliation is a statistical procedure which enables the adjustment of data so that weighted differences between calculated and measured values are minimized subject to the modeling equations. Selected measurements are used as input to the model. Differences are compared between the calculated values based on the model inputs and the measured values. Gross errors are detected, eliminated from the measurements and reported for operator attention. The objective is to determine the best set of adjusted process data which satisfies the energy and material balances for the process in conjunction with the thermodynamic and equilibrium laws of nature.

An exact or rigorous model of the process is absolutely necessary to accurately evaluate the multivariable interactions in a complex process plant. Short-cut models do not have the range of accuracy and are often based on engineering estimates or specific correlations that are valid only for a narrow range of conditions.
Rigorous modeling also provides for a true evaluation of the mechanical constraints of a process. Such items as distillation tower flooding, heat exchanger performance, compressor horsepower, etc. are continuously checked during the optimization calculations to limit the solution to a feasible action. The optimized solution will most always be constrained at one or more of the mechanical limits of the process. As the economics of the process change the optimization will move the operation to a new point which most likely will be constrained by a new mechanical limit. Approximate models by their intended nature are not capable of the exact analysis necessary for this type of evaluation.

The economic objective function used to optimize the operating conditions can be as unique as each company's operations. Factors which are normally included in the function are feed and product economics, utility expenses and any special contractual obligations such as take or pay arrangements. These factors may be structured as functions of market prices, transfer prices, fuel values or other influences. Whatever the combination of factors they should be consistent with the company's overall business plan.

After the process data are reconciled, the model confirmed, mechanical constraints verified and the components of the economic objective function established, the true power of the equation based simulator is revealed in the multivariable interaction analysis which seeks the conditions to maximize the economic objective function. Once the optimum

operating conditions are determined, they are reported to the operator to be implemented by the distributed control system (DCS). In most instances advanced process control (APC) techniques are required to permit manipulated variables to be controlled against mechanical constraints. The benefits inherent to advanced process control are enhanced by overall process unit optimization. Current APC's are capable of optimizing limited windows in a process through approximate models. Real time optimization with rigorous models allows, for the first time, an overall process perspective. Multivariable interaction throughout the process can be predicted by an exact model. The APC handles real time disturbances - rain shower, cold front, etc. - dynamic situations. "Real time" optimization projects steady state targets for maximizing profits.

We have discussed how the system manipulates operating variables to improve operations, but how does it perform on-line the steps necessary to produce the results?

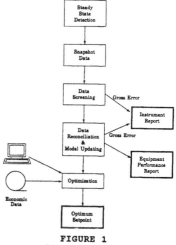

**FIGURE 1**
**BLOCK DIAGRAM**

An on-line run as shown in Figure 1, begins by monitoring key process variables as a function of time to determine steady state. Once steady state is determined, a time average set of process measurements is taken from the process. These initial values are screened for obvious gross errors and suspect measurements are identified and set aside. The screened data is then reconciled against the rigorous model of the process. Additional suspect measurements are identified and combined with the original list and reported as Bad Instruments. Data reconciliation also provides a list of equipment information pertaining to fouling factors, catalyst activities and others.

Once the data are reconciled they can be combined with current economic data to perform multi-variable optimization. Optimum conditions are reported to management for implementation by the control system.

Once a new steady state is reached the system is ready to begin again.

We have recently installed this system at a 240 million standard cubic foot per day (MM scfd) Natural Gas Liquids (NGL) - Nitrogen Recovery Unit (NRU).

The plant consists of two identical fractionation trains, requiring modeling a total of over 100 pieces of equipment. The heat pump and refrigeration fluids are common to both trains requiring the simultaneous solution of both trains to determine system interactions. Over 450 process measurements are reconciled to the rigorous model.

Each train consists of an NRU column, expander/compressor, demethanizer column and cold boxes. The cold units consisting of brazed aluminum plate-fin heat exchangers. All operations are rigorously modeled in ProCAM including thermal rating of the cold box exchangers.

The hardware configuration, Figure 2, for this installation consists of two process computers. A MicroVAX II operating in VMS and an Apollo DN 10000 operating in UNIX. Communication is accomplished through an ethernet system by Excelan.

Data is obtained from the DCS by a plant information package which resides on the MicroVAX. The ProCAM executive program communicates with the plant information package monitors for steady state and manipulates input and output files is also on the MicroVAX.

The Apollo DN 10000 is a 30 MIP machine which handles the actual modeling calculations. All models reside on the Apollo. Access is also provided via modem for headquarters and remote locations to check plant status. Results are displayed in summary tables on the CRT's with detailed hard copy

printouts available for trouble shooting and analysis.

While the system has only been on-line a few months it is clear that the payout will be less than one year. Data reconciliation alone has uncovered several operational inefficiencies that only rigorous modeling could detect.

An additional benefit of on-line optimization is the capability to do off-line "what if" studies using real-time data. The same tool which analyzes current operations can be used to analyze a feed stock change before it is purchased or the effect of anticipated product pricing changes.

There are additional implications for marketing, supply, maintenance, planning and just-in-time (JIT) production.

For the future, the simultaneous equations solving techniques used in on-line optimization make it an excellent foundation for building exact dynamic models for processes. Research studies have already demonstrated its viability for dynamics. When dynamics programs can be executed economically on high availability computers, exact dynamic models will provide the ultimate process control method.

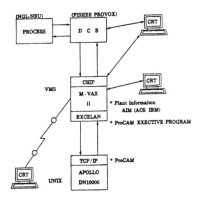

**FIGURE 2**
**HARDWARE CONFIGURATION**

# DISCUSSION

## Sesson 1: Production Planning

*Chairmen*
**C. McGreavy**
*Leeds University, UK*
**T. Ibaraki**
*Kyoto University, Japan*

This session covered a range of themes related to production planning which extended from broadly conceptual principles and structural relationships to particular methodologies and algorithms. It was useful particularly to have perspectives from Europe, the United States and Japan discussed in the three papers since they provided an interesting insight to some of the problems surrounding technical human interface which introduced the session. There was a general consensus that the hierarchal structure of the decisions and operations required a top down interpretation, although the evolutionary pattern which was responsible for developing this would inevitably be based on a bottom up approach. What was clear was that there was a crucial need to ensure that the appropriate level and form of information was effectively integrated into the system. In particular, the interface between the hardware and software systems and the operators and managers.

There was an appreciation that to assist the human in handling the increasingly complex problems required a comprehensive range of software and procedural tools which provided a framework, not only for implementing sophisticated algorithms, but would integrate the necessary decision support systems which enable the human participants to play a more active and decisive role. Integration of activities, whether computer based or management driven was seen as fundamental, but at the same time, there must be an evolutionary path forward which necessarily poses difficulties in maintaining integrity in highly bound systems.

Two specific issues which emerged and which provided the focus for much of the discussion related to human factors in system development and state of the art in application of expert systems.

Human factors such as communication between system engineers and operators, and the willingness to cooperate in a traditionally departmentalized organization, are regarded as vital for the successful production management. In order to achieve the necessary system integration and cooperation at all levels, considerable discussion is required to convince higher management levels of the importance of information technology. They should be encouraged to participate in a workshop with the managers responsible for production to try to find solutions jointly, rather than simply providing them at the production manager level so they feel that they not only own the problem, but also are finally the owner of the solution.

Expert systems were seen as playing an important role and examples were very actively discussed, although not everyone was convinced of their value. Most of them are still for small scale applications, especially in relation to production scheduling, which involve combinatorial optimization problems. Characteristically, production scheduling involves situations where there are many variables including integers as well as constraints. However, experience suggests that integer programming methods are difficult to apply to daily scheduling. Consequently, industry usually uses simulation methods and/or expert systems. Unfortunately, it is not known whether solutions obtained by these methods are optimal or not, so some exploration to evaluate the significance of the solutions is required. The discussion pointed to the fact that usually such an evaluation is not made. Generally the results obtained by the expert system are simply compared with those suggested by human experts. The application of integer programming methods to provide an algorithmic solution was suggested. After developing a prototype expert system it is considered essential to compare solutions after a long time interval. Unless this is done it is not possible to make an assessment as to whether the solutions are in any sense optimal.

The additional contributions dealt with a number of specific techniques and approaches in particular industries. These were useful case studies which gave a useful appraisal of the effectiveness of the methodologies in particular situations. In this sense, the discussion was largely concerned with examining how much generalization of this experience could be expected to apply to other situations. There was a clear message that resolving these issues was an important step forward in moving towards a fully computer integrated manufacturing system.

# ON-LINE SCHEDULING FOR
# A POLYMERIZATION PROCESS

## Y. Yamasaki,* T. Iio,* Y. Nojima* and H. Nishitani**

*Engineering Research Laboratory, Kanegafuchi Chemical Industry Co. Ltd.,
Takasago, Hyogo, Japan
**Department of Information and Computer Sciences, Osaka University, Toyonaka,
Osaka, Japan

**Abstract.** A total production control system has been developed for a polymerization plant. This system is composed of three subsystems, long-range planning, re-scheduling and real-time process control system. A hierarchical structure of these subsystems can treat data sampled at different intervals. Especially, re-scheduling system based on the rolling scheduling plays a key role for on-line scheduling. By using this system, the effects of uncertainties or unexpected changes in both the market and production system can be avoided over the planning horizon. This total production control system can be effectively used in both sales and production departments. This system is the first stage of computer integrated manufacturing and management system (CIM).

**Keywords.** Chemical industry; Production control; Production scheduling; On-line scheduling; Management system; Hierarchical system.

## INTRODUCTION

In a polymerization plant, many products are produced on several lines. So far, production plan has been determined weekly by a project engineer with his experiences. A plan determined in this way has been good for stable plant operation. But this may be not so good from the viewpoint of strategic production.

There are many kinds of uncertainties or unexpected changes in both market and production system such as changes in delivery due date and troubles in facilities. To cope with these uncertainties, a total production control system including information of market and production plant is necessary.

Necessary information depends on the person in charge. For example, a sales manager, a plant manager and a plant operator will ask for different accuracy of the information. Under these situations, a hierarchical structure of information based on the sampling period is useful (Cott and Macchietto, 1988). This also means a hierarchical structure of subsystems in the total production control system, which is composed of long-range planning, re-scheduling and real-time process control.

In a new total production control system, various information is transmitted at different time intervals. On-line scheduling based on the concept of the rolling scheduling (Baker, 1977,1979) is applied in the re-scheduling system to increase the real-time property of the system.

## A POLYMERIZATION PLANT

The schematic flowsheet of a polymerization plant considered in this paper is shown in Fig. 1. Each production line has several batch reactors, a dryer and tanks for intermediates and finished products. Usually, each reactor belongs to a specified line as shown in Fig. 1. However, some reactors such as reactor #10 and #14 may be used on different lines. The production rate in a reactor depends on the product.

Dozens of products are made in this plant. All the products can be grouped in accordance with product quality. Table 1 shows product family grouping and their assignment to production lines. Each product family is usually produced on a specified line as summarized in Table 1. However, product family g is also produced on line 2 when production volume of product family a and b is over the capacity of line 1.

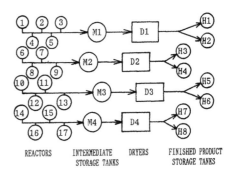

REACTORS  INTERMEDIATE  DRYERS  FINISHED PRODUCT
STORAGE TANKS      STORAGE TANKS

Fig. 1  The schematic flowsheet
of a polymerization plant.

For this polymerization plant, a total production control system is required, in which both sales and production information will be effectively used to manage the whole activities related to production. Figure 2 illustrates the information flow in the polymerization plant. There are mainly three subsystems, i.e., the physical distribution system, the planning/scheduling system and the real-time process control system. The downward information

TABLE 1    Product Family Grouping and its Assignment to Production Lines

| Product Family | Products | Production Volume | Production Line |
|---|---|---|---|
| a | A1,A2,A3,A4,A5,A6,A7,A8 | small | 1 |
| b | B1,B2,B3,B4,B5,B6,B7 | small | 1 |
| c | C1,C2,C3 | large | 4 |
| d | D1,D2 | large | 3 |
| e | E1,E2 | medium | 2 |
| f | F1,F2,F3 | medium | 2 |
| g | G1,G2,G3 | small | 1,(2) |
| h | H1,H2 | medium | 2 |

flows are commands from above such as shipment, production schedule, processing sequence and set-point of the process control. The upward information flows are collection of measured or processed data. They are sampled at different intervals. For strategic production, these three subsystems should be well organized.

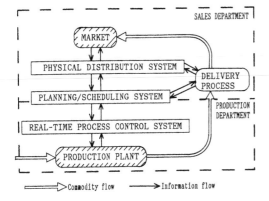

Fig. 2    Information flow in a polymerization plant.

Up to now, these three subsystems were treated in two departments. Sales department takes charge of physical distribution and data for planning and scheduling. Production department is responsible to make a production plan and a schedule and to implement them by using the real-time process control system. Strictly speaking, communication between these two departments was intermittent. But recently the concept of computer integrated manufacturing and management (CIM) recommends more close communication among these departments.

## PRODUCTION CONTROL SYSTEM IN A NARROW SENSE

As a production control system, we proposed three level structure shown in Fig. 3. This structure can coordinate long-range planning, re-scheduling at a short interval, and real-time process control.

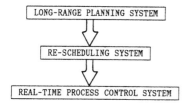

Fig. 3    Hierarchical structure of the production control system in a narrow sense.

In the long-range planning system, the production plan is determined for a long-range period at the same interval as the planning horizon, for example a month. In the re-scheduling system an enforcement schedule is determined at a short interval, for example a day. And in the real-time process control system the process states are sampled at an interval of seconds to minutes and these sampled data are used for real-time process control. In the following section the characteristics of each subsystem are described.

### Long-Range Planning System

This system is the conventional planning system. A production plan is made in a feedforward manner. This is a basic plan for other subsystems. For example, it is transmitted to the Material Requirements Planning (MRP) system and the Physical Distribution System (PDS). So, it is a basis for the activities of a delivery planner and a sales man.

The prediction data of sales, facilities, and resources over the horizon is necessary for this subsystem. For example, demand volume and production volume of each product, capacity of each facilities, productivity of each line and a schedule of shut-down for maintenance should be predicted over the corresponding horizon.

Based on these information, this system generates a basic production plan including the assignment of products and reactors to all lines. The predicted results of the changes in both the line load and inventory are presented to sales and production departments.

Generally, the planning horizon should be dependent on production volume. Product family c and d should have a short horizon of a month, and product family a, b, and g a long horizon of a few months. Especially, product B6 and B7 should have a horizon of a year, because these products are usually supplied only in winter. However, a month was selected as the planning horizon for all lines to avoided complicated treatment.

### Re-Scheduling System

This system is to realize the on-line scheduling. The re-scheduling system is very important under a lot of uncertainties in the manufacturing. Required information for re-scheduling is basically same as for the long-range planning system. Therefore, both systems can use the same algorithm. From the real-time process control system, we can get some information which indicates progress of batch sequence, actual amount of production volume, and process states such as volume in tanks and current capacity of facilities. More reliable information on the due date and volume of shipment are available from the updated data file.

In general, as the period of re-scheduling the shorter is the better in regard to attenuate the various kinds of disturbances. But some data have limits in the sampling interval, especially for market information. Also the larger computational power will be required for a shorter period of re-scheduling. In the case of the polymerization process, one day is adopted as the period of the re-scheduling.

The output of this system is a production schedule for a dryer on each line. This result is used in actual production operation.

A plant manager and an operator have communication to this system through a man-machine interface. Their decisions are made based on the enforcement schedule every day. A sales man will confirm the amount of inventory and will predict changes of inventory. As a result, he will have correcter information for sales. In some cases, he will claim modification of the production plan.

## Real-Time Process Control System

This system works as the executor of the commands from the re-scheduling system. Functions in this system are the conventional process control system such as batch sequence control, PID control and so on.

### PRODUCTION SCHEDULING METHOD

We developed a production scheduling method for the polymerization plant, which is composed of four stages (Yamasaki and co-workers, 1988).
1. line balancing
2. production volume planning
3. production line scheduling
4. operation scheduling

Figure 4 shows the flowchart of the production scheduling procedure.

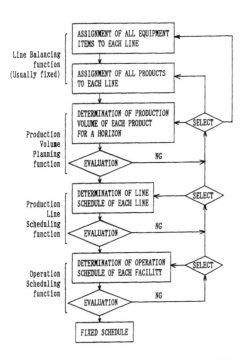

Fig. 4 Flowchart of the production scheduling procedure for a polymerization plant.

## Line Balancing Stage

All equipment and all products should be assigned by taking account of the load of each line. This line balancing is responsible for a plant manager and a planner.

## Production Volume Planning Stage

The accumulated production volume of product i, $P_{iT}$ over a horizon, T is calculated as follows:

$$P_{iT} = D_{iT} + [ X_i(T) - X_i(0) ] \qquad (1)$$

where
$D_{iT}$ : accumulated demand of product i at the end of horizon T
$X_i(k)$ : inventory of product i at the end of period k  ( k = 0,1,·····,T )

Once inventory at the end of the horizon, $X_i(T)$ is specified by taking into account a rational margin, the accumulated volume of production over the horizon, $P_{iT}$ is calculated. In the long-range planning system, the accumulated demand for i-th product over the horizon, $D_{iT}$ is given as a forecast by the sales department. In the re-scheduling system, the accuracy of this forecast will be improved by using the trend in the actual demand until the day before the re-scheduling date. Calculated $P_{iT}$ is checked for the line capacity and if necessary, the production volume planning will be modified or the line balancing is repeated.

## Production Line Scheduling Stage

Production period of each product is determined for each line. There are some scheduling method in the area of Operations Research. We have proposed a heuristic method by using the accumulation curves (Yamasaki and co-workers, 1988). This method is useful to evaluate the dynamic behavior of inventory for each product over the scheduling horizon. Therefore, tardiness of due date or short supply can be eliminated from the resulting schedule. A basic equation of the accumulation curve method is as follows:

$$P_i(k) = D_i(k) + [ X_i(k) - X_i(0) ] \qquad (2)$$

where
$P_i(k)$ : accumulated production of product i at the end of period k
$D_i(k)$ : accumulated demand of product i at the end of period k

A set of [$P_i(k)$,k=1,2,·····,T] indicates a production schedule for product i. A proposed schedule should be evaluated from multiple viewpoints such as inventory and changeover costs. If the schedule is not satisfactory, return to the upper stage.

## Operation Scheduling Stage

At this stage, the schedule of operation including equipment and personnel is determined and loading patterns of utilities and labour are evaluated. If some limitations are violated, the schedule of operation is changed under the fixed production sequences of products on the line, first. If there is no solution by shift of operation, the schedule of line should be changed.

### AN ON-LINE SCHEDULING SYSTEM

There are many kinds of uncertainties or unexpected changes in both market and production plant in Fig. 2. Factors of uncertainties in the polymerization plant are summarized in Table 2. Uncertainties in market include errors for the forecast of demand. Uncertainties in production plant include the changes in operating facilities and changes in the production capacity. So far, changes of due date or delivery volume and decrease of production capacity by accident of hardware often occurred in this polymerization plant. To cope with these obstructions, the re-scheduling system in the previous section is effective. This is based on a good real-time property of the system. By using this system, the effects of some disturbances can be attenuated over the planning period and bad effects in future can be avoided. The same concepts were found in some papers.

TABLE 2    Factors of Uncertainties or Unexpected Changes in a Polymerization Plant

1. UNCERTAIN FACTORS in MARKET

   (1) change of due date  ( postponement or shortening )
   (2) change of delivery volume  ( increase or decrease )
   (3) commercial or sample shipment to new customer

2. UNCERTAIN FACTORS in PRODUCTION PLANT

   (1) shut-down by accident of equipment
   (2) decrease of production capacity by accident of equipment
   (3) short of product volume by accident of product quality
   (4) addition of experimental production out of initial plan
   (5) shortening of line-up of new product
   (6) error between predictive capacity by model and actual one

Baker (1977,1979) proposed a rolling scheduling to solve a multiperiod model. A schedule for the multiperiod is solved at a small interval but its implementation is done for only the first period. One time period later, the model is updated by using new information and the same procedure is repeated. Onogi and co-workers (1986) also proposed an on-line scheduling system for a class of combined batch/semi-continuous process.

A TOTAL PRODUCTION CONTROL SYSTEM

Block diagram of the total production control system for the polymerization process is shown in Fig. 5. The long-range planning system makes a basis of production schedule. In the re-scheduling system an enforcement scheduling is made every day. The information flow makes a closed loop of the real-time process control system and the re-scheduling system.

The total production control system is structured hierarchically to treat data sampled at different intervals. The long-range planning system may be performed once a month. The re-scheduling system may be repeated at an interval of a day and the real-time process control system may be acting at an interval of seconds to minutes.

In the production scheduling system in the narrow sense, a few days for production volume planning, one day for production line scheduling and a few minutes for operation scheduling are suitable as the sampling periods in the polymerization process.

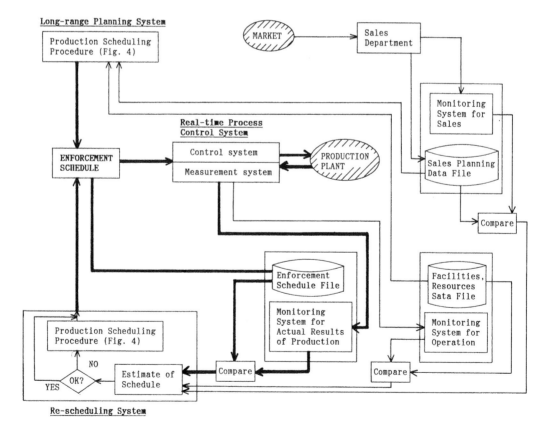

Fig. 5  Block diagram of the total production control system for a polymerization process.

On the other hand, uncertainties or unexpected
changes in Table 2 will also occur in different
frequency. Each uncertainty or unexpected change
should be treated in the suitable subsystem with an
appropriate sampling period. On-line scheduling
system with the hierarchical structure presented
here will be well adapted for this purpose.

### ILLUSTRATIVE EXAMPLE

A production volume plan in the long-range planning
for line 4 is presented in Table 3. As the horizon
for the long-range planning, a month (30 days) is
adopted. A production line schedule in the Gantt
chart for line 4, which is made by the
accumulation curve method, is shown in Fig. 6.
Let consider that we will get the information
about a change of delivery volume of product C3 on
the 18th day. It is an increasing of 100 tons
during 4 days from the 26th to the 30th day.

TABLE 3  Production Volume Planning
in the Long-range Planning
for Line 4 ( T=30 days )

| Product, i | C1 | C2 | C3 |
|---|---|---|---|
| Monthly production volume, $P_{i,T}$ [ton] | 2200 | 300 | 250 |
| Predicted demand for a week, $D_{i,T}$ [ton] | 2700 | 400 | 300 |
| Initial inventory volume, $X_i(0)$ [ton] | 750 | 180 | 100 |
| Required production period [day] | 23.1 | 3.4 | 3.0 |

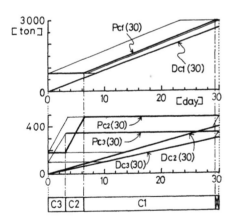

Fig. 6  Production line scheduling
in the long-range planning
for line 4.

So far in the conventional method, the schedule is
adjusted for this unexpected change on the 18th
day for the remaining 12 days. The result of re-
scheduling for 12 days is shown in Fig. 7. An
extra changeover between C1 and C3 and short
production of product C1 will be caused.

On the other hand, the re-scheduling system based
on the rolling scheduling will enable better
treatment for the disturbance on the 18th day.
Table 4 indicates the production volume planning
for the re-scheduling on the 18th day. The

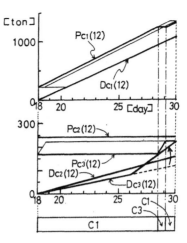

Fig. 7  Production line scheduling
in the re-scheduling for line 4
on the 18th day.
( conventional method )

TABLE 4  Production Volume Planning
in the Re-scheduling for Line 4
on the 18th day ( T=30 days )

| Product, i | C1 | C2 | C3 |
|---|---|---|---|
| Monthly production volume, $P_{i,T}$ [ton] | 2100 | 300 | 300 |
| Predicted demand for a week, $D_{i,T}$ [ton] | 2000 | 400 | 350 |
| Initial inventory volume, $X_i(0)$ [ton] | 230 | 240 | 170 |
| Required production period [day] | 22.1 | 3.4 | 3.6 |

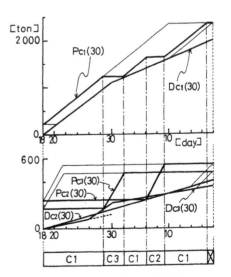

Fig. 8  Production line scheduling
in the re-scheduling for line 4
on the 18th day. ( new system )

resulting re-schedule is shown in Fig. 8. There is no short production and an excessive changeover in the line schedule can be avoided.

The reason of this result is that in the conventional method the period for the re-scheduling is limited by the date when the disturbances enter, but in the new re-scheduling system the period for the re-scheduling is the same as the long-range scheduling. Of course, it is necessary that the information over the horizon should have good accuracy.

### CONCLUSIONS

A total production control system has been developed for a polymerization plant. This system is composed of three subsystems, long-range planning, re-scheduling and real-time process control system. These subsystems are structured hierarchically. The re-scheduling system has a good on-line property based on the concept of the rolling scheduling. And this on-line property works effectively for uncertainties or unexpected changes in both market and production. Accordingly, this system is effectively used by a sales manager, a plant manager and a plant operator.

A same production scheduling procedure is adopted in both long-range planning and re-scheduling system. This procedure is composed of four functions which are (1) line balancing, (2) production volume planning, (3) production line scheduling and (4) operation scheduling. The first and second functions are closely related to both sales and production management. The third function is a key function for all sections. The forth function is related to actual plant operation.

The features of our system is a global application of the on-line concept to the whole production related system such as plant management and market. This total production control system aims at a CIM system. A real system for an existing polymerization plant is in the midst of development.

### ACKNOWLEDGEMENT

We dedicate this paper to Dr. Morikawa, to whom we look up as our preceptor, of glorious memory. He was a General Manager of Engineering Research Laboratory, Kanegafuchi Chemical Industry Co. Ltd. and died in February in 1989.

### REFERENCES

Baker, K. R. (1977). An experimental study of the effectiveness of rolling scheduling in production planning. Decision Sciences, 8, 19-27.

Baker, K. R., and D. W. Peterson (1979). An analytic framework for evaluating rolling schedules. Management Science, 25, 341-351.

Cott, B. J., and S. Macchietto (1988). An integrated approach to computer-aided operation of batch chemical plants. Proceedings of International Symposium of Process Systems Engineerings, PSE'88, Sydney, Australia, 243-249.

Onogi, K., Y. Nishimura, Y. Nakata, and T. Inomata (1986). An on-line operating control system for a class of combined batch/semi-continuous processes. J. Chem. Eng. Japan, 19, 542-548.

Yamasaki, Y., H. Morikawa, and H. Nishitani (1988). Production scheduling system for a batch process. Proceedings of International Symposium of Process Systems Engineerings, PSE'88, Sydney, Australia, 250-256.

# A GENERAL SIMULATION PROGRAMME FOR SCHEDULING OF BATCH PROCESSES

### S. Hasebe and I. Hashimoto

*Department of Chemical Engineering, Kyoto University, Kyoto 606, Japan*

Abstract.    Proposed here is a simulation algorithm which can be widely used in
determining the starting moments of operations for general batch processes.  In this study,
the processing order of jobs on every batch unit is assumed to have beed determined
before carrying out the simulation.  When we determine the starting moments of jobs on a
batch unit, many kinds of constraints such as the constraints on the storage policy and the
working pattern must be considered.  In order to take these constraints into account, the
operations to produce a batch of product are divided into a sequence of basic operations
such as filling, processing and discharging.  Then, the simulation algorithm to find the
earliest starting moments of these basic operations is developed.  In this study, a variety of
constraints are classified into four groups in order to make the characteristics of constraints
clearer.  The simulation algorithm proposed here consists of four stages, and one of the four
types of constraints is incorporated into each stage of the algorithm.  The proposed
simulation algorithm derives the strict optimal schedule if there are no constraints other
than those on the waiting period and the working patterns.

Keywords.    Production control; Scheduling system; Batch process; Multi-purpose plant,
Simulation;

## INTRODUCTION

In recent years, an increasing number of products
have been produced in batch processes in order to
satisfy diversified customers' needs.  The demand for
keeping deadlines for delivering products has also
been increasingly strict.  As a result, it has become
extremely difficult to manually generate operation
schedules.    The development of computer aided
scheduling systems is strongly desired in order to
quickly generate and revise appropriate operation
schedules.

One of the predominant characteristics of batch
processes is that the material leaving a batch unit is
fluid, and it is sometimes chemically unstable.
Therefore, the starting moments of operations must
be calculated by taking into account the storage policy
between two operations.  Furthermore, the operations
which can be carried out at night or over weekend
are limited and the simultaneous execution of some
operations may be prohibited.  So, even if the
processing order of jobs on each batch unit is fixed, it
is very difficult to determine the optimal starting
moments of jobs on each unit which satisfy these
constraints.

Recently, much effort has been devoted to develop
algorithms to determine the starting moments of
operations by taking into account several types of
interstage storage policies.  For a serial multiproduct
batch plant, Suhami and Mah (1981), Wiede and
Reklaitis (1987) and Rajagopalan and Karimi (1989)
proposed calculation algorithms in which the
completion time of the final job (makespan) is
minimized.    Egri and Rippin (1986) proposed a
simulation algorithm in which they considered the

due date of each job, the utility constraints, the
restrictions of the operations during the weekend and
several types of storage policies.

In this study, a variety of constraints are classified
into several groups in order to make the
characteristics of constraints clearer.  Then, a
simulation algorithm, which can be widely used in
determining the starting moments of operations for
general batch processes, is proposed.

## FORMULATION OF THE PROBLEM

### Scheduling System

The objective of scheduling systems is twofold: one is
to determine the sequence in which the products
should be produced (sequencing problem), and the
other is to determine the starting moments of various
operations such as charging, processing and
discharging on each unit (simulation problem).

There are two ways to solve the scheduling  problem.
One is to solve both the sequencing and the
simulation problems simultaneously.    Kondili,
Pantelides and Sargent (1988) formulated the
scheduling problem as an MILP and solved both
problems simultaneously.  They proposed an effective
branch and bound method but the problems which
can be treated by this formulation are still limited.

The other way is to solve the sequencing and the
simulation problems separately.  Figure 1 shows an
example of such a scheduling system.  Each $g_i$ in the
figure shows one of the processing orders of jobs on
every unit.  Hereafter, it is called "a production
sequence".  This system derives a good schedule by
the following steps:  At first, an initial production
sequence $g_0$ is generated.  Then, a set of production

sequences is generated by changing the production orders of some jobs for the production sequence prescribed by $g_0$. For each production sequence $g_i$, the starting moments of jobs and the performance index are calculated by using the simulation programme. The most preferable sequence in the generated production sequences is regarded as the initial sequence of the recalculation, and the modification of the production sequence is continued as far as the sequence can be improved.

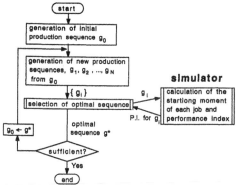

Fig. 1. Scheduling Algorithm of Batch Processes

The feature of this system is that the generation of production sequences and the calculation of the starting moments of jobs are completely separated. Therefore, we can develop each of these two algorithms independently without taking into account the contents of the other algorithm. In the previous paper (Hasebe, Hashimoto and Takamatsu ,1988), we proposed an effective method to generate a set of $g_i$.

## Assumptions

The aim of this paper is to propose a simulation algorithm which can generally be used in the simulator shown in Fig.1. It is assumed that the process satisfies the following constraints:
1) A production demand to make a batch of product is called "a job". Every job to be processed is already known before carrying out the simulation.
2) Jobs have different production paths, i.e. a multipurpose process is considered, and in some cases, each job can take one of several production paths. However, the production path of each job and the production order of jobs on every unit are determined by the sequencing programme and fixed before carrying out the simulation.
3) Operations of a job must satisfy many kinds of constraints to be explained in the following section.
4) The performance index is a monotonically increasing function of the completion time of each job. In other words, each job should be processed as soon as it satisfies the constraints.

## Basic Operations

The production on a batch unit consists of operations such as filling the unit, processing the materials, discharging and cleaning for the next batch. Each of these operations is hereafter called "a basic operation". In many cases, it is possible to insert a waiting period between two basic operations being successively operated. Therefore, in order to calculate the completion time of each job, the relationship among the starting moments of basic operations must be considered.

The procedure to produce a batch of product can be

expressed by a sequence of basic operations the order of which is given a priori. Therefore, the precedence relationship among all of the basic operations is determined by deciding the production sequence. Figure 2 shows an example of the processing order of jobs i and j on batch units m and n. A node and arrows in the figure indicates a basic operation and the processing order of basic operations, respectively.

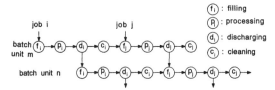

Fig.2 Processing Order of Basic Operations of Two Jobs

The problem discussed here is to determine the optimal starting moments of basic operations on every unit which have to satisfy many kinds of constraints.

## CLASSIFICATION OF THE CONSTRAINTS

In this section, we classify various constraints into four types, and explain the factors which should be considered in defining each type of constraint.

### Constraints on Waiting Period.

The processing order of basic operations can be expressed by a graph such as shown in Fig. 2. The arrow from basic operation i to basic operation j, $i \rightarrow j$, means that there exists some relationship between the starting moments of basic operations i and j. In some cases, operation j must be started as soon as operation i is completed, and it is sometimes possible to take some waiting time between two basic operations. These relationships can be expressed by the following inequalities:

$$t_i + h_{ij} \leq t_j \leq t_i + h_{ij} + h'_{ij} \qquad (1)$$

where

$t_i$ : starting moment of basic operation i,

$h_{ij}$ , $h'_{ij}$ : time determined as a function of basic operations i and j.

The value of $h'_{ij}$ depends on the chemical stability of the materials. If two basic operations must be executed successively without taking any waiting time, $h_{ij}$ is equal to the processing time of basic operation i and $h'_{ij} = 0$. If two basic operations must be started simultaneously, then $h_{ij} = h'_{ij} = 0$.

Four types of interstage storage policies have been discussed in the literatures (Suhami and Mah, 1981 ; Egri and Rippin, 1986 ; Wiede and Reklaitis, 1987 ; Rajagopalan and Karimi, 1989):
(a) Unlimited number of batches can be held in storage between two stages, (UIS).
(b) Only a finite number of batches can be held in storage between two stages, (FIS).
(c) There is no storage between stages, but a job can be held in a batch unit after the processing is completed, (NIS).
(d) A material must be transferred to the downstream unit as soon as the processing is completed, (ZW).

By assigning proper values to $h_{ij}$ and $h'_{ij}$ in Eq.(1), it is possible to express not only UIS, NIS and ZW storage policies but also some other storage policies.

When FIS storage policy is taken between two batch units, we must consider many factors such as the capacity, the name of currently stored material and its storage level of each tank in order to determine the filling time for the storage tank. In this case, it is very difficult to express the relationship between the starting moments of two basic operations by using simple inequality constraints. Therefore, FIS storage policy is dealt with separately and discussed in a following section as a different type of constraint.

### Constraint on Working Patterns

The second type of constraint is the restriction with respect to the processing of particular basic operations during a fixed time period. This type of constraint is hereafter called "the constraint on working patterns". In order to make this type of constraint clearer, we show several examples.

(a) The discharging operation cannot be executed during the night.
(b) Every kind of operation cannot be executed during the night. It is possible to interrupt the processing temporarily, and the remaining part of the processing can be executed next morning.
(c) Every kind of operation cannot be executed during the night. But it is possible to hold the material in a batch unit.
(d) Unit i cannot be used between $t_1$ and $t_2$ because of an overhaul.
(e) Processing operation of job k must be started between $t_1$ and $t_2$.

Figure 3 expresses the schedule which satisfies each of the above constraints. In this figure, the available starting moments of filling operation are all the same, but the scheduling results are completely different one another. In order to distinguish these examples from one another, the following 6 factors are required:
(i) Restricted period,
(ii) Cycle time of the restricted period. If the constraint is effective only once, this value is set equal to zero in order to denote that it is a temporary constraint.
(iii) Restricted batch units,
(iv) Restricted operations,
(v) Pattern of the restriction,
  s : starting of the operation is prohibited,
  p : processing of the operation is prohibited.
(vi) Availability of the storage
  y : material can be held in the unit,
  n : material cannot be held in the unit.
The availability of the division of a basic operation depends on the stability of the material. Therefore, the parameter which indicates the availability of the division of a basic operation should be assigned to each basic operation.

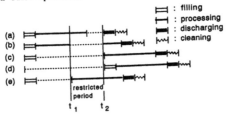

Fig. 3 Schedules for Various Types of Working Patterns

Constraints on working patterns can be expressed by using the above factors. For example, case (c) is

expressed as follows:
  { [ $t_1$ , $t_2$ ] , 24 hr , all , all , p , y }
The i-th element enclosed in the above braces corresponds to the parameter value of the i-th factor explained above.

### Utility Constraint

The third type of constraint is the restriction on simultaneous processing of several basic operations. If the maximum level of utilization of any utility or manpower is limited, the basic operations which use large amount of utility cannot be executed simultaneously. The feature of this constraint is that the restricted period is not fixed but depends on the starting moments of basic operations which are processed simultaneously. This type of constraint is hereafter called "the utility constraint".

It is assumed that the consuming rate of each utility during the processing of each basic operation is constant and it is given before carrying out the simulation. If the consuming rate of a utility changes during a processing, the basic operation is divided into several basic operations so that each basic operation has a constant consuming rate. In this study, it is assumed that each basic operation has data on the consuming rate of each utility, and the maximum level of utilization of each utility is given by a piecewise constant cyclic function.

### FIS Constraint

In an actual batch process, the capacity of each storage tank is finite. If FIS storage policy is taken between two batch units, we must adjust the starting moments of some basic operations so that each storage tank does not overflow. The constraint related to the finite storage is called "the FIS constraint". The FIS constraint has the following characteristics:
(a) If there are several storage tanks between batch units, we must decide a tank in which the material is stored.
(b) Many kinds of materials can be stored in a tank by sharing the period. In some cases, cleaning time and/or changeover cost are required to change the materials stored in the tank.

In order to consider these characteristics, we must prepare the following data for each storage tank: (i)the capacity of the tank, (ii)the names of the units before and after the tank, (iii)the names of the materials which can be stored in each storage tank, (iv)the cleaning time and/or the changeover cost between two materials.

### QUICK CALCULATION OF THE STARTING MOMENTS OF JOBS

When we use the scheduling system shown in Fig. 1, the calculation of the starting moments of jobs must be executed for a number of production sequences. Therefore, an effective simulation algorithm is required to find the plausible schedule within a reasonable computation time. In this section, we propose a quick calculation method to estimate the starting moments of jobs for a given production sequence.

### Grouping of the Units

As the materials treated in the process are fluid, the processing order of jobs cannot be changed if the process does not have any storage tanks between two units. In this section, the process is divided into several groups of units so that the production order of jobs on every unit is uniquely determined by deciding the production order of jobs on these unit groups. The following two rules are used to group the units.

(a) If there are some alternative units for the operation of a job, each of these units must belong to a different unit group.

(b) If every job processed on unit i is directly transferred to unit j, these units can belong to the same unit group. If every job processed on unit j is the job which is directly transferred from unit i, these units can belong to the same unit group.

### Relationship among the Starting Moments of Jobs

In this section, the relationships among the starting moments of jobs are derived under the following assumptions:

(i) For each job, ZW storage policy is taken within a unit group.

(ii) ZW or UIS storage policy is taken between two different unit groups.

(iii) The restrictions explained in the preceding section are not considered.

Figure 4 shows a part of a production sequence described by using the processing order of jobs on unit groups. The circle in the figure shows the processing of a job on a unit group and job number is shown in the circle. The processing order of jobs on a unit group is given by arrows, and a dotted arrow denotes the production path of a job.

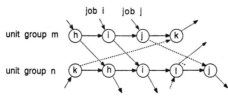

Fig. 4   Processing Order of Jobs on Two Unit Groups

The starting moment of job i on the first unit in unit group m is called "the starting moment of job i on unit group m", and expressed by $T_i^m$. Since ZW storage policy is employed in a unit group, the starting moment of every basic operation of a job is uniquely determined by deciding the starting moment of the job on every unit group.

We must consider two kinds of precedence relationships among the starting moments of jobs on unit groups. One is the relationship between the starting moments of a job on two unit groups. When job i use unit group m and unit group n successively, $T_i^m$ and $T_i^n$ satisfy one of the following relationships:

(a) If UIS storage policy is employed between unit groups m and n,

$$T_i^m + P_i^{mn} \leq T_i^n. \tag{2}$$

(b) If ZW storage policy is employed between unit groups m and n,

$$T_i^m + P_i^{mn} = T_i^n. \tag{3}$$

where $P_i^{mn}$ is the minimum time difference between $T_i^m$ and $T_i^n$, and it is calculated independently of the production sequence.

The other is the relationship between the starting moments of two jobs on a unit group. When job i and job j are successively processed on unit group m, the minimum time difference between $T_i^m$ and $T_j^m$ is called "the changeover time from job i to job j on unit

group m", and expressed by $C_{ij}^m$. As ZW storage policy is employed in a unit group, $C_{ij}^m$ is also calculated independently of the production sequence. Then, $T_i^m$ and $T_j^m$ must satisfy the following inequality:

$$T_i^m + C_{ij}^m \leq T_j^m \tag{4}$$

The relationship among $T_i^m$, $T_j^m$, $P_i^{mn}$ and $C_{ij}^m$ is graphically expressed in Fig. 5.

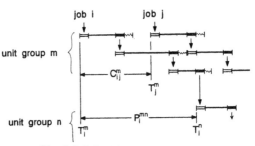

Fig. 5   Relationship among the Variavles

Each job i has the earliest starting moment, $Q_i$, which is determined before carrying out the simulation. Then, the following inequality must be satisfied:

$$Q_i \leq T_i^m \qquad \text{for } \forall m \in G_i \tag{5}$$

where

$G_i$ : set of unit groups which are used by job i.

### Simulation Algorithm

The problem is to find the earliest starting moment of each $T_i^m$ which satisfies Eqs.(2) to (5). We propose a simple and effective algorithm to find the optimal $T_i^m$ by using the characteristic in which each constraint expressed by Eqs.(2) to (5) has only one or two variables. The algorithm consists of the following steps:

&lt;Algorithm I&gt;

(1) $T_i^m \leftarrow Q_i$ for all $m \in G_i$ and every job i.  ($a \leftarrow b$ means that b is substituted for a.)

(2) Repeat steps (3) to (6) for every job.

(3) Select a job (job J), and repeat steps (4) to (6) for every unit group which is used by job J.

(4) Select a unit group (unit group M) which is used by job J, and find a unit group (unit group L) which precedes unit group M.

(5) Change the starting moments of job J on unit groups L and M if they satisfy the following conditions:

If $T_J^L + P_J^{LM} > T_J^M$, then substitute $T_J^L + P_J^{LM}$ for $T_J^M$.

If $T_J^L + P_J^{LM} < T_J^M$ and ZW storage policy is employed between unit groups L and M for job J, substitute $T_J^M - P_J^{LM}$ for $T_J^L$

(6) Find a job which precedes job J on unit group M. If job K selected at this step satisfies the condition that $T_K^M + C_{KJ}^M > T_J^M$ , then substitute $T_K^M + C_{KJ}^M$ for $T_J^M$ .

(7) If the starting moments of some jobs are changed during steps (2) to (6), return to step (2).

(8) Stop the calculation.

In this algorithm, the initial value of the starting moment of each job on a unit group is $Q_i$, and it is increased by using the following simple logics:

If $a \leq T_1$, $b \leq T_2$ and $T_1 + c \leq T_2$, then $a + c \leq T_2$.

If $a \leq T_1$, $b \leq T_2$ and $T_1 + c = T_2$, then $a + c \leq T_2$ and $b - c \leq T_1$.

At steps (5) and (6), the starting moment of a basic operation is delayed so that the basic operation is started as soon as the preceding operation is finished. Therefore, the derived starting moments are the earliest starting moments for the given production sequence. $P_J{}^{LM}$ and $C_{KJ}{}^M$ in steps (5) and (6) are calculated independently of the production sequence. The computation time of this algorithm is very short compared with the algorithm to be explained in the next section.

<Note 1> If $C_{ij}{}^m + C_{jk}{}^m < C_{ik}{}^m$ for some jobs, the result of algorithm I may be infeasible. In that case, the following condition must be added to step (6).

If $T_{ij}{}^m + C_{ij}{}^m + C_{jk}{}^m > T_{ik}{}^m$, then substitute $T_{ij}{}^m + C_{ij}{}^m + C_{jk}{}^m$ for $T_{ik}{}^m$.

<note 2> By assuming that every job is processed without taking any overlapping among the operations, we can calculate the upper bound of the completion time of the final job. If there are no feasible solutions for the given production sequence, the starting moments of some jobs become infinity by continuing the execution of algorithm I. Therefore, we are forced to judge that the given production sequence is infeasible when a starting moment of a job exceeds the precalculated upper bound.

## CALCULATION OF THE STARTING MOMENTS OF BASIC OPERATIONS

In this section, we propose a simulation algorithm to determine the earliest starting moments of the basic operations which satisfy every kind of constraint explained in a previous section.

It is advised not to calculate precisely the starting moments of jobs for a production sequence which has little possibility of being the optimal schedule. The proposed simulation algorithm consists of four stages. At the first stage, the starting moments of basic operations are calculated by taking into account the constraints on working period only. Then, the starting moments of the basic operations are recalculated by adding other types of constraints one by one. Algorithm consists of the following steps:

### Stage 1. Schedule which Satisfies the Constraints on Waiting Period

In this stage, the starting moments of the basic operations are calculated by taking into account only the constraints on waiting period between two basic operations. That is, the starting moments of the basic operations which satisfy inequality constraints expressed by Eq.(1) are derived. If FIS policy is employed between the two batch units, this FIS policy in this stage should be regarded as a UIS policy.

By deciding the production sequence, the precedence relationship among the basic operations is uniquely determined as shown in Fig. 2. The structure of Fig. 2 is similar to that of Fig. 4. Every relationship between two nodes in Fig. 2 and Fig. 4 is expressed by equality or inequality constraints, and a basic operation in Fig. 2 corresponds to a job on a unit group in Fig. 4. Therefore, we can generate a simulation algorithm which is very similar to algorithm I. The algorithm developed at this stage is hereafter called algorithm I'. The starting moments of basic operations derived here are the earliest starting moments which satisfy the constraints on waiting period.

### Stage 2. Schedule which Satisfies the Constraints on Working Patterns

If a schedule derived at stage 1 does not satisfy a constraint on working patterns, the starting moments of some basic operations must be delayed so as to satisfy the constraint. The modification of the starting moment of a basic operation affects the starting moments of other basic operations. So, after the modification of the starting moment of the basic operation, all steps of stage 1 must be executed again to derive the starting moments which satisfy the constraint on waiting period. By repeating the modifications, we can obtain the schedule which satisfies every constraint on working patterns. The algorithm consists of the following steps:

<Algorithm II>
(1) to (7): same as the steps (1) to (7) of algorithm I'
(8) Choose a constraint on working patterns.
(9) Find basic operations which do not satisfy the constraint selected at step 8. If every basic operation satisfies the constraint, go to step (12).
(10) For each of basic operations selected at step (9), delay the starting and ending moments so that the basic operation satisfies the constraint. Note that only the starting and ending moments of the selected basic operation are changed at this step.
(11) Go to step (2).
(12) If the schedule does not satisfy some constraints, go to step (8).
(13) Stop the calculation.

The order of selecting a constraint on working patterns does not affect the final schedule. But in order to reduce the calculation time, the selection of the constraints should be performed in a chronological order in terms of their restricted time periods.

### Stage 3. Schedule which Satisfies the Utility Constraints

Suppose the case where the three basic operations use the same utility and where the utility constraint is not satisfied during $s_2$ and $e_1$ as shown in Fig. 6. The feature of the utility constraints is that there are several alternative ways to satisfy the constraint. In this example, the starting moment of one of the three basic operations must be delayed to satisfy the constraint. That is, one of the following inequalities must be satisfied: $e_2 \leq s_1$ or $e_1 \leq s_2$ or $e_1 \leq s_3$.

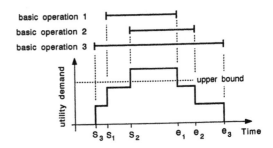

### Fig. 6 Schedule which does not Satisfy the Utility Constraint

We can consider many heuristics to determine a basic operation to be delayed. In the proposed algorithm, a performance index of the schedule is calculated for each of the alternative ways to satisfy the constraint, and the best alternative is chosen depending on the value of the performance index. The algorithm consists of the following steps:

<Algorithm III>
(1) to (12) : same as steps (1) to (12) of stage 2.
(13) Find utility constraints which are not satisfied during some time periods, and select one which has the earliest time period. If every constraint is satisfied, go to step (19).
(14) Enumerate the solution schemes and express them by using inequality constraints.
(15) Choose one solution scheme enumerated at step (14), and delay the starting moment of the basic operation so as to satisfy the selected constraint.
(16) Execute step (2) to step (12), and calculate the performance index.
(17) Repeat steps (15) and (16) for every solution scheme, and select the schedule which minimizes the performance index.
(18) The inequality constraint which is corresponding to the best solution scheme is incorporate to the calculation algorithm of the starting moments of basic operations, and return to step (13).
(19) Stop the calculation.

Stage 4. Schedule which Satisfies FIS constraints
For the schedule determined at stage 3, the storage level of each intermediate product during the scheduling period can be calculated. If the storage level exceeds the capacity of the tank or the number of stored materials becomes larger than that of storage tanks, we must delay the starting moment of the discharging of a certain job so that the job is transferred to next unit directly. The proposed algorithm consists of the following steps:

<Algorithm IV>
(1) to (18) : same as steps (1) to (18) of stage 3.
(19) Choose a discharging operation in a chronological order.
(20) If the starting moment of the selected discharging operation is equal to the starting moment of the filling operation to the next unit, go to step (19).
(21) Find the storage tanks in which the material can be stored. If it is possible to store the material to one of the tanks, change the storage level of the tank during the storing period and go to step (19). If there are no storage tanks which are able to store the material, delay the starting moment of the discharging operation till the starting moment of the filling operation to the next unit, and go to step (2).

By executing algorithm IV, we can derive a schedule which satisfies every kind of constraint explained in the previous section. Schedules derived at stages 1 and 2 are strictly the optimal schedules if only the constraints imposed at these stages are considered.

There are some alternative ways such that each of these alternatives satisfies one of the utility constraints or one of FIS constraints. In the algorithm explained at stages 3 and 4, we choose an alternative in order to satisfy a constraint. The selected alternative is fixed and not changed during the later calculation. We do not consider all of the combinations of alternatives which are generated by selecting an alternative from a set of alternatives each of which satisfies a constraint. Therefore, the results derived at stages 3 and 4 may not be optimal.

At the end of each stage, the performance index of the production sequence can be calculated, and it can be used as the lower bound of the final result. So, if the result of one of these stages is worse than the result of stage 4 which was calculated for other production sequence, we need not to continue the calculation for this production sequence.

CONCLUSION

A simulation algorithm, which can be widely used in determining the starting moments of operations for general batch processes, is proposed. When we determine the starting moments of jobs on a batch unit, many kinds of constraints must be considered. In this study, a variety of constraints are classified into four groups in order to make the characteristics of constraints clearer.

The production on a batch unit consists of basic operations such as filling, processing, discharging and cleaning. In many cases, it is possible to insert the waiting period between two successively processed basic operations, and the waiting time affects the completion time of each job. In this paper, the operations to produce a batch of product is divided into a sequence of basic operations, and an effective simulation algorithm which calculates the starting moments of those basic operations was proposed.

The simulation algorithm proposed here consists of four stages, and one of the four types of constraints is incorporated into each stage of the algorithm. The proposed simulation algorithm derives the strict optimal schedule if there are no constraints other than those on the waiting period and the working patterns.

REFERENCES

Egri, U. M. and D. W. T. Rippin (1986). Short-term Scheduling for Multiproduct Batch Chemical Plants. Comp. & Chem. Eng., 10, 303-325.

Hasebe, S., I. Hashimoto and T. Takamatsu (1988). Scheduling through Reordering Operations. Proc. of PSE'88, Sydney, 76-81.

Kondili, E., C. C. Pantelides and R. W. H. Sargent (1988). A General Algorithm for Scheduling Batch Operations. Proc. of PSE'88, Sydney, 62-75.

Rajagopalan, D. and I. A. Karimi (1989). Completion Times in Serial Mixed-Storage Multiproduct Processes with Transfer and Set-up Times. Comp. & Chem. Eng., 13, 175-186.

Suhami, I. and R. S. H. Mah (1981). An Implicit Enumeration Scheme for the Flowshop Problem with No Intermediate Storage. Comp. & Chem. Eng., 5, 83-91.

Wiede Jr, W. and G. V. Reklaitis (1987). Determination of Completion Times for Serial Multiproduct Processes-3. Mixed Intermediate Storage Systems. Comp. & Chem. Eng., 11, 357-368.

# ADAPTIVE DISCRETIZATION OF CONTINUOUS-TIME INVENTORY CONTROL PROBLEMS AND ITS APPLICATION TO PRODUCTION PLANNING AND SMOOTHING

## P. P. Monteiro and A. D. Correia

*Departamento de Engenhairia Electrotécnica, Largo Marquês de Pombal,*
*3000 Coimbra, Portugal*

**Abstract.** The problem of controlling a continuous-time production inventory system subject to deterministic demand and to upper and lower bounds in all production rates and inventory levels is addressed. Inventories are here to be exploited in order to minimize the number of production rate changes (NPRC). It is found that, under mild assumptions, it is feasible to keep production rates constant during discrete time intervals. An algorithm is develloped that maximizes their length. A discrete-time version of the continuous-time control problem is then formulated. This adaptive discretization is found to be usefull for: i) it eliminates the a priori choice of a (rather short) discretization period found in previous work; ii) it therefore leads to smaller discrete-time problems, and efficiency gains; iii) when the discrete smoothing problem does not explicitly minimize NPRC, which is a non-convex criterion, NPRC-better solutions should be obtained, because production rates are constrained to be constant for longer times. These results are illustrated by a numerical example.

**Keywords:** Production control;production scheduling; inventory control; production smoothing; production-inventory systems; process industries; pulp and paper industry; discrete parts manufacturing industries.

## INTRODUCTION

In the process industries there are often plant sections that suffer very much when production rate changes occur: efficiency depends frequently on space fields of process variables, and the ideal fields are only attainable in steady state. Changes are very costly as they disrupt the ideal steady state fields.

The sections are connected by means of inventories (Fig. 1) that allow a step production rate change in one section not to require a step in the "neighbor" sections - the inventories have a decoupling effect.

In the present study we aim at using the inventories in order to minimize the number of production rate changes (NPRC) necessary to obtain the forecasted production.

This study is based on our current experience with short-term production planning in the pulp and paper industry and may be seen at least as a complement to the studies on this subject mentioned in the references section. However, an effort was made to bring generality to the  problem statement and solution. It should be possible to model the system to be controlled as a production-inventory system as in Fig.1.Inventory holding costs should be small when compared to production rate change costs or, in broader terms, to set-up/shut-down costs.

## PROBLEM STATEMENT

All production rates and inventory levels refer to a production-inventory system (Fig. 1) in which the i-th section, whose production rate is $u_i(t)$, fills/empties the inventories $x_j$ to which it is connected at $b_{ji}u_i(t)$ rates. This leads to the dynamic equation

$$dx/dt = Bu(t) , x(0) = x^{(0)} \tag{1}$$

For a known (or forecasted) demand of end-products

$$u_i(t), t \in [t_0, t_f [, i \in E_p \tag{2}$$

(in Fig. 1., $E_p=\{7\}$ ) and a known initial inventory

$$x (t_0)= x^{(0)} \tag{3}$$

it is necessary to decide on

$$u_i(t), t \in [t_0, t_f [, i \notin E_p \tag{4}$$

such that

- the forecasted production is realized
- the capacity of the sections is not exceeded
- planned shut-downs are taken into account
- inventory  safety margins are respected
- the final inventory is statistically
  favorable.

55

These requirements are expressed by appropriate use of the inequality constraints

$$u_{min}(t) \leq u(t) \leq u_{max}(t) \qquad (5)$$
$$x_{min}(t) \leq x(t) \leq x_{max}(t) \qquad (6)$$

as table 1 clarifies. And, last but not least,

- unnecessary product rate changes should be avoided. $\qquad (7)$

This last requirement must be expressed in a performance criterion optimized under (5) and (6). Acording to (Edlund and Rigerl, 1978),(7) is the most important requirement, hence our interest in diminishing NPRC.

## ADAPTIVE DISCRETIZATION

It is convenient to introduce the following theorems:

### Theorem A

Suppose that there is a feasible solution $u^f(t)$ to the planning problem and that a time interval $\Delta = [t_1, t_2[$ exists such that

i) $u_{min}(t)$ and $u_{max}(t)$ are constant in $t \in \Delta$
ii)$x_{min_i}(t)$ and $-x_{max_i}(t)$ are convex in $t \in \Delta$ for all i
Then constraining the production rate function to be constant in $t \in \Delta$ does not lead to infeasibility.

### Proof

We shall show that $u^c$, as defined in (8), which is constant in $t \in \Delta$, is feasible.

$$u^c(t) = \begin{cases} \dfrac{1}{t_2 - t_1} \displaystyle\int_{t_1}^{t_2} u^f(\xi)\, d\xi \, , \ t \in \Delta \\[2mm] u^f(t), \qquad t \notin \Delta \end{cases} \qquad (8)$$

Because of i), $u^c(t)$ satisfies (5). On the other hand, it can be verified that the inventory trajectory $x^c$ generated by $u^c$ is

$$x^c(t) =$$

$$\begin{cases} \dfrac{t_2 - t}{t_2 - t_1} x^f(t_1) + \dfrac{t - t_1}{t_2 - t_1} x^f(t_2) \, , \ t \in \Delta \\[2mm] x^f(t) \, , \ t \notin \Delta \end{cases} \qquad (9)$$

By using

$$\theta = \frac{t_2 - t}{t_2 - t_1} \qquad (10)$$

we can rearrange the case "$t \in \Delta$" in (9) as

$$x^c(\theta t_1 + (1-\theta)t_2) = \theta x^f(t_1) + (1-\theta)x^f(t_2) \qquad (11)$$

Because of assumption ii),

$$x_{min}(\theta t_1 + (1-\theta)t_2) \leq \theta x_{min}(t_1) + (1-\theta)x_{min}(t_2) \qquad (12)$$

But, because $\theta \in [0,1[$, $\theta \geq 0$ and $(1-\theta) \geq 0$, from which

$$\theta x_{min}(t_1) + (1-\theta)x_{min}(t_2) \leq \theta x^f(t_1) + (1-\theta)x^f(t_2) \qquad (13)$$

By linking (12), (13) and (11) in this order and then using (10) to introduce t , we show that

$$x_{min}(t) \leq x^c(t) \, , \ t \in \Delta \qquad (14)$$

The proof that $x^c(t)$ also respects its upper bound is similar, and therefore ommitted. In view of (9), we have proved that $x^c$ satisfies (6). As $x^c$ is obtained by use of (8) in (11), (11) is also satisfied. $u^c$ is therefore feasible, q. e. d. .

### Theorem B

Under the assumptions of theorem A, every $(u^c, x^c)$ that satisfy (1), (5) and

a) $u^c(t)$ is constant in $t \in \Delta$
b) $x_{min}(t_1) \leq x^c(t_1) \leq x_{max}(t_1)$
c) $x_{min}(t_2) \leq x^c(t_2) \leq x_{max}(t_2)$
d) $x_{min}(t) \leq x^c(t) \leq x_{max}(t) \, , \ t \notin \Delta$
is feasible.

### Proof

Because of the assumptions, in order to prove the theorem, it is only necessary to show that

$$x_{min}(t) \leq x^c(t) \leq x_{max}(t) \, , \ t \in \Delta \qquad (15)$$

Using a) in (1) and then using (10) we may write

$$x^c(\theta t_1 + (1-\theta)t_2) = \theta x^c(t_1) + (1-\theta) x^c(t_2) \qquad (16)$$

This expression equals (11) if $x^f$ is replaced by $x^c$. The remainder of the proof follows (12),(13) and (14) with the above mentioned difference, and is therefore ommitted.

### The Adaptive Discretization Vector

Let us have the verification of the hypotheses of theorem A formulated as a Boolean function of $t_1$ and $t_2$ (the bound functions are ommitted because they are fixed in every particular problem)

$$hyp\,(t_1, t_2) = \begin{cases} 1 & \text{if the hypotheses of theorem A are} \\ & \text{verified in } \Delta = [t_1, t_2[ \\ 0 & \text{otherwise} \end{cases} \qquad (17)$$

Let us introduce a discretization vector $\tau$ and divide the total horizon $[t_0, t_f[$ in discrete intervals $\Delta_k$ according to

$$\begin{cases} \Delta_k = [\tau(k), \tau(k+1)[, & k=0,1,...,nk-1 \\ hip(\tau(k), \tau(k+1)) = 1, & k=0,1,...,nk-1 \\ \tau(0) = t_0 \\ \tau(nk) = t_f \end{cases} \quad (18)$$

By successive applications of theorem A, if the continuous-time problem is feasible, there is also a feasible plan given by

$$u(t) = u_k, \ t \in \Delta_k \qquad k=0,1,...,nk-1 \quad (19)$$

The constraint (5), in view of assumption i) of theorem A and (19), may be expressed by

$$umin(\tau(k)) \leq u_k \leq umax(\tau(k)), \qquad k=0,1,...,nk-1 \quad (20)$$

On the other hand, by successive applications of theorem B, we may replace (6) by

$$xmin(\tau(k)) \leq x(\tau(k)) \leq xmax(\tau(k)), k = 0,1,...,nk \quad (21)$$

Finally, using (19) in (1), we obtain

$$x(\tau(k+1)) = x(\tau(k)) + (\tau(k+1) - \tau(k))B \ u_k, k=0,1,..., \ nk-1,$$
$$x(\tau(0)) = x^0 \quad (22)$$

The constraints (20),(21) and (22) are in discrete-time and their satisfaction was shown to insure feasibility of the continuous-time problem. They are functions of $\tau$ and, together, constitute an adaptive discretization.

## The (Optimal) Adaptive Discretization

Because we aim at minimizing NPRC, (19) is a very fortunate result: it insures that the number of production rate changes per section will be less than nk-1. The best discretization within this paper's context, which we define as *the* adaptative discretization, is therefore

$$\min_{\tau} \ nk \ under \ (18) \quad (23)$$

The following results hold:

$$hyp(t_1, t_2) = 1 \Rightarrow hyp(\gamma, \delta) = 1 \ for \ \gamma \geq t_1, \ \delta \leq t_2 \ and \ t_1 \leq t_2 \quad (24)$$
$$hyp(t_1, t_2) = 0 \Rightarrow hyp(\gamma, \delta) = 0 \ for \ \gamma \leq t_1, \ \delta \geq t_2 \ and \ t_1 \leq t_2 \quad (25)$$

From these two results it is possible to infer that the region $\rho$ defined by

$$\rho = \{ (t_1, t_2) \mid hyp(t_1, t_2) = 1 \} \quad (26)$$

is as depicted in Fig. 2. On the other hand, every $\tau$ may be represented on a two-dimensional plane by line($\tau$), defined by

line ($\tau$) = the broken line that connects the points $(\tau(0), \tau(0))$, $(\tau(0), \tau(1))$ , $(\tau(1), \tau(1))$, $(\tau(1), \tau(2))$,..., $(\tau(nk-1), \tau(nk))$, $(\tau(nk), \tau(nk))$ $\quad (27)$

It is the thick line of Fig. 2. Because (18) requires $hyp(\tau(k), \tau(k+1)) = 1$ and $\tau(k) \leq \tau(k+1)$, by looking at (24), (18)$\Rightarrow$line($\tau$) $\in \rho$. But, by the very definitions of $\rho$ and line($\tau$), it follows that (line($\tau$) $\in \rho$ and $\tau(0) = t_0$ and $\tau(nk) = t_f$ ) $\Rightarrow$ (18), from which

$$(18) \Leftrightarrow (line(\tau) \in \rho \ and \ \tau(0) = t_0 \ and \ \tau(nk) = t_f) \quad (28)$$

The optimization (23) may now be graphically paraphrased as "find a line($\tau$)that goes from $t_0$ to $t_f$ without leaving $\rho$ with a minimum number of angles." This is graphically seen to be equivalent to making line($\tau$) the upper boundary of $\rho$.

## Optimal Discretization Calculus

Because of $\rho$'s shape, it may be easily seen that algorithm (29) finds the upper boundary of $\rho$.

1. $\tau(0) = \tau(1) = t_0$; $k=0$; $nk=0$
2. max $\tau(k+1)$ under $hyp(\tau(k), \tau(k+1)) = 1$ and $\tau(k+1) \leq t_f$
3. $nk = k+1$
4. if $\tau(k+1) = tf$ then stop
5. $k=k+1$; go to 2 $\quad (29)$

The assumptions under which the adaptive discretization is valid are that (18) should allow $\tau(k+1) > \tau(k)$, for all $k = 0,1,...,nk-1$, and (tf-t0) to be finite. In fact, under these assumptions, (29) terminates in a finite number of steps (nk is clearly finite).

## NUMERICAL TEST

### Continuous-Time Problem

The integrated pulp and paper mill of Fig. 1 and the continuous-time problem defined in Table 1 were used. The problem shows a shut-down in section 5. xmin and xmax express the initial inventory x(0), allow for decoupling effects in the middle of the planning horizon, and constrain the final inventory to be statistically favorable (Yorke 1984).

### Adaptive Smoothing Criterion and its Optimization

A smoothing criterion should be used together with our adaptive discretization (20), (21), (22) to constitute a discrete-time optimal production planning problem.

Tamura's algorithm (Tamura 1975) with strong quadratic production rate weighting was tested. This approach, described in detail in (Leiviskä 1982), is considered in (Leiviskä 1980) as of "high performance" and also adopted in (Ruiz, Muratore, Ayral and Durand, 1986).

We have modified the performance criterion in order to adapt it to the generic discretization expressed by $\tau$. The general $\tau$-problem is thus

$$\min_{u,x} J(u,x,\tau) \text{ subject to } (20),(21),(22) \qquad (30)$$

where

$$J(u,x,\tau) =$$

$$= \sum_{k=0}^{nk-1} \frac{1}{2} (\tau(k+1) - \tau(k)) \parallel u_{\tau(k)} - u^0_{\tau(k)} \parallel^2_R +$$

$$+ \sum_{k=1}^{nk} \parallel x_{\tau(k)} - x^0_{\tau(k)} \parallel^2_Q$$

$$(31)$$

with

$R = diag\{ r_i \}$, $Q = diag\{q_i\}$, $r_i = r$, $q_i = q = 1$.

Tamura's algorithm solves (30) by means of a dual

$$\max_{\lambda} \ \phi(\lambda) \qquad (32)$$

The maximization (32) was done by the steepest ascent method; a composite gradient and feasibility tolerance rule

$$\parallel \partial \phi / \partial \lambda \parallel_\infty \leq 2 \qquad (33)$$
$$xmin_k - 2 \leq x(u)_k \leq xmax_k + 2 \qquad (34)$$

was used as stopping criterion in (32) [$x_k(u)$ are the inventories calculated by using u in (22) ]

## Test Targets

Leiviskä(1982) found that, with an homogenous discretization, r/q should be high in order to reduce NPRC, and that computation time increased with r/q. We tested how the adaptive discretization (Table 2) interacts with this key finding. The results are presented in tables 3, 4 and 5.

## Test Results: Computation Time

A program was coded in FORTRAN and compiled and run in a μVAX II computer. Adaptive discretization reduces the discrete horizon very much (tables 2 and 3). As a result, there are less iterations and less CPU time per iteration. As its calculus [algorithm (29)] is very quick, the total CPU time is always very much reduced by using it (table 4).

## Test Results: NPRC

The optimal plans (table 5) using a "high" r/q (r/q=20) are all of equivalent (high) merit: the production rate changes that occur are necessary in order to keep the inventories in their allowable level (table 6). The similarity of these plans shows that the general $\tau$-problem (31) is indeed well chosen. As shown, all yield NPRC=3.

If the fractional part of the plans were included, however, NPRC would be higher. In fact, consecutive discrete production rates $u_{ki}$ and $u_{k+1i}$ that appear to be equal are, in general, only *approximately* equal. The results obtained with r/q=5 are not truncated up to integer and show clearly this fact. Small differences between consecutive production rates are not, however, of much importance – they may be eliminated by truncating or, better, by averaging nearly equal consecutive production rates without changing the plan and the inventory trajectory significantly.

Note that even truncating the production rate's fractional part, with r/q=5 and T=4, NPRC is much higher than 3 -- this plan is clearly worse than the plans obtained with r/q=20. However, with r/q=5, adaptive discretization, and truncation up to integer, NPRC is still 3. Within this reasoning, adaptive discretization yielded, at "low" r/q=5, a plan whose merits are comparable to those of the constant discretizations with r/q=20. The efficiency gain is indeed very large.

## CONCLUSIONS

Adaptive discretization reduced the horizon of the discrete problem very much, and this always led to very large computation time savings (table 4).

It also led to NPRC reduction. This reflects that (30) does not minimize NPRC, a non-convex function of u.

Step 2. of algorithm (29) may be interpreted as an effort to maintain u(t) constant for as long as possible, and,then , indirectly, diminish NPRC.

Adaptive discretization would certainly improve the ranking of the quadratic programming approach in (Leiviskä and Uronen,1980), and probably the ranking of the other approaches as well.

To sum up with, the test results back up the claims introduced in the abstract and in the introduction. The author's current experience with adaptive discretization is therefore very positive.

## AKNOWLEDGEMENTS

This work is finantially supported by PORTUCEL, EP: the authors express here their recognition to this company in the person of Mr. Eng. Rui Ribeiro and Mr. Eng. José Luis Amaral.

## REFERENCES

Drew, S.A.W. (1975). The application of hierarchical control methods to a managerical problem. Int. J. Systems Sci, 6, nº4, 371-375.

Drew, S.A.W. (1982). A study in the application of large scale control methods to a practical industrial problem. 8th IFAC Trienual World Congress, Kyoto, Japan.

Edlund, S.G. and K. H. Rigerl (1978). A computer based production control system for the coordination of operations in a pulp and paper mill. IFAC World Congress, Helsinquia.

Leiviskä, K. and P. Uronen (1979). Dynamic optimization of a.sulphate mill pulp line. IFAC/IFORS Symposium, Toulouse, France.

Leiviskä, K. and P. Uronen (1980). Different approaches for the production control of a pulp mill. IFAC PRP4, Automation, Ghent, Belgium.

Leiviskä, K. (1982). Short term production scheduling of the pulp mill. Acta Polytechnica Scandinavica, Math. Comp. Sci, nº 36, Helsinquia.

Ruiz, J., E. Muratore, A. Ayral and D. Durand (1986). Optimill®:optimal management of pulp mill production departments and storage tanks. Proceedings of the 6th International IFAC/IMEKO Conference.

Tamura, H. (1975). Decentralized optimization for distributed-lag models of discrete systems. Automatica, 11, 593-602.

Yorke, G.L. (1984). The utilization of a mill operations model for kraft mill operation. 70th Annual Meeting of CPPA.

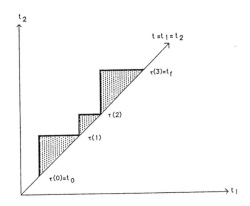

Fig. 2. The region $\rho$ where $hyp(t_1,t_2)=1$ (shown dotted) and $line(\tau)$ (the thick line) for a given discretization vector $\tau$

## TABLE 1 Continuous-Time Problem Parameters

number of sections and inventories
$nu=7$, $nx=8$

continuous-time horizon
$t_0 =0$, $t_f =72$ hours

shut-down in section 5
$Umax_5(t)= 0$, $t \in [t_0, 8[$
$Umax_5(t)=100$, $t \in [8, t_f[$

required production
$Umin_7(t)=Umax_7(t)=80$, $t \in [t_0, t_f[$

other production rate bounds
(section capacities)
$Umin_i(t)= 0$, $t \in [t_0, t_f[$ , $i=1,2,3,4,5,6$
$Umax_i(t)=100$, $t \in [t_0, t_f[$ , $i=1,2,3,4,6$

initial inventory
$x^{(0)}_i = Xmin_i(t_0) = Xmax_i(t_0) = 50$, $i=1,2,...,nx$

allowable final inventory levels
$Xmin_i(t_f)=45$, $i=1,2,...,nx$
$Xmax_i(t_f)=55$, $i=1,2,...,nx$

inventory level bounds during the time horizon
$Xmin_i(t)=30$, $t \in ]t_0, t_f[$ , $i=1,2,...,nx$
$Xmax_i(t)=70$, $t \in ]t_0, t_f[$ , $i=1,2,...,nx$

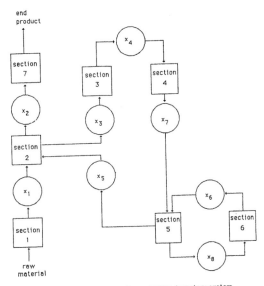

Fig. 1. Graphical representation of a production-inventory system. Productive sections and inventories are represented respectively by rectangulars and circles. The arrows show the way the materials flow

### TABLE 2 Discretization Vectors Used In The Numerical Test

adaptive discretization vector:

$\tau(0)=0, \tau(1)=8, \tau(2)=72.$

constant discretization vector of period T:

$\tau(k)= kT, k=0,1,..., (t_f-t_0)/T-1$

### TABLE 3 Influence of Discretization on CPU Time per Tamura's 2-Level Algorithm Iteration

| discretization | discrete-time horizon nk | CPU time per iteration |
|---|---|---|
| adaptive | 2 | 15 ms |
| constant (T=8) | 9 | 65 ms |
| constant (T=4) | 18 | 125 ms |

### TABLE 4 Influence of Discretization and r/q on Total CPU Time

r/q=5

| discretization[1] | iterations[2] | totalCPUtime |
|---|---|---|
| Adaptive | 1730 | 26 s |
| T=8 hours | 4720 | 320 s |
| T=4 hours | 9540 | 1182 s |

r/q= 10

| discretization[1] | iterations[2] | totalCPUtime |
|---|---|---|
| Adaptive | 7220 | 108 s |
| T=8 hours | 9380 | 610 s |
| T=4 hours | 18880 | 2360 s |

r/q=20

| discretization[1] | iterations[2] | totalCPUtime |
|---|---|---|
| Adaptive | 1920 | 28 s |
| T=8 hours | 18380 | 1176 s |
| T=4 hours | 37880 | 4697 s |

[1]Adaptive discretization calculus CPU time was 80 ms
[2]Number of Tamura's 2-level algorithm iterations

### TABLE 5 Influence of Discretization and r/q on the Optimal Production Plan

r/q = 20, adaptive discretization (*)

| t \ $u_i(t)$ | 1 | 2 | 3 | 4 | 5 | 6 | 7 |
|---|---|---|---|---|---|---|---|
| [0,8[ | 0 | 78 | 77 | 60 | 0 | 28 | 80 |
| [8,72[ | 0 | 78 | 77 | 78 | 85 | 80 | 80 |

r/q = 20, T=4 or T=8   (*)

| t \ $u_i(t)$ | 1 | 2 | 3 | 4 | 5 | 6 | 7 |
|---|---|---|---|---|---|---|---|
| [0,8[ | 0 | 78 | 77 | 54 | 0 | 27 | 80 |
| [8,72[ | 0 | 78 | 77 | 78 | 85 | 80 | 80 |

r/q=5, adaptive discretization (**)

| t \ $u_i(t)$ | 1 | 2 | 3 | 4 | 5 | 6 | 7 |
|---|---|---|---|---|---|---|---|
| [0,8[ | 0.1 | 78.7 | 77.2 | 54.8 | 0.0 | 27.4 | 80 |
| [8,72[ | 0.1 | 78.8 | 77.4 | 78.9 | 85.8 | 81.5 | 80 |

r/q=5,  T=4  (**)

| t \ $u_i(t)$ | 1 | 2 | 3 | 4 | 5 | 6 | 7 |
|---|---|---|---|---|---|---|---|
| [0,4[ | 0.1 | 77.8 | 76.1 | 54.4 | 0.0 | 27.3 | 80 |
| [4,8[ | 0.1 | 77.8 | 76.2 | 54.4 | 0.0 | 27.6 | 80 |
| [8,12[ | 0.1 | 77.9 | 76.3 | 78.3 | 87.7 | 78.4 | 80 |
| [12,16[ | 0.1 | 78.0 | 76.4 | 78.3 | 87.0 | 79.8 | 80 |
| [16,20[ | 0.1 | 78.1 | 76.5 | 78.4 | 86.3 | 79.3 | 80 |
| [20,24[ | 0.1 | 78.2 | 76.7 | 78.4 | 85.8 | 79.6 | 80 |
| [24,28[ | 0.1 | 78.3 | 76.7 | 78.4 | 85.3 | 79.9 | 80 |
| [28,32[ | 0.1 | 78.3 | 76.8 | 78.4 | 84.9 | 80.2 | 80 |
| [32,36[ | 0.1 | 78.4 | 76.9 | 78.4 | 84.5 | 80.4 | 80 |
| [36,40[ | 0.1 | 78.4 | 77.0 | 78.4 | 84.3 | 80.5 | 80 |
| [40,44[ | 0.1 | 78.5 | 77.0 | 78.4 | 84.0 | 80.7 | 80 |
| [44,48[ | 0.1 | 78.5 | 77.1 | 78.4 | 83.8 | 80.8 | 80 |
| [48,52[ | 0.1 | 78.6 | 77.1 | 78.4 | 83.6 | 80.9 | 80 |
| [52,56[ | 0.1 | 78.6 | 77.2 | 78.4 | 83.5 | 81.0 | 80 |
| [56,60[ | 0.1 | 78.6 | 77.2 | 78.4 | 83.4 | 81.0 | 80 |
| [60,64[ | 0.1 | 78.6 | 77.2 | 78.4 | 83.4 | 81.1 | 80 |
| [64,68[ | 0.1 | 78.6 | 77.2 | 78.3 | 83.3 | 81.1 | 80 |
| [68,72[ | 0.1 | 78.6 | 77.2 | 78.3 | 83.3 | 81.1 | 80 |

(*) truncated to integer
(**) rounded decimal part

### TABLE 6 Inventory Trajectories

r/q= 20, adaptive discretization (*)

| t \ $x_i$ | 1 | 2 | 3 | 4 | 5 | 6 | 7 | 8 |
|---|---|---|---|---|---|---|---|---|
| 0 | 50 | 50 | 50 | 50 | 50 | 50 | 50 | 50 |
| 8 | 49 | 49 | 50 | 59 | 31 | 64 | 72 | 29 |
| 72 | 45 | 43 | 55 | 55 | 43 | 47 | 52 | 55 |

(*) The trajectories have constant gradient between the times that appear in the table. Other discretizations and r/q yield approximately the same trajectories.

# INTEGRATION OF PRODUCTION PLANNING
# AND REAL-TIME PROCESS CONTROL

## H. Nishitani, I. Tamura and E. Kunugita

*Department of Information & Computer Sciences, Osaka University,*
*1-1 Machikaneyama, Toyonaka, Osaka 560, Japan*

**Abstract**. Some computer control algorithms known as the model predictive control can treat the future information of desired trajectories of the controlled variables and measurable disturbances. Consequently, it can include both the feedback and feedforward control elements. This implies that the algorithm can be used as an interface between the non-real-time planning system of which emphasis is on the feedforward effects and the real-time control system of which emphasis is on the feedback effects.

**Keywords**. Production control, hierarchical system, non-real-time planning, real-time control, interface, model predictive control algorithm

## INTRODUCTION

Plan-do-see are the fundamental functions in the management of a production system. So far, the planning function is placed at the management level in the head office. This is a non-real-time function. On the other hand, the do-see functions are placed at the actual manufacturing site in the factories. They are achieved by the real-time control system. Recently most companies in the process industry are looking for an effective integration of these functions because they intend to read the market and produce their products more timely. For this strategic production both non-real-time planning and real-time control should be coordinated closely.

The integration of planning, scheduling, and process control has been studied by a few researchers (Lasdon & Baker, 1986; Lasdon, Waren & Sakar, 1988). An integrated system should have the adaptive capabilities of a feedback control system in that its primary purpose is to respond to disturbances at all functional levels. But their interests were mainly in the utilization of the information at the regulatory control level to the steady state optimization at the supervisory control level and in the development of effective optimization algorithms. Their paper contain no mention of integration methods of non-real-time and real-time functions. For an effective integration the inherent differences between real-time and non-real-time applications should be considered with the role of the data sampled at different intervals (Gore, 1987).

On-line optimization is a closely related practical issue to the integration problem (Darby & White, 1988). However, no general principle for structuring an integrated system has been extracted from a bunch of know-how yet.

Generally an optimal plan can be transmitted to the lower level control system which lays stress on the feedback properties such as attenuation of unmeasurable disturbances and measurement noise and robust stability. The real time control system is housed in the modern distributed control system (DCS). But even today most control loops are composed of PID control elements. Therefore,

only the present values of the optimal plan can be utilized as the set-points of control loops. Unfortunately, the production plan in future can not be put to practical use. This is a great loss of planning information. In this paper, the necessity of an interface between planning and real-time control is proposed.

## OPTIMAL PLANNING

An optimal planning over a fixed time horizon in future is formulated as a problem of calculus of variations. For this purpose a mathematical model of the production system must be built based on the information from do-see functions at the actual production plant. The accumulated data of production activities should be reflected in the model. The principal reason for the use of finite-horizon planning is that forecasts or prediction for the more distant future tend to be increasingly unreliable.

The following optimization problem is formulated at time $t_k$ to determine an optimal plan or schedule over a finite time horizon of $\theta_1$.

$$\text{Minimize} \int_{t_k}^{t_k+\theta_1} g[y(t),x(t)]dt \tag{1}$$

where $y(t)$ and $x(t)$ are vector functions of time-dependent decision variables and state variables, respectively. The relationships between these two vector functions are represented by a set of ordinary differential equations, i.e.

$$d[x(t)]/dt=f[x(t),y(t),p(t)] \tag{2}$$

where $p(t)$ is a vector function of time-dependent parameters. The above equations represent the interactions of cause and effect in the corresponding plant. Since this model usually includes uncertainties, the model should be updated. As a simple method to cope with uncertainties the state variables are measured directly or indirectly at the same interval as the planning period. The measured values are used as the initial values of the above ordinary differential equations.

$$x(t_k)=x_0(\text{measured}) \tag{3}$$

A problem to determine the decision variable vector $[y(t): t_k \leq t \leq t_k+\theta_1]$ which minimizes the objective function eq.(1) subject to eq.(2) is a typical problem of calculus of variations. However, when the planning horizon is divided into M small equal periods at an interval of $\theta_2$ for which the optimal decision variables are constant, the variational problem becomes a parametric optimization problem of which solution is represented by a stair-step like function. It approximates the solution of the variational problem.

The ordinary differential equations can be also discretized by a small time interval of $\Delta t$ as follows:

$$x(t_{k+i})=f'[x(t_{k+i-1}),y(t_{k+i-1}),p(t_{k+i-1}),\Delta t] \quad (4)$$

where the differential is approximated by a difference.

$$dx(t_{k+i})/dt=[x(t_{k+i})-x(t_{k+i-1})]/\Delta t \quad (5)$$

If $\Delta t$ is chosen as $\theta_2$, which is the interval of the discretization of the solution of the variational problem, then the optimal planning problem becomes a multi-period problem.

$$\text{Minimize} \sum_{i=0}^{M-1} g[y(t_{k+i}),x(t_{k+i})] \quad (6)$$

subject to

$$x(t_{k+i})=f'[x(t_{k+i-1}),y(t_{k+i-1}),p(t_{k+i-1}),\theta_2]$$

$$(i=1,\ldots,M) \text{ and } x(t_k)=x_0 \quad (7)$$

In most on-line optimizing control system for the processes with slow state dynamics a single-period optimization problem, i.e. a steady-state optimization problem, is solved at a supervisory control level.

REAL-TIME CONTROL

The real-time control system is concerned with regulation despite unpredictable disturbances by the minor control loops. Implementation of the optimal plan, which is determined at the planning level, is also an important objective of the control system. For this purpose the decision variables at the planning level are selected as the controlled variables. It is assumed that the controlled variables can be measured at any interval of sampling period. To control these variables some variables which can affect the controlled variables should be selected as the manipulated variables. Usually dynamic elements such as time delay and/or dead time are included in the relationships between the manipulated and the controlled variables. The dynamics can be represented by some methods.

In the design of the conventional control system the transfer function in the Laplace transform domain is usually used to represent the dynamics. For a multiple input- multiple output process it is represented as follows:

$$Y(s)=G(s)U(s) \quad (8)$$

where $Y(s)$ and $U(s)$ are the variables in the s-domain and $G(s)$ is the transfer function matrix.

The control system usually drives at an interval of $\theta_3$ based on the error between the desired set-points and the actual outputs at present and in the past. It only uses the set-point at the present time. This is due to an over-emphasis on the feedback control aspect of the real-time control system. Although the optimal plan determined at the upper level includes the future information, the conventional control system can not use the available information which will be valuable.

In the conventional control system the follow-up property to the set-point will be displeased when a plant has a slow dynamics. It is not the feedback property. Therefore, some feedforward control elements added to the real time control system can improve the property considerably. In the next section a computer control algorithm known as the model predictive control is applied to coordinate the non-real-time planning and real-time=control system.

MODEL PREDICTIVE CONTROL AS AN INTERFACE BETWEEN PLANNING AND REAL-TIME CONTROL

The model predictive control is a digital computer control algorithm which is widely used in the process control area. Especially, there are many applications to the processes with long time delays and/or dead times.

In the design of the model predictive control the data set of responses of the output to the impulse or step changes in the input is used to represent the process dynamics. They are called the impulse response model or the step response model, respectively. Since these two models are equivalent each other, the step response model is used in this paper. Hereafter, the time $t_{k+j}$ is denoted simply by k+j.

The multiple outputs at sampling period k+j are represented by the following model.

$$y_m(k+j)=\sum_{i=1}^{N} A_i\Delta u(k+j-i)+A_N u(k+j-N-1) \quad (9)$$

where

$$\Delta u(k+j-i)=u(k+j-i)-u(k+j-i-1) \quad (10)$$

and the $A_i$'s $(i=1,\ldots,N)$ are the coefficient matrix of the step responses of the outputs. The outputs at sampling period k+j in future are predicted by the following equation at the present sampling period k:

$$y_f(k+j)=y_m(k+j)+d(k+j) \quad (11)$$

where $d(k+j)$ is the unmeasurable disturbances which will enter the process at future sampling period k+j. Usually it is assumed that the disturbances in future is equal to the discrepancy between the actual process output, $y(k)$ and the model output, $y_m(k)$ at the present sampling period k.

$$d(k+j)=y(k)-y_m(k) \quad (12)$$

Consequently, the outputs in future are predicted as follows:

$$y_f(k+j)=y(k)+[y_m(k+j)-y_m(k)] \quad (13)$$

In the case that the numbers of the manipulated variable vector ($u_1$) and measurable disturbance vector ($u_2$) are $q_1$ and $q_2$ respectively, the set of outputs predicted over the horizon, i.e. from j=1 to j=P, is represented by a set of linear equations with respect to the vector of the input increments as follows:

$$YF=Y+A_{f1}\Delta U_1+A_{p1}\Delta V_1+A_{f2}\Delta U_2+A_{p2}\Delta V_2 \quad (14)$$

where

$$YF=(y_f(k+1)^t, \ldots ,y_f(k+P)^t)^t$$

$$Y=(y(k)^t, \ldots ,y(k)^t)^t$$

$$\Delta U_1=(\Delta u_1(k)^t, \ldots ,\Delta u_1(k+P-1)^t)^t$$

$$\Delta V_1=(\Delta u_1(k-1)^t, \ldots ,\Delta u_1(k-N+1)^t)^t$$

$$\Delta U_2=(\Delta u_2(k)^t, \ldots ,\Delta u_2(k+P-1)^t)^t$$

$$\Delta V_2=(\Delta u_2(k-1)^t, \ldots ,\Delta u_2(k-N+1)^t)^t$$

superscript t: transpose

$$A_{f1}=\begin{bmatrix} A_1 & : & 0 \\ : & : & : \\ A_P & : & A_1 \end{bmatrix}$$

$$A_{p1}=\begin{bmatrix} A_2-A_1 & : & A_N-A_{N-1} \\ : & : & : \\ A_{P+1}-A_1 & : & A_N-A_{N-1} \end{bmatrix}$$

$A_{f2}$ and $A_{p2}$ are the same as those obtained by substituting the coefficient of step response to the measurable disturbances for the coefficient of step response to the manipulated variables in $A_{f1}$ and $A_{p1}$, respectively.

Usually, the reference $y_d(k+i)$ at sampling period $k+i$ is made from the original optimal values $s(k+i)$ and the present measured outputs as follows:

$$y_d(k+i)=(I-C^i)s(k+i)+C^i y(k) \qquad (15)$$

where

$$C=\text{diag}(c_1, \ldots ,c_{q_1})$$

This reference model is a command converter. The parameter matrix C is chosen to be able to operate the plant stably. Over the whole predictive horizon the reference trajectory is represented as follows:

$$YD=(I-C^*)S+C^* Y \qquad (16)$$

where

$$YD=(y_d(k+1)^t, \ldots ,y_d(k+P)^t)^t$$

$$S=(s(k+1)^t, \ldots , s(k+P)^t)^t$$

$$C^*=\begin{bmatrix} C^1 & : & 0 \\ : & : & : \\ 0 & : & C^P \end{bmatrix}$$

In the model predictive control a set of values of the manipulated variables for the future $\Delta U_1$ is determined by minimizing the following objective function.

$$F=[[YD-YF]]^2_Q{}^t_Q + [[\Delta U_1]]^2_R{}^t_R \qquad (17)$$

where

$$Q^t Q=\begin{bmatrix} Q_1{}^t Q_1 & : & 0 \\ : & : & : \\ 0 & : & Q_P{}^t Q_P \end{bmatrix}$$

$$R^t R=\begin{bmatrix} R_1{}^t R_1 & : & 0 \\ : & : & : \\ 0 & : & R_P{}^t R_P \end{bmatrix}$$

$$[[X]]^2_B=X^t BX$$

Sometimes the following conditions are added to suppress the changes in the manipulated variables.

$$\Delta u_{1i}(k+M_i)= \ldots =\Delta u_{1i}(k+P-1)=0 \qquad (18)$$
$$(i=1, \ldots , q_1)$$

Under the above conditions the set of unknown variables $\Delta U_M$ is represented as follows:

$$T_M \Delta U_M=\Delta U_1 \qquad (19)$$

where $T_M$ is a $[q_1 P]\times[\sum_{i=1}^{q_1} M_i]$ dimensional matrix.

Minimization of the quadratic objective function of eq.(17) subject to linear equations of eqs.(14),(16),&(19) with respect to $\Delta U_M$ gives the following optimal solution.

$$\Delta U_M=G^{-1}T_M{}^t A_{f1}{}^t Q^t Q\ E \qquad (20)$$

where

$$G=(T_M{}^t A_{f1}{}^t Q^t Q A_{f1} T_M+T_M{}^t R^t R T_M)$$

$$E=YD-[YF:\Delta U_1=0]$$

$$=YD-[Y+A_{p1}\Delta V_1+A_{f2}\Delta U_2+A_{p2}\Delta V_2]$$

Driving error vector, E is the discrepancy between the reference trajectory, YD and the predicted output trajectory under no changes in the manipulated values, $[YF: \Delta U_1=0]$. The manipulated values to be implemented at the present time, k is represented as follows:

$$\Delta u_1(k)= \sum_{i=1}^{P}[E_{1i}(I-C^i)[s(k+i)-y(k)]- \sum_{i=1}^{N-1}D_{1i}\Delta u_1(k-i)$$
$$- \sum_{i=1}^{P}E_{2i}\Delta u_2(k+i-1)- \sum_{i=1}^{N-1}D_{2i}\Delta u_2(k-i) \qquad (21)$$

where

$$[E_{11} \ldots E_{1P}]=bG^{-1}T_M{}^t A_{f1}{}^t Q^t Q$$

$$[E_{21} \ldots E_{2P}]=bG^{-1}T_M{}^t A_{f1}{}^t Q^t Q A_{f2}$$

$$[D_{11} \ldots D_{1N-1}]=bG^{-1}T_M{}^t A_{f1}{}^t Q^t Q A_{p1}$$

$$[D_{21} \ldots D_{2N-1}]=bG^{-1}T_M{}^t A_{f1}{}^t Q^t Q A_{p2}$$

$$b=[I\ 0 \ldots 0]$$

By using the z transformation the above control algorithm is rewritten as follows:

$$U_1(z)=[1/(1-z^{-1})][Dc(z)^{-1}Nc(1)]$$
$$\times[Nc(1)^{-1}Nc(z)]S(z)-Nc(1)^{-1}F1(z)-Y(z)$$
$$-[Nc(1)^{-1}Rc(z)](1-z^{-1})U_2(z)] \qquad (22)$$

where

$$Dc(z)=I+D_{11}z^{-1}+ \ldots +D_{1N-1}z^{-(N-1)}$$

$$Nc(z)=E_{11}(I-C)z^1+ \ldots +E_{1P}(I-C^P)z^P$$

$$Rc(z)=E_{21}+D_{21}z^{-1}+ \ldots +D_{2N-1}z^{-(N-1)}$$

$$F1(z)= \sum_{i=1}^{P}E_{1i} \sum_{j=1}^{i-1}s(j)z^{i-j}$$

The structure of the control law is shown by the block diagram in Fig.1. This control law includes both feedback and feedforward control elements.

(1) As the feedback control elements an integral element $[1/(1-z^{-1})]$ and a controller $[Dc(Z)^{-1}Nc(1)]$ are included. Therefore, no offset occurs for step changes in the set-point or disturbances.
(2) As the feedforward control elements a compensator $[Nc(1)^{-1}Nc(Z)]$ for the set-point change and a compensator $[Nc(1)^{-1}Rc(Z)]$ for the measurable disturbances are included.

HIERARCHICAL STRUCTURE OF THE INTEGRATED SYSTEM

## Two Level System

So far a two level system as shown in Fig.2, which
has no interface between the planning and control
levels, has been used. At the upper level with
the optimizer an optimal planning over a specified
time horizon or in most cases a steady-state
optimization is performed in a feedforward manner
at an interval of $\theta_1$. The process measurements
are used periodically to update the planning or
optimization model.

The transmission of the reference must be
synchronized with the actual time in the control
system, because the control system accepts only
the set-point at the present. In this two level
control system the process dynamics between the
manipulated variables and the process outputs can
not be taken into account to determine the set-
points of the control system.

## Three Level System

In a three level control system shown in Fig.3
the same optimal planning problem as the two level
system is solved at the optimizer. In this case
the resultant whole information of the optimal
plan can be utilized to determine more suitable
set-points to the lower level control system. For
this purpose the model predictive control
algorithm is adopted at an interval of $\theta_2$. The
algorithm can treat the dynamics of the augmented
process including the process and the minor
control loops and the reference trajectories in
future, which are determined at the planning
level.

As a result, three feedback control loops are
included in the new hierarchical control system.
The sampling periods for the optimizer, the model
predictive control and the real-time minor control
system are $\theta_1$, $\theta_2$, and $\theta_3$, respectively.

## ILLUSTRATIVE EXAMPLE

In an experimental fixed-bed reactor in which a
reforming reaction of methane by carbon-dioxide
occurs, catalyst deactivates rapidly by coking
under some operating conditions. The schematic
diagram of the experimental reactor system is
shown in Fig.4. The rate of the reforming
reaction on deactivated catalyst was represented
as follows (Kunugita & co-workers, 1989):

$$R=k(T)f(X) A \qquad (23)$$

where

$$k(T)=\bar{k} \exp(-E/T) \qquad (24)$$

and $f(X)$ is a function of the conversion of
methane at the reactor exit. The deactivation
factor, A depends on the history of the reaction
temperature.

$$dA/dt=-K(T) A \qquad (25)$$

It is assumed that both profiles of the
temperature and the deactivation factor are
uniform along the flow direction because of a
small amount of catalyst in the reactor.

At the optimizer the optimal schedule of the
reactor temperature , which gives the maximum
yield of the product during the whole operation,
is determined. It is known that under some
reasonable conditions the optimal policy of the
reaction temperature is equivalent to the
temperature which keeps the conversion constant at

the reactor exit. To cope with the uncertainties
in the deactivation model for planning, the
deactivation factor A is measured indirectly every
20 minutes, at which interval the composition of
the products are measured by gas chromatographs.
The measured value is used as the initial value of
the ordinary differential equation of the
deactivation model of eq.(25). This is a simple
way to update the model for planning by using the
sampled data.

The reactor temperature at time t, which gives the
same value of conversion at the starting time 0,
can be obtained by solving the following equation.

$$A(t)k[T(t)]=A(0)k[T(0)] \qquad (26)$$

In the experiment a discrete optimal temperature
schedule at an interval of 1 minute was
determined. The planning horizon was selected as
20+P, where P is the predictive horizon for the
model predictive control.

As a minor control system a PI controller of the
velocity form was used at an interval of 5
seconds. The dynamic behavior between the heater
current and the reactor temperature was
approximated by a transfer function model
(Kunugita & co-workers, 1989).

$$G(s)=T(s)/U(s)$$

$$=Kp[(t_3s+1)/[(t_1s+1)(t_2s+1)]]e^{-t_4s} \qquad (27)$$

where Kp and $t_i$ (i=1,...,4) are parameters, of
which values were determined from experimental
data. These values depend on the extent of
deactivation and operating conditions. Therefore,
a robust tuning of the controller parameters
should be considered. The following two
specifications were adopted for robust tuning.

"The damping ratio for the command response should
satisfy the following condition under three
representative extent of deactivation."

Specification 1:    $0.6 \le \xi \le 0.8$
                                                    (28)
Specification 2:    $0.8 \le \xi \le 1.0$

The control system with the former specification
is superior in the speed of response than the
control system with the latter specification.

The dynamic behavior of the augmented process
including the reactor and the PI controller was
represented by the step response model to design a
model predictive control system. The set of
sampled values of the step response model at an
interval of 1 minute are summarized for both
tunings in TABLE 1.

## Two Level System

The set-point of the PI control system is given
from the data set of the optimal temperature
schedule. In this case the tuning specification 1
was chosen for a fast follow-up to the set-point
changes. Figures 5 & 6 shows the reactor
performance during the whole operation and the
follow-up property for the most rapid deactivation
period.

## Three Level System

The data set of the optimal temperature schedule
at an interval of 1 minute is transmitted to the
model predictive control system as the reference
for the future. The model predictive control
algorithm calculates the modified set-point to the
PI controller by taking into account of the
dynamics of the augmented process. In this case

the tuning specification 2 is chosen because the follow-up property will be improved by the preview action in the control law. Figures 7 & 8 shows the reactor performance during the whole operation and the follow-up property for the most rapid deactivation period.

## Comparison Between Two Hierarchical Systems

The three level system with an interface is superior in the follow-up property than the two level system without the interface. As a result, the reactor performance was improved considerably. This is an evidence for the effectiveness of the proposed integrated system of the non-real-time planning and real-time control.

## CONCLUSION

In this paper a new three level system was proposed for an integration form of non-real-time planning and real-time control. An interface between the planning and control systems should have both the feedforward and feedback elements, which are emphasized at the planning and control functions, respectively. From this reason the model predictive control algorithm is an appropriate candidate for an interface. An application of the integrated system to a catalytic reactor system with catalyst decay illustrated the effectiveness of the proposed integration scheme.

## REFERENCE

Darby, M.L. & D.C. White (1988). On-line optimization of complex process units. Chem. Eng. Prog., 59, October, 51-59.

Gore F.E. (1987). Plantwide computerization: where are we now?. HydroCarbon Processing, October, 26-A-26-D.

Kunugita, E., H. Nishitani, & X.W. Tao (1989). Dynamic model of a fixed-bed reactor with catalyst deactivation. J. of Chem. Eng., Japan, 22, 258-263.

Lasdon L.S. & T.E. Baker (1986). The integration of planning, scheduling, and process control. In M. Morari & T.J. McAvoy (Ed.), Chemical Process Control 3, CACHE-ELSEVIER, pp.579-620.

Lasdon L.S., A.D. Waren, & S. Sarker (1988). Interfacing optimizers with planning languages and process simulation, In G. Mitra (Ed.), Mathematical Models for Decision Support, Springer-Verlag, pp.239-262.

Nishitani, H. (1989). Applications of the model predictive control (In Japanese). Keisoku to Seigyo, In Press.

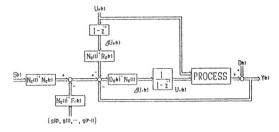

Fig. 1. Structure of the model predictive control.

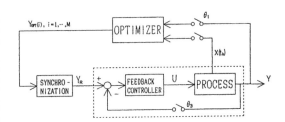

Fig. 2. Conventional two level system.

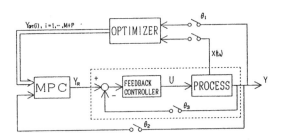

Fig. 3. New three level system.

## TABLE 1 Sampled Values of the Step Response Model

Tuning Specification 1: Sampling Period= 1 min.
N=16

| | | | | | |
|---|---|---|---|---|---|
| 0.1575 | 0.7874 | 1.2598 | 1.3386 | 1.1911 | 0.8976 |
| 0.8346 | 0.8976 | 1.0394 | 1.0551 | 1.0236 | 0.9921 |
| 0.9764 | 0.9843 | 0.9921 | 1.0000 | | |

Tuning Specification 2: Sampling Period= 1 min.
N=10

| | | | | | |
|---|---|---|---|---|---|
| 0.0222 | 0.4436 | 0.8095 | 0.8871 | 0.8760 | 0.9204 |
| 0.9758 | 0.9980 | 1.0202 | 1.0000 | | |

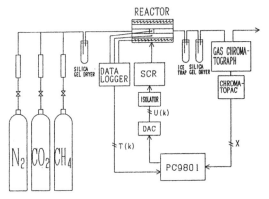

Fig. 4. Schematic diagram of the experimental fixed-bed reactor system.

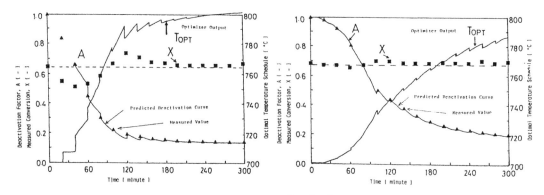

Fig. 5.   Reactor performance in the two level
          system

Fig. 7.   Reactor performance in the three level
          system.

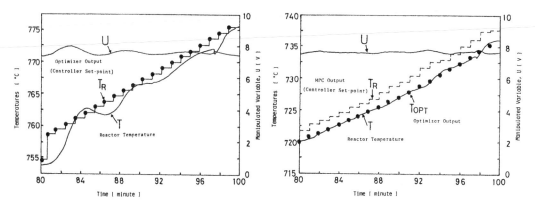

Fig. 6.   Follow-up property for the most rapid
          deactivation period in the two level
          system.

Fig. 8.   Follow-up property for the most rapid
          deactivation period in the three level
          system.

# OPTIMAL PRODUCTION OF GLUTATHIONE BY CONTROLLING THE SPECIFIC GROWTH RATE OF YEAST IN FED-BATCH CULTURE

## H. Shimizu,* K. Araki,* S. Shioya,* K. Suga and E. Sada**

*Department of Ferm. Technology, Osaka University, Osaka 565, Japan*
**Department of Chemical Engineering, Kyoto University, Kyoto 606, Japan*

**Abstract** The optimal profile of the specific growth rate was obtained utilizing simple mathematical model in the yeast fed-batch culture. The model was built based on the mass balance around the fed-batch system and relationship between the specific growth rate, $\mu$, and the specific production rate of glutathione, $\rho_G$. Optimal profile of $\mu$ was calculated as a bang-bang type one. That is, $\mu$ should start from the maximum value, $\mu_{max}$ and should be kept at $\mu_{max}$. And then $\mu$ should be switched to $\mu_C$, that gives a maximum value of $\rho_G$. It has been proved from the maximum principle that switching was once and switching time from $\mu_{max}$ to $\mu_C$ depended on the final required glutathione concentration in the cell. Finally, this ideal profile of $\mu$ for the maximum production of glutathione was realized by manipulating the substrate feed rate, and $\mu$ could be controlled at the optimal profile given in advance, using the extended Kalman filter and PF(Programmed Controller/ Feedback Compensator) System. As the result, the maximum production of glutathione was accomplished successfully. The control strategy employed here can be applied to the other batch reaction processes.

**Key Words** Maximum principle; Bang-bang control; Extended Kalman filter; PF System; Glutathione production from the yeast

## INTRODUCTION

Glutathione (GSH) is a kind of useful tripeptide consisting of L-glutamate, L-cysteine and glycine, which affects widely the oxidation and reduction in vivo. Recently glutathione has been reported as a medicine for the lever and a scavenger of the toxic compounds. It is well known that glutathione content in a certain strain of the yeast is high. It can assimilate the glucose and also produce ethanol due to the Crabtree effect when the glucose concentration in the fermentor is very high. And it has been reported that the productivity of glutathione depended upon the kinds of carbon sources (Sakato et al., 1985).

The aim of this paper is to attain the optimal production of glutathione by controlling the specific growth rate of the yeast. First, the effect of the concentration of glucose on the specific growth rate, $\mu$, and the specific production rate of glutathione, $\rho_G$, which are defined as the growth rate per unit cell and production rate of glutathione per unit cell respectively, were investigated. And simple model of the glutathione production system was constructed. This model was based on the mass balance around the fermentor and the relationship between $\mu$ and $\rho_G$. Second, the optimal profile of $\mu$ for the glutathione production was obtained by the maximum principle. Finally, this ideal profile of $\mu$ for the maximum production of glutathione was realized by manipulating the feed rate of the substrate. Here, $\mu$ could be controlled at the optimal profile given in advance, using extended Kalman filter and PF (Programmed Controller/ Feedback Compensator) System.

## GLUTATHIONE PRODUCTION RATE

Metabolic pathway of glutathione in the yeast is shown in Fig.1. When excess amount of the glucose is supplied, the yeast produces ethanol due to the Crabtree effect. Pyruvate and Acetyl-CoA enter TCA cycle and energy is released. Glutamate is made from α-Ketoglutarate which is a member of TCA cycle. And glutathione is made from glutamate and other amino acids, such as cysteine and glycine enzymatically. That is, these precursours are important for glutathione production. In the previous report it has been shown that when glucose was used as a sole carbon source, both of $\mu$ and $\rho_G$ became larger than that in case using ethanol as a carbon source(Shimizu et al., 1988). In this

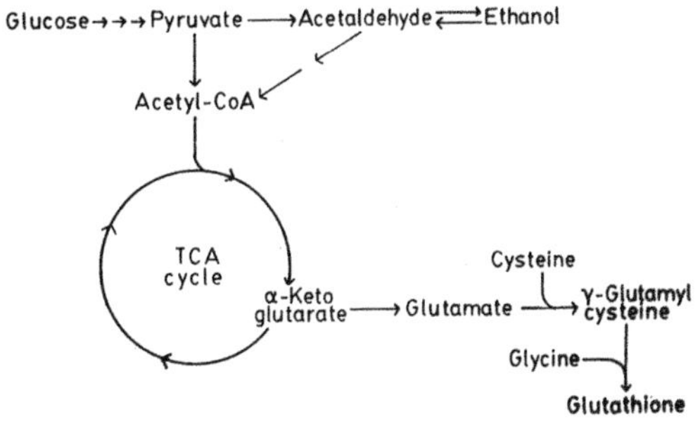

Fig.1 Metabolic pathway of glutathione in the yeast.

paper glucose was utilized as a sole carbon source and relationship between specific growth rate $\mu$ and specific production rate of glutathione $\rho_G$ was investigated in the fed-batch culture. Experimental apparatus is shown in Fig.2. The main facilities used are; 5L-jar fermentor with temperature controller (Mitsuwa Co.Ltd, Japan), microtube pump (Eyla Co.Ltd, Japan) used for feeding substrate, pH and DO controller (Mituswa Co.Ltd, Japan) connected to the computer, microprossesor $\mu$-MAC 5000 (Analog Devices Co.Ltd, USA), which was used as an A/D, D/A converter and first step data processing, and personal computer PC-9801 (NEC Co.Ltd, Japan), which is a main computer.

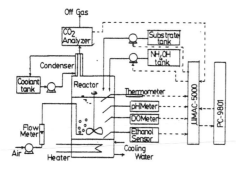

Fig.2 Schematic diagram of the experimental apparatus.

Relationship between $\mu$ and $\rho_G$ is shown in Fig.3. The relationship between $\mu$ and the specific production rate of ethanol $\rho_E$, which is defined as the production rate of ethanol per unit cell is also shown in Fig.3. It should be notified that $\rho_G$ had the maximum value at 0.30 (hr$^{-1}$) of $\mu$. Ethanol was produced at higher $\mu$ than 0.30 (hr$^{-1}$) by the Crabtree effect. This critical value of $\mu$ coincided completely with the value at which $\rho_G$ had the maximum. That is, when the glucose concentration was too high and ethanol was produced, $\rho_G$ decreased. It was obviously lower than the highest value in the case of no production of ethanol.

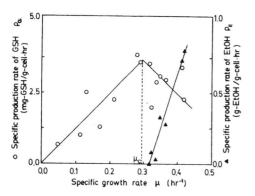

Fig.3 Relationship between $\mu$ and $\rho_G$, $\rho_E$.

## MODEL OF GLUTATHIONE PRODUCTION IN FED-BATCH CULTURE

The model of the glutathione production in the fed-batch system can be represented as follows.

(Cell growth)
$$\frac{d(VX)}{dt} = \mu(VX) \qquad (1)$$

(Glutathione production)
$$\frac{dp}{dt} = \rho_G - \mu p \qquad (2)$$

Where, V, X, and p are liquid volume($\ell$), cell concentration (g/$\ell$) and glutathione concentration in the cell (mg-GSH/g-cell), respectively. When the total cell mass, VX, was defined as Z, Eq.(1) can be rewritten as,

$$\frac{dZ}{dt} = \mu Z . \qquad (3)$$

The relationship between $\mu$ and $\rho_G$ shown in Fig.3 can be approximately formulated as two straight lines as,

$$\rho_G(\mu) = a_i \mu + b_i \qquad (4)$$
$$(i=1; \ 0 \le \mu \le \mu_C)$$
$$(i=2; \ \mu_C < \mu) .$$

Each parameters were estimated as,

$a_1 = 11.78, \quad b_1 = 0,$
$a_2 = -9.903, \quad b_2 = 6.359,$

using the least square method.

## FORMULATION OF OPTIMIZATION PROBLEM

When the glutathione is produced using the yeast fed-batch culture, two objectives should be evaluated.

(1) The amount of the total glutathione, that is the product of cell mass, Z, and the glutathione concentration in the cell, p, should be maximized.

(2) Glutathione concentration in the cell, p, should be kept at a high value at the final time of the fermentation because of the economic factor of the down stream processing.

Considering the above two objectives, an optimization problem can be stated as: to find the optimal profile of the specific growth rate, $\mu$, so as to maximize the total amount of glutathione and to keep the glutathione concentration at the high value at the fixed final time. Therefore, the optimization problem can be formulated mathematically as follows.

<Optimization problem of GSH production>
The problem is to find the optimal profile of $\mu$ so as to maximize the following performance index,

$$J = pZ(t_f)$$
$$= pZ(0) + \int_0^{t_f} \rho_G Z \, dt , \qquad (5)$$

under the constraint at the final time as,

$$p(t_f) = p_f \ . \tag{6}$$

in the glutathione fed-batch culture described by Eq.(2), (3) and (4). Where, $t_f$ and $p_f$ are the fixed final time of the fermentation and the concentration of the glutathione in the cell at the final time, respectively. Further, the following inequality condition of $\mu$ and the initial conditions were imposed.

$$\mu_{min} \leq \mu \leq \mu_{max} \tag{7}$$

$$\begin{aligned} Z(0) &= Z_0 \\ p(0) &= p_0 \end{aligned} \tag{8}$$

Where, $\mu_{min}$, $\mu_{max}$, $Z_0$ and $p_0$ are the maximum and minimum values of $\mu$ to be manipulated freely in the fed-batch culture, and initial values of cell mass and concentration of glutathione in the cell, respectively.

## OPTIMUM PROFILE OF THE SPECIFIC GROWTH RATE

The optimal profile of the specific growth rate, $\mu$, could be obtained by the maximum principle and was a bang-bang profile. That is, $\mu$ should start at the maximum value, $\mu_{max}$, and should be kept at $\mu_{max}$. And then, $\mu$ should be switched to $\mu_C$, which gave a maximum value of $\rho_G$. Switching time $t_b$ depended on the final required value of glutathione concentration in the cell, $p(t_f)$. It has been proved using the theory of the maximum principle that the switching is once or zero times and there exists no singular control under any initial conditions, operation time, $t_f$, and any constraint of $p(t_f)$.

One of the examples of the optimal profile of $\mu$ is shown in Fig.4. The final time of the fermentation, $t_f$, the final required concentration of glutathione, $p_f$, the initial cell mass, $Z_0$ and the initial concentration of the glutathione in the cell, $p_0$ were 10(hr), 11(mg-GSH/g-cell), 1(g), 5.24(mg-GSH/g-cell), respectively. The switching time $t_b$ was 3.5(hr) as shown in the figure. Glutathione concentration in the cell, p, increased after switching $\mu$ from $\mu_{max}$ to $\mu_C$, because $\mu_C$ gave the maximum value of $\rho_G$. The solid line shows the total amount of GSH in the broth. For comparison, the total amount of glutathione in the case of the constant control of $\mu$ is also shown as the chain line in Fig.4. When $\mu$ was kept at $\mu_C$ from initial time to the final time, $\rho_G$ was kept at a maximum value. However, when $\mu$ was controlled to the optimal profile, cell mass increased much more than the case of controlling $\mu$ at $\mu_C$ constantly, although $\rho_G$ was not always maximum. That is, in the optimal strategy, the cell mass was requested to increase quickly in the period controlling $\mu$ at $\mu_{max}$ and the glutathione concentration was requested to increase largely after switching $\mu$ from $\mu_{max}$ to $\mu_C$. Finally, the total amount of glutathione at the final time according to the optimal strategy was larger than that using constant control of $\mu$ as shown in the figure.

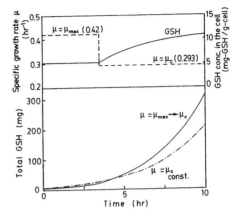

Fig.4 Optimal profile of $\mu$ and the comparison of the total amount of glutathione between optimal control and constant control of $\mu$.

Fig.5 shows the relationship between glutathione concentration in the cell at the final time $p(t_f)$ and switching time $t_b$ when the operating time $t_f$ is fixed as 10(hr). And the total amount of GSH vs $p(t_f)$ is also shown. When required $p(t_f)$ was high, $t_b$ should be small and possible amount of total GSH must be also small. The value of $t_b$ and total amount of GSH increased according to decreasing $p(t_f)$ and the amount of the total GSH had the maximum value at 8.19(mg-GSH/g-cell) of $p(t_f)$. In that case, switching time $t_b$ was 7.95(hr). From another viewpoint, this was shown to be the solution of the maximizing GSH production problem without any constraint for the final glutathione concentration in the cell. That is, possible maximum value of the amount of GSH with 10 hours operation, $p_M$, is

421.4 (mg-GSH).

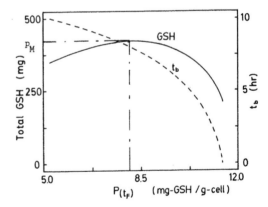

Fig.5 Relationships between glutathione concentration in the cell at the final time $p(t_f)$ and switching time $t_b$, and total GSH vs $p(t_f)$.

The effect of the uncertainties included in the initial value of the glutathione concentration, $p_0$, was studied. Initial value of the glutathione concentration in the cell cannot be measured on-line because glutathione is produced in the cell and there is no available sensor for glutathione concentration measurement. Therefore, if there are some errors in the pre-set initial concentration in the cell, the optimal profile of $\mu$, especially switching time $t_b$ should be changed. In Fig.4 and Fig.5, the initial value of the glutathione concentration in the cell was set at 5.24(mg-GSH/g-cell), that is obtained in the steady state keeping $\mu$ at $\mu_{max}$ and that is the smallest value of $p$. If there are some errors in $p_0$, then $p(t_f)$ and the total amount of glutathione are shifted to the larger values than the expected values in the optimal solution.

Even if $p_0$ includes some errors or changes to the different value from the pre-set value, $\mu$ must be switched from $\mu_{max}$ to $\mu_C$. correctly. Fig.6 shows the time course of $p$ according to each optimal strategy, that is, in which the optimal switching time, $t_b$, is taken correctly for various $p_0$. It should be notified that the dynamics of $p$ doesn't depend on the cell mass, $Z$, as seen in Eq.(2). When $\mu$ is kept at $\mu_{max}$, $p$ goes to 5.24(mg-GSh/g-cell), which is a value in the steady state keeping $\mu$ at $\mu_{max}$ and that is the minimum value of $p(t_f)$. And when $\mu$ is switched to $\mu_C$ and kept constant, $p$ goes up to 11.78(mg-GSH/g-cell), which is the value in the steady state keeping $\mu$ at $\mu_C$ and that is the maximum value of $p$. The break points in every lines indicate the switching of $\mu$ from $\mu_{max}$ to $\mu_C$. The differences of $t_b$ were in the range of about 30(min). Fig.7 shows the time courses of $p$ for various $p_0$ when the switching time is 3.5(hr). Four lines indicate $p$ at time t, when $p_0$ are 10, 8, 6, 5.24(mg-GSH/g-cell), respectively. All values of $p(t_f)$ are almost the same. According these results, it is shown that the sensitivity of the uncertainties of $p_0$ for the optimal trajectory of $\mu$ is very small. So it is true that there is no problem about uncertainties of $p_0$ in the actual operation.

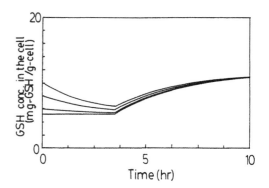

Fig.7 Time course of concentration of glutathione for various initial concentrations of GSH.

## REALIZATION OF OPTIMAL PRODUCTION OF GSH

The optimal profile of $\mu$ for the maximum production of glutathione was actually realized by controlling the feed rate of the substrate. At first, $\mu$ was indirectly controlled by exponential feeding plus ethanol concentration control. One of the control results is shown in Fig.8. The experimental conditions are same as those of experiment shown in Fig.4. Dotted line in the figure of $\mu$ is the desired trajectory of $\mu$ and the solid line is the real value estimated by the extended Kalman filter(Shimizu et al., 1989a). Exponential feeding of the substrate was accomplished for the initial 2.5(hr). Because of the delay of the response of $\mu$ against feeding flow rate change, the desired $\mu$ for the exponential feeding was changed to the new one at 2.5(hr).

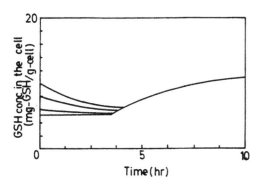

Fig.6 Optimal trajectories of $p$ for various initial concentration of the glutathione in the cell, $p_0$. $p_0$ are: 10, 8, 6, 5.24 (mg-GSH/g-cell).

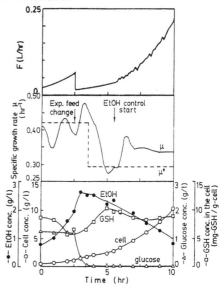

Fig.8 Time course of the state variables in the fermentor when $\mu$ was controlled to the optimal profile using the exponential feeding plus ethanol concentration control.

After 5(hr), ethanol concentration was controlled by feeding the substrate. The yeast shouldn't produce nor assimilate ethanol at $\mu_C$. Then, ethanol concentration should be constant at $\mu_C$, if the dilution by the feeding of the fresh medium is negligible. However, this experiment was not such case. On the contrary, ethanol concentration should be decreased according to the dilution due to the feeding, if $\mu$ is controlled at $\mu_C$. Then, the objective value of ethanol concentration for the control was given as the decreasing curve by the dilution. If ethanol concentration is controlled to the desired decreasing curve, $\mu$ and $\rho_G$ can be controlled at $\mu_C$ and the possible maximum value, respectively. This control strategy was realized by PF(Programmed controller/Feedback Compensator) System (Takamatsu et al., 1985), (Shimizu et al., 1989b). PF System consists of programmed controller and pre-compensator. Programmed controller gives the desired value to the controlled variable if there is no disturbance. Pre-compensator can compensate the undesired disturbances. Ethanol concentration was measured by the ethanol sensor using teflon tube and FID detector (Dairaku et al., 1981). Indirectly, $\mu$ was controlled at $\mu_C$ successfully as seen in the figure. Glutathione concentration in the cell, p, increased after switching $\mu$ from $\mu_{max}$ to $\mu_C$ as expected from the model.

The comparison of the time courses of the amount of the total GSH between constant $\mu$ and bang-bang control of $\mu$ was performed as shown in Fig.9. Open circle indicates the time course of total amount of GSH obtained by the optimal profile control of $\mu$. On the other hand, closed triangle indicates the amount of total GSH obtained by constant control of $\mu$. The constant value of $\mu$ was 0.305(hr$^{-1}$), which is very close to $\mu_C$. As shown in Fig.9, total GSH obtained by the optimal strategy, was about 1.41 times larger than that by constant control of $\mu$.

The maximum production of glutathione was also accomplished by controlling $\mu$ to the optimal profile directly, with the extended Kalman filter and PF System. Estimation and control system of $\mu$ is shown in Fig.10. $\mu$ could be estimated as follows (Shioya, 1988). The nitrogen uptake rate by the cell, $R_{NH3}$ could be calculated from the ammonia feeding rate, $F_{NH3}$, for pH control of the fermentor, considering the material balance of the nitrogen. Growth rate $\mu X$ could be estimated from the relationship between $F_{NH3}$ and $\mu X$. Moreover $\mu$ and cell concentration X were estimated by the extended Kalman filter using $\mu X$ as the measured variable. Estimated $\mu$ could be controlled by PF System. The standard feed rate $\bar{F}$ that was calculated as exponential feeding was given by programmed controller and actual feed rate F was given as a sum of $\bar{F}$ and $\Delta F$ that is calculated by the pre-compensator.

Estimation and control result for the maximum production of glutathione under 6(hr) operation is shown in Fig.11. Dotted line in the figure of $\mu$ is the desired profile of $\mu$. For the maximum production of GSH without any restriction for $p(t_f)$, the switching time $t_b$ was given as the time 2.05(hr) before the final time $t_f$ as mentioned before. The solid line indicates the estimated values by the extended Kalman filter. Open circle and solid line in the figure of cell concentration were off-line data and estimated values by the extended Kalman filter, respectively. Estimated values of cell concentration coincided with the measured one accurately and it was confirmed that $\mu$ and cell concentration could be estimated by extended Kalman filter satisfactorily. And $\mu$ was controlled at the optimal trajectory by PF System.

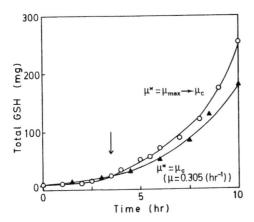

Fig.9 Comparison of the total GSH by the optimal control of $\mu$ with that by the constant $\mu$.

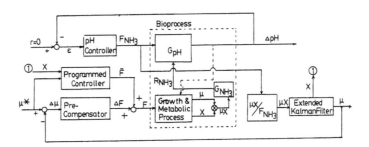

Fig.10 Structure of the estimation and control system of $\mu$.

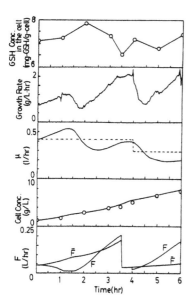

Fig.11 Result of estimation and optimal
profile control of μ.

The comparison of the total amount of
glutathione for both control strategies, was
performed as shown in Fig.12. Open circle is
the time course of the total GSH obtained when
μ was controlled to the optimal profile. On
the other hand, closed triangle indicates the
time course of the total GSH by constant
control of μ. When μ was controlled at the
constant value, 0.305 $(hr^{-1})$, $p(t_f)$ was higher
than that of the optimal control of μ.
However, when μ was controlled to the bang-
bang profile, the cell mass at the final time
increased much more than that by the constant
control of μ. As the result, the maximum GSH
production could be accomplished by
controlling μ to the optimal bang-bang profile
of μ.

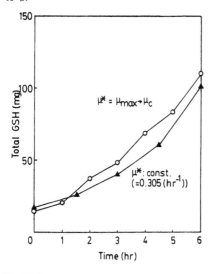

Fig.12 Comparison of the total GSH obtained by
the optimal control with that obtained
by the constant μ.

## CONCLUSIONS

Optimal profile of μ for the glutathione
production in the fed-batch culture was
determined as a bang-bang profile. The
optimal profile of μ could be realized by
feeding the fresh medium based on the PF
System. And the maximum production of
glutathione could be also realized
experimentally, using the extended Kalman
filter and PF System. The control strategy
employed here can be applied to the other
batch chemical and biochemical reaction
processes.

## ACKNOWLEDGEMENT

The authors would like to thank
Mr. A.Kanda and Mr. C.G.Alfafara for their
assistance of these works and also thank Kyowa
Hakko Kogyo Co.Ltd which kindly offers us the
yeast strain used in the experiment.

## References

Dairaku, K., Y.Yamasaki, K.Kuki, S.Shioya and
T.Takamatsu, (1981). Biotechnol. &
Bioeng., , vol. 23, 2069.
Sakato, K., H.Tanaka, and S.Hamada, (1985).
Proc. of the annual meeting of Ferm.
Technol., Japan, 49 Japanese
Shimizu, H., K.Araki, M.Ogura, S.Shioya,
K.Suga and E.Sada, (1988). Proc. of the
4th International Cong. on Computer
Applications in Ferm. Technol.,
Cambridge.
Shimizu, H., S.Shioya, K.Suga and T.Takamatsu,
(1989a). Biotechnol. & Bioeng., vol. 33,
354.
Shimizu, H., S.Shioya, K.Suga and T.Takamatsu,
(1989b). Appl. Microbiol. & Biotechnol.,
vol. 30, 276.
Shioya, S., (1988). Proc. of the 4th
International Cong. on Computer
Applications in Ferm. Technol.,
Cambridge.
Takamatsu, T., S.Shioya, Y.Okada and M.Kanda,
(1985). Biotechnol. & Bioeng., vol. 27,
1675.

# DESIGN OF THE INTELLIGENT ALARM
# SYSTEM FOR CHEMICAL PROCESSES

## Rong Gang, Wang Shu-Qing and Wang Ji-Cheng

*Research Institute for Industrial Process Modelling and Control, Zhejiang University,
Hangzhou, 310027, PRC*

**Abstract**. In large process plants , alarm systems play a very important role in aiding
operator in his primary tasks of detecting and interrupting the progression of a fault,
and providing corrective actions for fault conditions. In order to improve the success
likelihood of the operator in receiving and processing alarms, an intelligent alarm
system is required. The concept of intelligent alarm means that the information given
by the process alarms and the manner in which this information is provided are based
on the process knowledge and helpful for improving operator's fault diagnostic efficiency. In this paper, the intelligent alarm system of a chemical reaction process has
been described. The fault propagating model used in design is a knowledge base with a
hierarchical structure, in which, the high level is a semantic network describing the
causal relationships among faults and the lower levels are functional components and
unit operations. Various conditions in which alarms can be activated are adopted so as
to indicate different fault situations in different operational mode of the plant. The
alarm-placement problem for optimal fault information output has also been solved, and
two kind of constraints considered. Alarm handling and reduction is implemented in two
steps, firstly, a mixed-direction reasoning technique is employed to acknowledge as
many activated alarms as possible and detect spurious alarms, and then, the alarm information is displayed in a prescribed order. The designed alarm system is implemented
with a personal computer although both knowledge-based techniques and conventional
numerical algorithms involved. A test case is discussed in detail.

**Keywords**. Alarm systems; system failure and recovery; artificial intelligent; chemical
industry; computer application; optimization.

## INTRODUCTION

Modern chemical plants tend to be large, complex
and highly instrumented. When a plant was partly
or completely stopped under fault conditions, it
will be expensive for the community,and the maintenance costs may be considerable. Moreover, some
of the ignored faults propagate in the plants and
result in a serious accident. Then the study of
how to assist human operator in his tasks of efficient detection, diagnosis and compensation of
the process faults takes on a great important.

There are mainly two kinds of methodology proposed
to aid human operator in this task,one is oriented
to automated fault diagnosis systems and the other
to diagnostic decision support systems or training
systems. Many research efforts on automation of
fault diagnosis have been reported and reviewed (
Lees, 1983; Isermann, 1984; Kokawa and co-workers,
1983; Kramer and Palowitch, 1987). In these methods, process computers serve as the diagnostic decision makers on behalf of human operator. On the
other hand, human operator makes decisions on detecting and diagnosing faults when the computer
decision aid system is adopted. The investigation
in this area includes alarm management and alarm
system analysis (Lees, 1983; Krigman,1985; Corsberg, 1987; Modarres and Cadman, 1986), decision
support for the operation at the fault conditions
(Yufik and Sheridan, 1986; Rasmussen, 1987), and
fault daignosis training simulator (Su and Govindaraj, 1986).

The selection of different diagnostic approaches
depends on the processes to be diagnosed. In the
complex chemical process plants, it is almost
impossible to use only one diagnostic technique to

deal with all of the faults automatically and effectively. Some of the malfunctions can be detected
and located by process computers in real-time, the
others probably require analysed by human operator
with the help of computers. Combination of the two
kinds of methodology will be helpful and necessary
if one tried to improve the efficiency of the daignostic system. To meet the informational need of
such combination, the design of alarm systems needs
to be updated.

In this paper, an intelligent alarm system has been
set up and the concept of intelligent alarms was
defined as a set of fault information that is based
on process knowledge and provided in a knowledge-
based manner. The knowledge base used in design of
the alarm system is a fault propagation model with
the hierarchical structure in which the high level
takes the form of a semantic network representing
causal relationships among faults of the nodes that
are functional components and unit operations defined in the lower levels (Rong and co-workers,1988).
Because of the structure of the proposed alarm system, both numerical and knowledgr-based methods can
be used to handle alarms, i.e., a bidirected reasoning technique is employed to detect if there are
any spurious alarms in the activated alarms, and
then the alarm messages will be displayed in a prescribed order. Besides, an informational entropy
measure of alarm system has been proposed to solve
optimal alarm-placement problem so as to provide as
much informatiom as possible,and two constraints
considered. A test case is discussed in detail.

### THE INTELLIGENT ALARM SYSTEM

#### The Structure of the Alarm System

Fig. 1 shows the flowsheet of the knowledge-based
alarm system built up in this work. The function
of the alarm system is to provide fault information about the process plant to a fault diagnostic
system where both computer and human operator can
be the decision maker for fault location and fault
correction, and different kinds of fault diagnostic methodology can be adopted.

There are six typical modules in the alarm system
desinged. They can be implemented with differen
algorithms in computer. The modules of DATA SAMPL-
ING, DATA BASE and FAULT DETECTION are constructed
by utilizing conventional numerical techniques and
can form a traditional alarm analyses system (Lees,
1983). In the module of FAULT DETECTION, real-time
data and off-line data as well as operational mode
of the plant have been employed to detect all of
the potential malfunctions in the plant. A set of
algorithms helpful for detecting fault, such as
filters, estimators and observers, have been built
in the module and can be put into use if needed.
Moreover, various conditions in which alarms will
be activated have also been considered according
to the definition of the plant.

### A Model of Fault Propagation

A hierarchical model of fault propagation which
depends on the structure of the plant, the functional specifications of the components and the
cause-and-effect relationships between variables
of the process is required to represent fault behavior and even to diagnose them. Given suitable
definitions of the nodes, directed digraph (Kokawa
and co-workers, 1983) can be used to represent the
high level fault propagation through the entire
plant. The definition of the node in a directed
digraph depends on the way in which the plant was
decomposed. In this paper, four categories typical
node have been classified:
·primary node, a node needs no decomposition.
·feedback control loop, local SISO loop.
·node that needs particular approach for fault
detection
·node in which a well-established diagnostic algorithm works properly.

### The Knowledge Base

The content of the knowledge base shown in Fig. 1
consists of the following parts:
KB1. knowledge about fault propagation represented
as a model with hierarchical structure.
KB2. importance of faults, priority of alarms and
cost of detecting faults in a node.
KB3. rules for generating diagnostic problems or
alarms to be displayed.
KB4. availability of the fault diagnostic problem
solvers (computer algorithms and human operator).

KB1 takes the form of a semantic network in which
the nodes are the same as those defined in directed digraph of the plant, and the meaning assigned
to the links between nodes represent the nature of
fault progressing. The semantic network is convenient to convert into other representations, for
instance, into a rule base in this work.

From rules in KB2-KB4, we are able to make inference of how to present alarms and allocate diagnostic tasks among computer diagnostic system and
human operator.

### Inference Engine

DIPAG (DIagnostic Problem and Alarm Generator), as
illustrated in Fig. 1, is the main module in the
intelligent alarm system. Once FAULT DETECTION
reports one or more abnormal measurements, DIPAG
tries to acknowledge activated alarms and find out

spurious alarms by using a bidirected reasoning
technique, and then decides to whom the current
diagnostic task will be assigned in terms of KB3
and KB4. If DIPAG believes that the problem must
be submitted to human operator, it has access to
KB2 and handles alarms and provides suitable alarm
messages for display in the man-machin interface.

### AN EXAMPLE IN CHEMICAL REACTOR DOMAIN

### Description of the Process Plant

The simplified flowsheet of a chemical reaction
plant is illustrated in Fig. 2. It is a extended
version of the example in the work of Rong and co-
workers (1988). The typical unit operation is a
continous-stirred tank reactor CSTR where the
irreversible first order reactions

$$A \xrightarrow{k1} B \xrightarrow{k2} C$$

take place. It is required to control $c_A$ and $c_B$ as
close as possible to the desired set point $c_{AS}$ and
$c_{BS}$ by adjusting $c_{AF}$ and $c_{BF}$ which are, in turn,
controlled by diluting prime feed A and B with the
concentration $c_{AO}$ and $c_{BO}$ in the TANK 1 and TANK 2
of Fig. 2. Output of the reactor was pumpped into
TANK 3 in which a level alarm was equipped,and the
concentration of byproduct C, was monitored with
an off-line measurement periodically.

### Decomposition of the Plant

A semantic network representation of the plant is
illustrated in Fig. 3, where each node represents
a functional subsystem which has one or more characteristic variable(s). One can detect if a node
is failure in terms of the value of its character-
ristic variables. For instance, the pressure difference between outlet and inlet of the pump is a
characteristic variable reflecting the performance
of the pump, if it dose not take a positive value,
one will infer that the pump may be fail to work.

According to the decomposition method proposed in
previous part of the paper, the reactor process is
decomposed into 12 nodes (/subsystems). The composition of each node in Fig. 3 is listed in Table 1
and the task of locating fault in Fig. 3 is defined as the global fault diagnostic problem, i.e.,
high level diagnosis. In turn, the tasks of locating original fault causes within a failure node is
called lower level diagnosis or local diagnosis.
The hierarchical structure proposed here makes it
possible to solve different diagnostic problem
with different diagnostic strategies.

### The Alarm System

We define an alarm as a characteristic variable in
a node. The value of these variables can be known
in three ways:
·directly measured by on-line sensors
·estimated with a mathematical model
·observed or checked by human operator.
If the characteristic variable takes an abnormal
value, the alarm is activated. If one of the alarms of a node is activated, the fault status of the
node has a binary value "1", otherwise, it takse
"0". Assum only the fault status of a node propag-
ates through a semantic network, meaning assigned
to links between nodes simply takes "AFFECT", and
the network is simplified into a directed digraph.

In the computer alarm system proposed in Fig. 1,
the module FAULT DETECTION will activated alarms
in terms of available information. We introduce a
cost function of fault detection that evaluates if
it is easy or not to find out a malfunction in a
node.

TABLE 1  Decomposition of the Plant

| Node Number | Composition of the Node | Candidate Alarms |
|---|---|---|
| $s_1$ | reactant A | $c_{AO}$, $F_A$ |
| $s_2$ | reactant B | $c_{BO}$, $F_B$ |
| $s_3$ | VA3, TANK1, PIPE1 | $c_{AF}$ |
| $s_4$ | VA4, TANK2, PIPE3 | $c_{BF}$ |
| $s_5$ | FC1, VA1, SF1, PIPE2 | $F_1$ |
| $s_6$ | FC2, VA2, SF2, PIPE4 | $F_2$ |
| $s_7$ | TC, ST, VA6, Fw | T |
| $s_8$ | CSTR, HE, AGIT A, B, C | T, L, $c_A$ $c_A$, $c_B$ |
| $s_9$ | PIPE5, PUMP, PIPE6 | $P_6$-$P_5$ |
| $s_{10}$ | SL1, LC, VA5 | L |
| $s_{11}$ | TANK3, PIPE8, VA7, SL2 | L2, $c_C$ |
| $s_{12}$ | PIPE7 | / |

The cost function is defined as

$$C(i) = w_{1i}d(i) + w_{2i}o(i) + w_{3i}m(i) \qquad (1)$$

where, C(i) - cost of fault detection in node i,
d(i) - cost of equipment for detecting fault,
o(i) - computational cost of model-based fault
      detection,
$w_{1i}$, $w_{2i}$, $w_{3i}$ - wighting parameters.
m(i) - man power needed for check and observation.
The candidate alarms in every node are also listed
in Table 1. We attempt to select alarms that have
to involve much information and lower costs.

OPTIMAL ALARM-PLACEMENT

A Brife Introduction of the Problem

The goal of fault diagnosis is to distinguish as
many fault modes from each other as possible, with
the least computation, in terms of available measurements. Fault diagnosability is a quantitative
index to evaluate the potential performance of a
fault diagnostic system. Given a measurement scheme
of the plant, its diagnosability is determined by
the structure of the plant. Ishida and co-workers
(1985) had presented a topological approach to diagnosability analysis, however, it cannot be applied
to general fault diagnosis model subjected to propagation constraints.

The counterpart of the diagnosability problem is
the problem of alarm-placement, that is to design a
alarm-placement scheme for the process plant so as
to provide maximum information about faults of the
plant with a given cost.

In this paper, we propose a quantitative index of
alarm systems. This index represents the information that an alarm system can provide and can be
used to evaluate performance of the alarm system.
It also affects diagnosability of the fault diagnostic system based on the alarm system designed, but
it is independent of the daignostic methods adopted, so that it is a general performance index of
alarm systems. This index is called informational
entropy measure in the sense of Shannon's entropy

(Shannon, 1948).

Informational Entropy Measure

Assum that the system $S=(s_1,\cdots,s_n)$ has N fault
states

$$A = \{A_i\}, \; i = 1, 2, \ldots, N \qquad (2)$$

and the informational entropy of the system is
written as

$$H(S) = H(A) = -\sum_{i=1}^{N} p(A_i)\log p(A_i) \qquad (3)$$

where, H(S) is the informational entropy of the
system and $p(A_i)$ is the propability of fault state
$A_i$.

Suppose the alarm system $Y=(y_1,y_2,\ldots,y_h)$ will
generate a set of activated alrms when the system
S is in fault state $A_i$, and each activated alarm
set is called fault symptom $B_j$

$$B = \{B_j\}, \; j = 1, 2, \ldots, K \qquad (4)$$

Because the alarm system is connected with system
S, some of information of system S can be obtained
from the alarm system Y. We define part of system
S's information which is involved in the alarm
system Y as

$$J_S(Y) = J_A(B) = H(A) - H(A/B) \qquad (5)$$

where, H(A/B) is the conditional entropy of S when
Y is attached to it.

Definition 1: The informational entropy measure
I(Y) is a dimensionless index given in Eq. (6)

$$I(Y) = (J_A(B)/H(A))\cdot100\% \qquad (6)$$

It is easy to see that $0 \leq I(Y) \leq 1$, and
1. if Y is independent of S, then H(A/B)=H(A) and
I(Y)=0
2. if Y and S have definite relationships, then
H(A/B)=0 and I(Y)=100%

Definition 2: Given the cost of fault detection $C_f$,
the alarm-placement problem is to maximize I(Y)

$$\begin{cases} \text{Max } I(Y) = (H(A)-H(A/B)) \, / \, H(A) \\ \sum_{i=1}^{k} C(i) \leq C_f \\ Y \subset \{y_1, y_2,\ldots, y_n\} \end{cases} \qquad (7)$$

Definition 3: Given Io, an entropy measure of the
alarm system, the alarm-placement problem is to
minimize the cost of fault detection in the alarm
system

$$\begin{cases} \text{Min } C_{as} = \sum_{i=1}^{k} C(i) \\ I(Y) \geq Io \\ Y \subset \{y_1, y_2,\ldots, y_n\} \end{cases} \qquad (8)$$

Algorithms for Optimization

Let us set up an alarm system with only one alarm
$y_1$ and the corresponding fault symptom is $B_1$, and

$$B_1 = \begin{cases} B_{11} & \text{when } A_i \text{ can be detected} \\ & \text{by } y_1 \\ B_{10} & \text{otherwise} \end{cases}$$

The informational entropy of $y_1$ is

$$\begin{aligned} J_A(B_1) &= H(A) - H(A/B_1) \\ &= H(A) - p(B_{11})H(A/B_{11}) - p(B_{10})H(A/B_{10}) \end{aligned}$$
$$(9)$$

The optimization of $y_1$ is to find out $y_{\overline{1}}$ from $\{y_1, y_2, \cdots, y_n\}$ so that

$$J_A(B_{\overline{1}}) = \text{Max} \qquad (10)$$

Secondly, we select $y_{\overline{2}}$ from $\{y_1, y_2, \cdots, y_n\}$ but $y_{\overline{2}} \neq y_{\overline{1}}$, so that

$$J_A(B_{\overline{2}}/B_{\overline{1}}) = \text{Max} \qquad (11)$$

In this way, we set up alarm system step by step and obtain an alarm system with k alarms so that

$$\begin{aligned} J_S(Y) &= J_A(B_{\overline{1}}) + J_A(B_{\overline{2}}/B_{\overline{1}}) + \cdots + \\ &\quad J_A(B_{\overline{k}}/B_{\overline{1}}\cdots B_{\overline{k-1}}) \\ &= \text{Max} \end{aligned} \qquad (12)$$

Theorem 1: If we build up an alarm system with k alarms, the alarm-placement scheme $Y=\{y_{\overline{1}},y_{\overline{2}},\ldots,y_{\overline{k}}\}$ from Eq. (10) to Eq. (12) will involve maximum information about system S, i.e., $J_S(Y)=\text{Max}$.

And if we need an alarm system $Y=\{y_1, \ldots, y_h\}$ whose $J_S(Y) \geq J_0$ (a given entropy index), then the alarm scheme from Eq. (10) to Eq. (12) will have the least number of alarms, i.e., h=Min.

In order to solve both optimization problem written in Eq. (7) and Eq. (8), a heuristic optimization algorithm is proposed. After define a heuristic factor

$$F_k(y_j) = \frac{1}{C(j)} \cdot J_A(B_j/B_{\overline{1}} \cdots B_{\overline{k-1}})$$

the algorithm is written as

STEP 1. let k=1
    for all $y_j \in Y_a(1)$, calculate $F_1(y_j)$
    select $y_{\overline{1}}$ from $Y_a(1)$ such that $F_1(y_{\overline{1}})=\text{Max}$
where, $Y_a(1)=\{y_1, y_2, \cdots, y_n\}$
STEP 2. let k=k+1
    for all $y_j \in Y_a(k-1)-y_{\overline{k-1}}$
    calculate $F_k(y_j)$, and select $y_{\overline{k}}$
    such that $F_k(y_{\overline{k}})=\text{Max}$
STEP 3.
  case 1 given by Eq. (7):
    (1) if

$$\sum_{i=1}^{k} C(i) \geqslant C_f$$

       then go to STEP 4.
    (2) if $J_S(Y) = H(A)$
       then go to STEP 4
  case 2 given by Eq. (8):
    (1) if

$$J_A(B_{\overline{1}}) + \cdots + J_A(B_{\overline{k}}/B_{\overline{1}}\cdots B_{\overline{k-1}})$$
$$\geqslant I_0 \cdot H(A)$$

       then go to STEP 4
    (3) if $k \geqslant h$   ( $h=|Y_a(1)|$ )
       then go to STEP 4
    (4) otherwise
       go to STEP 2
STEP 4. end and
    the resulted alarm system is

$$Y = \{y_{\overline{1}}, y_{\overline{2}}, \ldots, y_{\overline{k}}\}$$

with the fault detection cost

$$C_{as} = C(\overline{1}) + C(\overline{2}) + \cdots + C(\overline{k})$$

## An Example

Each candidate alarm listes in Table 1 is different from the others in its fault detection cost and its informational contribution to alarm system designed. It is easy to see that the alarm-placement problems given by Eq. (7) and (8) are the simplified multigoal optimization problem. In the algorithm proposed, multigoal problem bas been solved by considering two kind of constraints and introducing a heuristic selection factor. To show the usfulness of the algorithm, it has been used to select alarms for the plant illustrated in Fig. 2. The computational results are listed in Table 2 and the desired solutions have been got.

TABLE 2   Optimal Alarm-Placement

| case | constraints | optimal results | number of node need alarm |
|---|---|---|---|
| single goal optim. | $C_f = 100$ | $I_{max}=84\%$ $C_{as}=76$ | 9,3,4,5,6,7 |
| | $I_0 = 90\%$ | $C_{min}=118$ $I=91\%$ | 9,3,4,5,6,7,8 |
| multi-goal optim. | $C_f = 76$ | $I_{max}=84\%$ $C_{as}=59$ | 6,5,10,7,4,12, 2 |
| | $I_0 = 95\%$ | $C_{min}=106$ $I=95.3\%$ | 6,5,10,7,4,12, 2,9,3 |

## CONCLUSIONS

The proposed alarm system has been implemented in a personal computer, as part of the intelligent supervisory control system (Rong, 1989), it has satisfactory performance in simulation run. The structure of the proposed alarm system shows great flexibility in that it is able to support both automated diagnostic system and computer based decision aid for operator, to meet the increasing demand of system safty and productivity. And the optimal alarm placement makes the resulted alarm system not only efficient but also economical.

## REFERENCES

Corsberg, D. (1987). Alarm filtering: practical control room upgrade using expert systems concept. InTech., 34, 4, 39-42.
Isermann, R. (1984). Process fault detection based on modelling and estimation methods - a survey Automatica., 20, 387-404.
Ishida, Y. and co-workers (1985). A topological approach to failure diagnosis of large-scale systems. IEEE Trans. Syst. Man, Cybernet., SMC-15, 327-333.
Kokawa, M., S. Miyazaki and S. Shingai (1983). Fault location using digragh and inverse direction search with application. Automatica., 19, 729-735.
Kramer, M. A. and B. L. Palowitch (1987). A rule-based approach to fault diagnosis using the signed directed graph. AIChE J., 33,1067-1078.
Krigman, A. (1985). Alarms, operators, and other nuisances: cope... or court catastrophe. InTech., 32, 12, 33-40.
Lees, F. P. (1983). Process computer alarm and disturbance analysis: review of the state of the art. Comput. Chem. Eng., 7, 669-694.
Modarres, M. and T. Cadman (1986). A method of alarm system analysis for process plants. Comput. Chem. Eng., 10, 557-565.
Rasmussen, J. and L. P. goodstein (1987). Decision support in supervisory control of high-risk industrial systems. Automatica., 23, 663-671.
Rong, G., S. Q. Wang and J. C. Wang (1988). Building a knowledge base of a fault diagnosis system for chemical processes. Preprint of

IFAC Workshop on RAM., Bruges, Belgium, 67-74.

Rong, G. (1989). Design of intelligent supervisory control systems for chemical processes. Ph.D. dissertation of Zhejiang University, Hangzhou, China, 1989.

Shannon, C. E. (1948). A mathematical theory of communication. Bell Syst. Tech. J., 27, p379.

Su, Y. L. and T. govindaraj (1986). Fault diagnosis in a large dynamic system: experiments on a training simulator. IEEE Trans. Syst. Man, Cybernet., SMC-16, 129-141.

Yufik, Y. M. and B. Sheridan (1986). Hybrid knowledge-based decision aid for operator of large scale systems. Large Scale Syst., 10, 133-146.

Acknowledgement - this work was supported partly by National Natural Science Foundation of China.

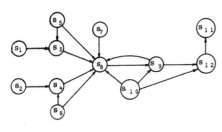

AFFECT

Fig. 3  Global Fault Propagating Model

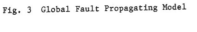

DIAGNOSTIC SYSTEM

HUMAN OPERATOR

DIP

ALARMS

MAN-MACHINE INTERFACE

KNOWLEDGE BASE

DIPAG (inference engine)

Observation / Operation Mode

ALARM SYSTEM

DATA BASE

FAULT DETECTION

DATA SAMPLING

PROCESS PLANT

Fig. 1  Structure of Alarm System

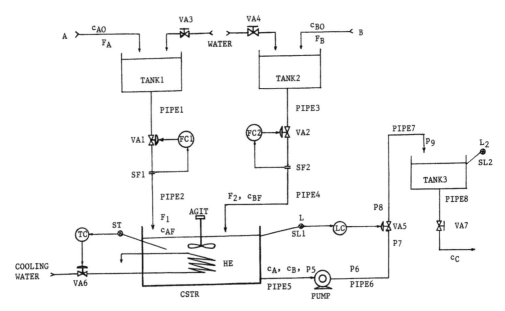

VA - VALVE; HE - HEAT EXCHANGER; AGIT - AGITATOR

FC, LC, TC - FLOWRATE, LEVEL, and TEMPERATURE CONTROLLER

SF, SL, ST - SENSOR OF FLOWRATE, LEVEL and TEMPERATURE

p - PRESSURE;  c - CONCENTRATION

Fig. 2  Flowsheet of CSTR Process

# PRACTICAL ASPECTS OF A MILL WIDE QUALITY CONTROL SYSTEM IN A MODERN NEWSPRINT MILL

## A. Arjas, I. Tiitinen and E. Härkönen

*Kajaani Oy, Kajaani, Finland*

**Abstract.** The paper describes the normal type of quality control that is used in modern newsprint mills. We discuss the present state of quality control and compare it with what is desirable. With the aid of EDP and modern sensors it will be possible to improve quality control greatly and at the same time omit some impractical procedures and standards. This needs new thinking and freedom from trade-defined analysis methods.

**Keywords.** Quality control; paper industry; pulp industry; automatic testing; data handling; data processing; sensors, standards.

## INTRODUCTION

A highly developed quality control system is certainly one of the key functions in any quality conscious industry. Its purpose, tasks and way of operation vary from industry to industry and even from factory to factory and mill to mill.

Traditionally, quality control represents the customers' interests at the mill. It monitors the quality level with the recommendation to reject products which don't meet the customers' requirements. This simple approach is no longer the only reason for an industry to operate such a system. In fact, a much wider use of the quality function is necessary for an industry that wants to be modern and competitive.

Unfortunately, there often are traditional trade practices which require the use of old and impractical methods. This refers to quality standards which are defined by and tied to old measuring procedures and instruments.

The rapid development of measuring systems, instruments, data processing and process control provide a possibility and a challenge which should be fully utilized.

### TASKS OF THE QUALITY CONTROL SYSTEM IN A NEWSPRINT MILL

Newsprint is usually regarded as a cheap and simple paper grade of low quality, neither glossy nor durable and used in a short-life product. However it does have to satisfy the demands of the newspaper industry, which are for a high quality of consistency, uniformity and reliability - and at the same time be very competitive in price. This is because the cost of newsprint is the major factor in pricing newspapers. In addition, recent developments in the print quality of major newspapers have meant that the printing properties of newsprint have also become and are becoming more important than they used to be. This is happening especially in countries where newspapers compete with commercial TV channels.

Seen from the customer's point of view, an advanced quality control system of a newsprint mill therefore secures excellent runnability, the agreed printability and, above all, the consistency and uniformity of both of these aspects.

From the manufacturer's point of view, the scope is much wider than the customers' satisfaction alone. Firstly, the papermaking process is an integral chain of operations where quality has to be monitored and necessary control actions taken between each consecutive section. Each control loop has to be as short as possible so that corrections are made where errors are detected, instead of trying to compensate for them in the succeeding phases. Only this can guarantee the smooth operation of the integrated process and lead to a consistent and uniform end product.

Another manufacturers' aspect is the delicate balance between quality and cost. There are obviously paper properties, where from the customer's point of view only maxima or minima should be specified. It is unlikely that any printer would be dissatisfied with an unusually high strength or opacity level. However, a quality level which is essentially better than the specification indicates that the costs tend to be out of control. Such a phenomenon leads to low margins with known consequences in the long run.

A further task of the quality control system is to provide data for product development. Customer feed-back has to be related to mill data that define the measured properties of the paper and the corresponding state of the process variables. Such information is extremely important regardless of whether customer experiences were positive or negative, whether they led to repeated orders or to complaints.

### PRESENT STATUS OF PAPER MILL CONTROL SYSTEMS

The systems used nowadays are hybrids consisting of old and new. The traditional parts of the system are based on samples which are measured in a laboratory. Samples are collected systematically.

The systems used nowadays are hybrids consisting of old and new. The traditional parts of the system are based on samples which are measured in a laboratory. Samples are collected systematically from the process and the results of the measurements are displayed on TV screens in various parts of the mill. Besides being the only quality information in dealings with customers, such measurements also form the back bone of the whole integrated mill system. There are two reasons for this. Firstly, various in-line measurements are calibrated against such conventional data, and secondly, the non-standardized in-line systems aim at a quality which meets traditionally defined standards.

The rest of the systems comprise automatic control loops, on-line measurements with manual control and fully manual procedures. As such information and control methods are used internally only, there is no need to standardize them. Such data need never to be accepted by anybody else but the producer.

## DESCRIPTION OF ONE EXISTING SYSTEM

Without a carefully planned and controlled production and transportation chain a morning paper would never reach a reader in time. The most critical links are the newsprint mill and the printing plant where both the schedule and the quality must be under continuous control. The system introduced here is not meant to be an exemplary one but it gives a general idea about common measures in paper mills. (Fig. 1)

### Checkpoint No. 1: Wood yard

Fresh wood with its natural moisture content is the best rawmaterial for mechanical pulps. To utilize this the period of time wood from harvesting to pulping must be minimized. Therefore a close link with the forest department is necessary so that supply and consumption are in balance. In seasons with the temperature above freezing, the wood yard is also hosed to prevent drying. If this first stage is out of control, serious quality problems can be expected.

### Checkpoint No. 2: Pulp mill

These days newsprint often only contains 5 - 10 % chemical pulp, so that the quality of the mechanical furnish component is even more important than in the past. The fast 8 - 9 meters wide paper machines have also become very sensitive to variations in mass flow. Though various feedback information from the process is available on the display screen the pulping equipmet, i.e. refiners and the grinders, are still controlled manually. The necessary feedback for adjustments is available through the laboratory and through the on-line quality monitor, the so called "freeness" testers installed in each of the main lines.

Besides degree of refining, which still seems to be the most popular characteristic for a pulp, distribution of fiber length, shive content and strength properties such as tear, tensile and

burst, are tested by the laboratory. As the test are done by hand the number of tests is limited.

### Checkpoint No. 3: Paper machine

The basic characteristics such as weight (g/m$^2$), moisture and thickness were already changed to computerised testing and controlling years ago. As to the weight, particularly, it is also checked per reel by measuring its weight, metreage and width.

Though newsprint is often considered a low-grade bulky product quite a number of different tests are still made on it in the laboratory, where samples are sent of every machine reel, i.e. at 30 - 60 min intervals. The testing instruments are widely used all over the world so there is no reason for going into details. It is, however, worth mentioning that we have recently installed a new, automated system, the "Autoline 2000". It measures weight, thickness, roughness, porosity and gloss for a cross-machine sample and feeds the data into the computer automatically. The other routinely measured properties are tensile, tear and burst strength, elongation, oil absorption, brightness, Y-value, opacity and shade. These are still both tested manually. Additional tests for surface strength, ink requirement, print through and some other properties characterizing the printability are carried out occasionally.

The methods are in accordance with Scan/ISO-recommendations.

The number of samples and the frequency of measurements are certainly statistically too small. That often makes it difficult for a machine crew to draw the right conclusions and even causes lack of confidence in the laboratory work.

### Checkpoint No. 4: Winder

Winding is one of the most critical operations in the production line. Both man and machine may destroy excellent quality paper by producing reels which look beautiful but have mechanical faults such as wrinkles, bursts etc. inside. If anything doubtful is found, reels are taken to a rewinder to be checked there. Winding is a stage where the importance of care can never be overemphasized.

### Checkpoint No. 5: Wrapping and handling

A wrapped reel with various information printed on it is like a mill's visiting card. Therefore it is everyones' responsibility to pay attention to any defect observed in the reels, in or outside the mill.

### Checkpoint No 6: Printing plant

Despite all tests and checks in the mill printer from time to time rejects a product. That is partly due to the fact that a control model without gaps is an impracticability. This is caused by the weak correlation between the tests and the printing process. For these reasons continuous communication with customers is important.

## SHORTCOMINGS OF THE PRESENT SYSTEM

Like any paper, newsprint is made of natural fibre pulp (Fig. 2, 3 and 4) which is non homogeneous and the properties of which vary depending on the raw material. Over the years scientists have attempted to characterize pulps in an unambigous way to assess their paper making potential and to give basic information for the process control. The results have not been convincing. Measurements that can be and are made correlate with important functional properties, such as strength and optical properties, but they are seldom exactly what should be measured and known. Another shortcoming is that many of the measurable properties are interrelated.

The picture is further complicated by the fact that the desired properties of most paper grades are a compromise of various desirable functional properties. Additionally, the process control parameters that influence such properties are never specific. The challenge for the paper maker is therefore to aim at an optimum combination of real functional paper properties by using non-specific control parameters and measuring the result with methods that characterize but don't define the quality.

An additional difficulty comes from the fact that some of the paper properties, or rather laboratory measurements, were developed to simulate the end use of the paper. They tend therefore to correlate with several physical, chemical and structural properties without being an exact measure of any one of them. However they are still being used because this is traditional practice. Fortunately, a number of other properties have direct and meaningful connection with the physics or chemistry of the paper web, although none of them alone can define the functional behaviour of the paper.

## IMPACT OF AUTOMATION AND ELECTRONIC DATA PROCESSING ON QUALITY CONTROL

The general tendency is to increase the number of measurements in an analysis. One figure or the average of several figures is no longer enough to describe the result of the analysis. We might describe the deviation of measured values or utilize the time history of the measured values.

Since the target is to help to control the paper machine to produce paper which has an even and desired quality, it would be logical if all these measurements were done on the paper web in the running paper machine. This indicates where we are going. The future analysis will be done on line at very high sampling frequency both in machine direction and cross direction and the numerical material will be transformed to an easily understandable form with the help of effective high-speed computers.

Where are we today?
Methods applied for paper testing are classified into destructive and nondestructive. Destructive methods, such as tensile and tear tests require a paper sample and therefore it is not possible to apply these methods as such on line. They are limited to the laboratory and to a quite small number of measurements. There have been some attempts to measure on line paper properties which

would correlate with the results of the destructive methods. Nondestructive methods don`t set any absolute limitations to applying them on line. Today, weight, thickness, moisture and colour are commonly measured on line. In most cases these on line measurements are used to control the paper machine. To control the quality of the paper an off-line anlysis made in a separate paper laboratory is needed today and probably also in the future, even if its role might change towards calibration and R & D purposes with developing on line measurements.

Computers and their memory units have been developed very quickly during the past few years. A typical computer application today is a laboratory computer where all numerical data, which is created in this computer, is stored. All meters are connected to this computer and the data is stored in an immediately available form for 1 or 2 years. The sampling frequency of these measurements is quite low, one sample per machine reel and maybe 10 analysis of cross machine directions. Right now the statistical analyses of this collected data is coming to the paper mills and there is a major task to study what the most convenient way to utilize this material would be (Jarmo Häyrynen, personal communication). If we were to measure on line periodic variations of different paper properties, we would have a real challenge. Let us take an example. The measurement we wish to make fluctuates approximately every 5 mm. The speed of the paper machine is 20 m/s. Therefore we will need to have a sampling frequency of 8000 Hz to recognise this detail. On the other hand, if we want to have good resolution between periodic changes having a small difference in their frequencies, say 0.005 Hz, we must have a paper sample of a length 4000 m. A large amount of data is needed to manage this task. Today, analysis of this type is not possible to do on line because the speeds of modern computers are not high enough to manage with the situation. This type of analysis is limited presently to the laboratory (Hannu Makkonen, personal communication).

Analysis of the above type is closely connected to the geometry and speed of different parts of the paper machine. A Comprehensive EDP-system on a paper mill also contains the control variables and running statistics of all separate paper machines. Such systems in different variations are quite common today. The target in the whole EDP application is to help to run the mill in a more profitable way. It is easy to collect data with the present systems, maybe too easy, and one of the main areas in the future will be to study the optimal way to utilize this data.

## FUTURE DEVELOPMENT POSSIBILITIES

Ideally the complete and real needs of a quality control system should be analysed carefully so that a totally modern control philosophy could be designed regardless of how the system is brought into being. It could be based on a widely applicable approach. This consists of two parts, namely a product analysis and a prosess analysis.

The product analysis goes back to what was described earlier in this paper. It lists carefully the most important functional properties that govern the usefulness of the given newsprint grade when used for a given application. These properties have to be unambiguous, precise, measurable and such that the desired compination

of them can be repeatedly achieved in a well controlled process. It is often necessary to start
with general terms such as runnability and printability and then to proceed to more precise aspects such as tensile strength or back scattering
power. When performing such an anlysis it will be
necessary to drop a few traditional measurements
from the specification since they bring no additional information and rather increase the complexity of the model.

The next step is to prepare a process analysis
which is the second part of the approach. The aim
of this analysis is to arrive at a reliable control model which defines the necessary control
variables of the paper making process that are
relevant for achieving the desired quality and
correspond to the essential paper properties defined in the product analysis.

Also the measuring devices and systems will develop. Traditionally the instruments have been relatively simple and manual. The first step towards
automation was to computerize the feeding of the
results. The next step was to automate a full
testing line while the measuring heads still functional in the original standardized way or were
nearly standard. Such systems are gaining ground
right now.

If traditional standards can be neglected, the
next step will be off-machine-off-standard automatic and computerized measuring systems which are
based on the above described product anlysis. If
this is succesfully done and the customers are
willing to approve it, the final development
stage will be on-line-off-standard measuring
systems. They will give the necessary feed-back
for the process control and this way guarantee a
good, consistent and uniform product. At the same
time information collected this way serves the
other purposes of quality control, that is product
development, technical customer service and control of production costs.

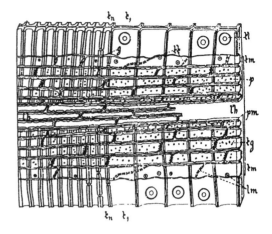

Fig. 2    Cross section/Picea Abies

Fig. 3    Mechanical pulp

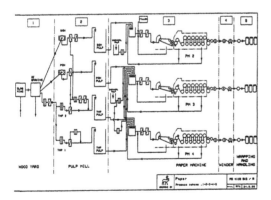

Fig. 1    Process scheme

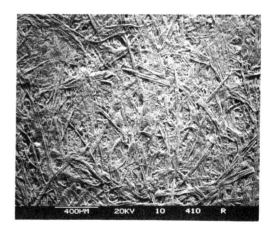

Fig. 4    Structure of paper surface,
          MF-newsprint

# ON THE QUALITY MANAGEMENT AND TECHNOLOGY OF PROCESS INDUSTRY

## Y. Arai, K. Kochiyama, and T. Shinohara

*Department of Engineering, Showa Denko K.K., Tokyo, Japan*

Abstract. For this decade,the chemical industry in Japan has strived
to reform their business structure to the specialty chemicals,where
the product quality is emerging out as the crucial technology and
management problems.This paper presents ,from the concept of quality,
the quality assurance,and its recent state in chemical industry.
Meanwhile,we discuss the quality technology in the process engineering
activities, and finally,try to place it to the Process Systems Engineering
as a key connotation for its next extent of this discipline.

Keywords:Quality control;quality assurance;speciality chemicals;quality
technology;process systems engineering;quality modelings.

## 0  Introduction

The recent chemical industry has been striving
to reform their business structures.
For the new speciality chemicals,where the more
quality oriented product is required.So,they
make their endeavor to satisfy the customers
through the production management system and
technology.
We present in this paper,the following items:
1 The recent Aspects of Chemical Industry in
Japan
2 On the Features of Process Technologies
3 Quality and its Connotations
4 On the Quality Efforts in Chemical Industry
   [CASE STUDY A]  Actual Examples of QA.
5 Quality technology  in Process Systems
Engineering
6 Process Quality Modelings
   [CASE STUDY B]  Application Examples are
                   illustrated.
7 Conclusions

## 1  The Recent Aspects Of Chemical Industry In Japan

We recognize that the chemical industry,
general speaking,  has a ditinguished features
in comparison with others :
   (1) Diversty of products
   (2) Short life cycle of a product
   (3) Feed material and process innovations
   (4) Impact of finding a market needs of a ne
material use.

After facing of twice of the energy crisis,the
chemical industry in Japan has continouly made
their efforts in  the followings:
   (1) for the mass commodity chemical fields,
       thoroughly seeking of energy savings  and
   its related investments to production
   process and
       boosting of  products development and
   release
   from their own unique techincal background
   and market fields needs,and

   (2) towards the shift to the specialty
       chemical field, which is to be used  for a
       particular  market,like as the computer
       makers.

Thus,even  though  they  have  enjoyed  the
preferable  business  atomosphere for the time
being,the chemical  industries  in   Japan has
proceeded to shift  their own business structure
to   so called "specialy chemistry"  field
which is commonly  taken now,as a long term
business strategy.(Watanabe(1988),and
           Kanenari(1988),Kojima(1986))

The research and development investments has
very consequently  been carried out.
(Fig.l)   shows  the  recent  trends  of  the
investments in Japan.(Statistic Bureau(1989))

## 2  On The Features Of Process Technologies

In  TABLE 1    are  listed  the differnces of
poduction features between   the commodity and
the speciality chemicals,which is from our own
perspectives.
   As pointed from Kojima(1984),as far as the
technonogy of chemical industry fundamentally
consitutes of reactions and separations, the
chemical engineering analysis and the process
system engineering approaches is to take
invariably important rolls,all throughout
   (1) the systematic analysis of a physico-
chemical phenomena,
   (2) the systematic process scheme buildings,
   (3) the commercial process developments,
   (4) the process designs,
   (5) the constructions  of the plants/the
equipements,
       and,
   (6) the production managenents.

Even though metioned above, new techonological
requiements have been given a rise from the
particular customers,who order  the particular
levels of the matericial quality.

For example,the extremally higher level of component purity, as compared with those of the convetional commodity, is required through the whole over productions of IC chip susbstrata material and its further manufacturing steps.

Two of technical points should be taken to overcome for this problems:
(1) innovations of the new unit
    operation process and its control
    designs  (Umeda(1988),Takamatsu(1988),and
             Kojima(1986))
The most of these fields are ones from applications of the fundamental physco-chemical transport phenomena and physical properties of the materials,which are requiring the more precise experimetal and the more theoreical analysis ,spanning over from a level of the electrons and the molecules (Quantum

Chemistry)to the macro physical chemistry (Thermodynamics,Kinetics,etc.) ,and
(2) the quality managements and technologies
             (Ishiguro(1984),Kusaki and
              Kobayashi(1985))
Supposedly,the quite adequate operations were provided,still cumbersome and difficult problems will remain to satisfy the customer's quality levels.

The production lines of customers are often in prematured state, where they are proceeding under a strong innovations and making their product for adding a higher quality levels from their products.

A chemical industury side as their material supplier is contantly requiring the cooperations ,especially in the material quality.

Fig.2  shows differeces of technical features between commodity and special chemistry from our perspecteves.

From this context,so far,it is easily recognized that the quality management and technology is becoming the most crucial points of the chemical industry.

The figure explains proximity of activities in the speciality ( Sudo(1986)) between D(Development) and P(Production),a framework of a process system engineering is to be very consequently different in the this field, rather than in the conventional commodity.

The commodity is mostly in the operation within a range of specified conditions given at the process designs.This means the process engineering objectives are rather definite in the commodity.

## 3     Quality And Its Connotations

### 3-1  Quality  and Related Terminology

quality
"The totality of features and characteristics of a product,process or service that bear on its ability to satisfy stated or implied needs."                    (ISO(1985))

quality control
"The operational techniques and activities are used to satisfy quality requirements.
                            (ISO(1985))

quality assurance
"All those planned and systematic actions necessary to provide adequate confidence that a product,process or service will satisfy given quality requirements.
                            (ISO(1985))

TABLE 2  shows the corresponding definitions of Japan Industrial Standard(JIS) and ISO.

### 3-2  Quality Control,Its Connotations

As for the concept of quality,it has been used in a broad connotation.(Ishikawa(1989), Hashimoto(1989))
(1)  quality (in narrow definition)
   such as physical properties(function, characteristics,size,weight,strength,etc.) ,chemical properties(concentrations , purity,etc.),reliabilities,outlook,taste, easiness to use,service,etc.

(2)  quality (in broader definition)
  i Q;  quality (in narrow definition)
  ii C;  performances or efficencies in the relation with cost/profit.
        such as cost throughput,profit management:
           yields,product cost,rate of loss, sales price,etc.
  iii D;  performances/characteristics in relation with quantity and delivery, such as production rate,sales rate, switching losses,inventory rate, consumption rate,delivery date-line, production scheduling,etc.
  iv S;  performances/characteristics at market/consumer sites, such as  safety,service,guarantee, reliability,product liability,etc.
  v E;  performances/characteristics in the relation with safety,and enviroment protection at the plant site.

Meanwhile,the concept of control is considerred as following connotations.(Ishikawa(1989),

Hashimoto(1989))
It connotes the loop  activities of "Plan", "Do","Check",and "Act",so called Deming's PD CA CIRCLE,which is widely recognized in Japanese industry in commemoration of Dr.W.E. Deming's contribution to nation wide spread of quality technology after the War II. Ishikawa(1989) presented PDCA CYCLE as ;

  PLAN
      i) Determine a target,
      ii) Setting up the measures to get the target,
  DO
      iii) Learning the methods(statistical tools,etc.),
      iv) Work and apply the methods to their problem,
  CHECK
      v) Check and confirm their work done,
  ACTION
      vi) Analyze how to make their readjustment and provide a adequate treatment to the objects.

The connotation of quality control ,as so far,implies always   seeking what consumer's substantial  quality is,and how to realize it,both in activity and technology.

## 3-3 Quality Assurance

We, so far ,stressed the connotation of quality control (broader) both in activity and technology.
In this section,is briefly taken the quality assurance as the systematic activity of quality.
(Historical Aspects)
   i) Product Inspection oriented
  ii) Process Management(Control) Oriented
 iii) New Product Design(Development) Oriented

In the progress,from i) to iii),the famous phrase;"Realization of quality at the each step of     process,and the best at the design step", is considerred, was emerging out.(Hashimoto(1989))
(Fundamental Procedures)
1  The organization sytem design with explicite form as:
   i) Who in where is in charge for whom
      in where,
        (;Top Management,Head Quarter,Lab,
        Factory[Technology,Testing,Production
        Planning,Prduction,Invetory,Delivery,
        Account,Quality System],Customer,etc.)

  ii) of What,  (;Policy Making,Surveying,

        Planning, ,Evaluation, Decision,Test
        and Inspection,  Production,Delivery,
        Order,Approvement,Measure, Reporting,
        etc.)

 iii) when   (;Market survey,Planning
        Research,Development,Engineering,
        Trial Production,Initial Production
        Tuning,Commercial Production,
        Inspection,Sales,Customer service,etc.
        )
  iv) how   (;Legal restrictions,and  guides,
        the company rules,and codes for
        Material ,Product,Prduction Technology,
        Inspection,Package Material, QC Flow
        Chart,Product Handling,Claim  ,etc.).

## 4   On the Quality Efforts In Chemical Industry

In the context of mentioned above,we are going to review the quality concepts of Japanese chemical industry.
Remembering a column (Management Factors) of TABLE 1,and if a concept frame is,hereafter, drawn in the broader definition of quality, the quality  is consequently to means a whole balance of  performances of five of management factors, itself;S,C,Q,D,and E.
For the commodity,where the concepts  and technology of unit operations    and process control are mostly established,  and through the  steps of scale-up development, the product  quality (;Q) is mostly guaranteed.
In this meaning,it is a very typical design oriented quality field,and we can say that the quality here  means   "cost"(C).

Throughout the efforts after twice of the energy crisis,we understand  two of quality problems are emerging out to our minds;

  i) Seeking the process quality(C)
        process improvements and operation optim-
        ization under various operation load,for
        saving the energy,for feed material
        conversions,etc.
 ii)Seeking the segragation  to distinctive
    quality products (Q).

During the time,they have made a shift toward two directions(Kojima(1986):
  i) Upstreams( Speciality chemicals)
        ex. high purity,high homogeity,
            high affinity and high composite
 ii) Downstreams( Manufacturings)
        ex. multi-grade,composite material,
            CAE and CAD,computer softwares,etc.
In above directions,the material quality is emerging out as the most crucial business points.
As mention before,the chemical industry is taking a participation to the customer's side of  technical  innovations, they have faced the more   different problems in  the quality management and technology than before.

## [CASE STUDY A]
  Since around 1985,almost every chemical comapany has had an unusual experience ,which they ever had,of quality assurance surveilances to the production lines  by their customers and would-be customers,so called "Quality Rushes". The most of customers,has requests more to guarantee the product quality at every step of a production line;
   i) Request of submit of the documents (from suppliers side) on the quality assurance systems and the operations conditions specified along
the processing.
  ii) On site surveilance by the customer,who
      checks the operation and its management
      in accordance to the submitted documents.
 iii) Score to the grade in the quality levels,
      and request to the further improvement,
      if necessary.
As for our case of Showa Denko,in 1986, the quality special committee,who is in charge of investigation of
   i) how the present quality system goes to
      working on, and
  ii) what problems they are facing ,and how we
      resolve them as the system.
Their report pointed 8 items:
   (1) Quality  at each of production stage ,
   (2) Clean workshop enviroment,
   (3) Quality assurance system for
       the new products(quality specifications,
       and its evaluations),
   (4) Contaminant proof,
   (5) Delivery oriented claim,
   (6) Segregated status of quality promoter
       from  production lines,
   (7) Production commissions(QA Contract,
       Guidance,and Training),
      and,
   (8) QA promoting in the head quarter.
TABLE 3 shows ,from our surveys, the leading chemical companies in Japan  how they are going to rebuild their own quality systems.

## 5   Quality Technology  In Process Systems Engineering(PSE)

As far as taking the images on a system,we can find  our minds,which are to be directed toward two of behaivioral ways:
   i)  as the methodology for recongition
        to the objects
        ;(Science like)
  ii)  as the methodogy for action
        to the Objects
        ;(Engineering like)

Fig. 4  shows a concept model for the relation of recognition  to action .
A frame placed at the top shows the difinition of sytem.

As the terminology,
  i) "OPERATION" means any operations working
     a object(not only for plant operation) ,
  ii) "FUNCTION" means as output from the
     object,
  iii) "INFORMATION" means as the same as ii),
     and
  iv) "STRUCTURE"means,here,the logical
     expression of the "OBJECT" to be taken as
     consideration,either in theoretical or
     in statistical modeling.
Supposedly,a process would be well
established  one,our technolgical activities
is mostly in the lower side of the frame
(;Frame of "MODELING FOR ACTION")
While,if  we take the technolgical field,
such as speciality chemistals,which we
dicussed in  the previous sections,the upper
frame of "MODELING FOR RECOGNITION" is
strongly  interrelated each other.
"ANALYSIS","DEVELOPMENT",and "IMPROVEMENT"
 brigde between both of frames.
For instace, the more precise  analysis for
understanding of the  mechanism  or the
behaivior of object is so often required
beyond the convetional engineering disciplines.
We stress,here,
  i) modeling work for the object  recognition
     is becoming more important in PSE.
                         (;Quality modelings)
and
  ii) a cyclic  activities bridging two frame
     is to mean the PDCA   cycles in  quality
     control.             (;Quality technology)
                                       F109

6  Process Quality Modelings

Since the history of the quality control
started from  statistics as SQC,we inclines
to take Statistics as a modeling tool.
We,however,take quality modeling tools in PSE
as two of the followings:
  i) Structural Model
  ii) Statistical Model.


(Fig. 3  )

[CASE STUDY B]
We categorized our modeling applications
from the mathematical tools;
  i) Differential Equations
                    (TABLE 4  )
  ii) Algebraic Equations
  iii) Statistics
  iv) Time series (Spectra) Analysis
                    (TABLE 5  )
In each of table,the examples of modeling,are
listed which selected from our work of this
decade.
Several examples among them is presented below:


(1) Polymethylene Grade Designs,
      (Kochiyama,Arai,Miyake,and Moteki(1983.
      and ,Moteki and Arai(1986))
This is our long period technical activities to
design of quality of polyethylene to make
matching  of multi-user's product requirement.
As the analytical metodology,the followings are
used:
  i) Multi-variate statistical analysis
By the use of this method,quantitative
relationships between the plant operation
factors(ex.monomer feed rates,catalyst rate,
temperature,etc.) and properties of resins
(ex.molecular ditribution,molecular groups,
viscosities,density,haze,clarity,etc.) are
obtained,.

and
  ii) Polymer kinetic modelings.
  By the use of this modeling simulation,the
molecular distribution are calculated,and  the
results of data analysis ,from i),were properly
evaluated.
.Fig.5-1   shows the typical steps of the
statistical modelings for i).

(2) Electrolysis Cells, Analysis and
Improvement through Electic Current Leakage
Distributions  (Arai,and Shibusa(1987).
               Arai,Shibusa,and Shinohara(1989))
Electrolysis reaction process has so often
problems to save  the energy consumptions.
Fig.5-2  shows FEM(Finite Element Method)
application to design  cell dimensions and
material of a multi-polar cell.
The model is taken as a surface reaction with
a combination of exess voltage and disociation
volatage over a boundary condition.

(3) Thin Multi-layered Composite Material,
Development,Analysis of Combinated links of
Phenomeana (Shibusa(1986),
CVD (Chemical  apor Deposition) is a typical
case of a coposite modeling problem with
  i)  thermo chemical surface reaction,
  ii)  thermo fluid dydnamics,
  iii)  wall radiations,
  iv)  mass transfer to substrate,

and
  v)  composite stress distributions.

(4)  Hydraurec Power Prediction,and
the Weather Forecasting (Arai,Shibusa,
and Miyazaki(1985),and Arai,and Shibusa(1985b))
The statistical time series models for
predicting water flow amount to a dam,
by the use of daily record more than 10 years.,
As canditate factors;rate of rainfall,
snowfall, day average temperature,hours
weighted average temperature,remained snow
amount.
Each factor are evaluate by the criteria of
Akaike(AIC).
Fig.5-3 shows the results of model fittings.


7 Conclusions

Reminding us back of the  PSE'82(Kyoto),where
the difinition of PSE were proposed
by Takamatsu(1982):
"In short,Process Systems Engineering is
an academic and technological field related
to methodologies for chemical engineering
decisions.
Such methodologies should be responsible for
indicating (i) how to plan,
          (ii) how to design,
          (iii) how to operate, and
          (iv) how to control
any kind of unit operation,chemical and other
production processes and chemical industries
themselves".

In MacGregor(1988),he stessed that ,in many
companies,process control groups have become
too isolated from the end user to relate
the quality problems that customers experience
back to the operation and control of the
process.
He gave the reasons as
  the chemical engineering eduction program
  placed in the haeavy emphasis on the
  petrochemical operation where quality has
  been less of concern than it is in special
  chemitry,etc.,and they are ,therefore,
  inadequate background in statistics.

Fig.6 shows that we trially put the academic
desciplines related to PSE and QC,where
each other,we think,has close in proximity.
We feel that the points are in how we can
answer the following quetions:
  i) whether is our  control domain
     (;frame of the system), either in unit
     operation,production  process,factory,
     company wide,or industry wide,etc.?
 ii) whether is our problem start
     in recognition mind,in action mind,or in
     both of them?
   and,
.iii) whether could we become a "super engineer"
     like Fransis Bacon and Issac Newton?
 iv) whether was Osborne Reynolds,
     a "super quality engineer"?


(Machine manufacturer) (Plastic Molder)
-------------------------------------
             (Chemical industry)

      = Re


REFERENCES

Arai,Y.,Y.Shibusa,and T.Miyazawa(1985 ).
   Jikoko Kaiki Moderu wo Tsukatta Suiryou
   Yosoku eno Pasokon Katsuyou.Keiso,28 ,
   pp45-49
Arai,Y.,(1985b).Seisan Kanri niokeru
   Hinshitsu Gijutsu.In JSPS No.143.July.
Arai,Y.,and Y.Shibusa(1989).Kagaku Purosesu
   niokeru Jikeiretsu Kaiseki no Tekiyou
   nituite.
   In T.Nakagawa(Ed.),Enjiniya notameno
   Purosesu  Kaiseki to Seigyo,
   Shisutemu Sogou,Tokyo,pp.224- 237.
Arai,Y.,T.Shinohara,T.Kato, O.Miyauchi,
   M.Inaba, T.Mizuo,N.Takagi,A.Kodama(1989),
   Application of ANSYS to Development of
   Plastic Products.
   In ANSYS CONFERENCE 89,Pittsb.Penn.
Hashimoto,I.(1989).Hinshitu Kanri to
   Statistical Process Control.
   In JSPS  No.143,July.
ISO.ISO 8402 -1985(E/F)(1985);Quality
   Assurance Vocabulary
Ishikawa,K (1989) Nihon Teki Hinshitsu Kanri.
   JUSE.
JIS.JIS  Z8101 (1981).
   In Japan Standard Association (Ed.),
   JIS HANDBOOK 1989 QUALITY CONTROL
Ishiguro,M.(1984).Porima no Hinshitsu
   Sekkei niokeru Kino,Kiko Tenkai,
   HINSHITSU KANRI,36, pp.1804      -1809
Kanenari,H.(1988).KAGAKU KEIZAI,Aug.pp.7-11
Kochiyama,K.,Y.Arai,Y.Miyake,and Y.Moteki
   (1983).On the Application of Multivariate
   Statistical Methods to a Polymer
   Reaction Process.
   KAGAKUKOGAKU,47,pp150-154
Kojima,Y.(1984).KAGAKUKOGAKU,48,1,pp17-19
Kojima,Y.(1987).Saikin no Kagaku Kogyo to
   Purosesu
   Kogaku,In T.Umeda(Ed.),Saikin no
   Purosesu Sekkei Gijutu Koenkai,
   SCEJ Kanto Shibu,pp1-3
Kojima,Y.(1986).New Trend of Process
   Engineering, KAGAKUKOGAKU,50,pp85-87
Kusaki,E.,and T.Kobayashi,(1985).Kotei Kanri
   notameno Kokoromi,HINSHITSU KANRI,36,
   pp607-613
MacGregor,F,J.,On-line Statistical Process
   Control  CEP,Oct.1988,pp21-31
Makabe,H.(1984).HINSHITSU HOSHO TO SHINRAISEI,
   Nikkagiren.

Moteki,Y.,and  Y. Arai (1986).Operation
   Planning and Quality Design of a Polymer
   Process.
   in proceeding of DYCORD(1986),Bournmouth,
   U.K.,
Shibusa,Y.(1986).Kagaku Purosesu niokeru
   Yugen-Yoso Ho no Riyo nitsuite.
   In JSPS No.143.Dec.
Sudo,G. (1986).Kogyo Zairyo no Hinshitu Hosho.


HINSHITSU,16 ,pp229-235
Statistic Beurau,(1989).Report on the Survey
   of Research and Development (1988),
   Management and Coordination Agency,pp.46
Takamatsu,T.,(1982).The Nature and Role of
   Process System Engineering,In Proceedings
   of PSE'82, Keynote Sessions,pp.3-17
Takamatsu,T.,(1988).Purosesu Kogaku no
   Rekishi to Kongo no Tenbo.
   In DAI 3 KAI PUROSESU SHISUTEMU
   KOGAKU SOGO SHIMPJIUMU,pp1-10
Yokoyama,T.,and Y.Arai (1980).Kagakusangyo
   niokeru Seisan Gijutsu Bumon no Yakuwari,
   HINSHITSU KANRI .31,pp.1094-1099
Umeda,T.(1988).Purosesu Sekkei Gijutsu :
   Genjou to Kadai,In DAI 3 KAI PUROSESU
   SHISUTEMU KOGAKU SOGO SHIMPOJIUMU,
   pp.35-44
Watanabe,K.(1988).21Seiki heno Chosen,
   KAGAKU KEIZAI, July,pp.8-15

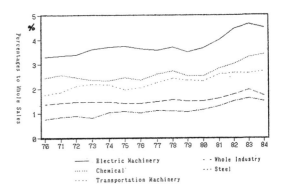

Fig.1  R and D Investments of Japanese Industries
        (Statistics Bureau (1989))

[COMMODITY]

Fig.2   Technical Development Modes

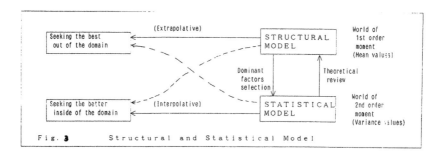

Fig. 3        Structural and Statistical Model

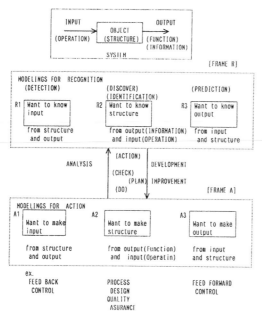

Fig. 4   ATTITUDES FOR MODELING

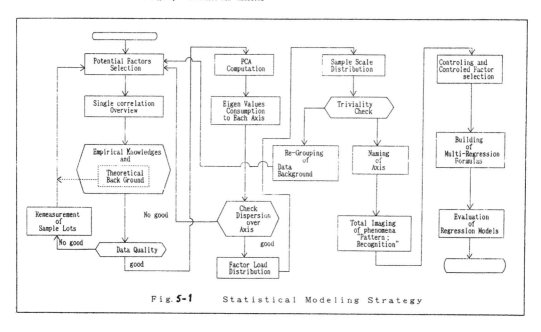

Fig. 5-1        Statistical Modeling Strategy

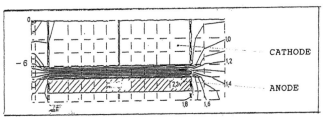

Fig.5-2   Electric Potential Distribution
In Electrolysis Cells

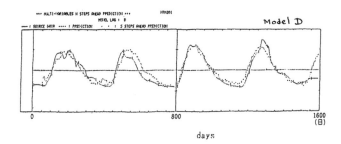

days

Fig.5-3   Hydrauric Power Forecasting
:solid line   for realization
dotted line for 50 days prediction

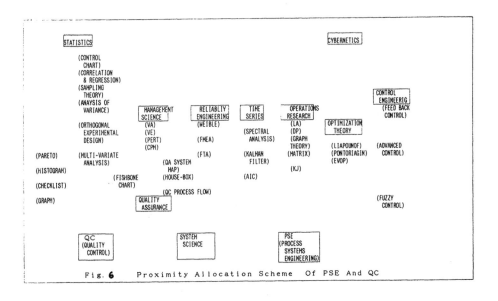

Fig. 6     Proximity Allocation Scheme  Of PSE And QC

TABLE **1**    Process Features(COMMODITY AND SPECIALITY)

| Main material | Process | Product | Management Factors (Potential priority) |
|---|---|---|---|
| [Commodity] Naphtha<br><br>Coal | large scale and Amount Continuous process Single purpose | Organics Inorganics Polymers | (Operation site)<br><br>Safety(Cost) Prodution cost Delivery(Cost) Quality(Cost) Enviroment(Cost) |
| [Specility] (Various) | Small scale and amount Batch process<br>Multi purpose | Pharmacy line chemicals Fine ceramics Electronics Biochemicals Composites | Safety Quality Delivery,Cost Enviroment |

TABLE **2**    Quality, and its related Terms

|  | JIS Z 8101 | ISO 8402-1985 |
|---|---|---|
| QUALITY | The totality of proper chararistics or performance which are the objects of estimation to determine whether a product or service satisfies the purpose of use or not... | The totality of features and characteritics of a product,process or service that bear on its ability to satisfy stated or implied needs. |
| QUALITY CONTROL | A system of means whereby the qualities of products or services are produced economically to meet the requirements of the purchaser. | The operational techniques and activies are used to satisfy quality requirements. NOTE<br>1 In order to avoid confusion,care should be taken to include a modifying term when referring to a sub-set of quality control, such as "manufacturing quality control",or when referring to a broader concept,such as "company-wide quality control".<br>2 Quality control involves operational techniques and activities aimed at both monitoring a process and eliminating causes of unsatisfactory perfomance at relevant stages of the the quality loop in order to result in econmic effectiveness. |
| QUALITY ASSURANCE | The systematic activities carried out by a producer to guaranty that the quality required by the consumer is fully satisfied. | All those planned and systematic actions necessary to provide adequate confidence that a product,process or service will satisfy given quality requirements. |

TABLE **3**    Quality Assurance Activities Surveys(1988)
( 16 chemical companies,how do they manage to do)

| QA Activities | Companies | From Distictives |
|---|---|---|
| 1 Company-wide QA Promotion In Headquarter | 15 | Productions segregated QA sytems with |
| 1-1 QA Division and its equivalences | (6) | charge,support, and promotion functions. |
| 1-2 Project Committee | (9) | 3 of assurances; product specification, quality,and function. |
| 2 Under Establishing or Reviewing of QA Rules and Codes | 5 | |
| 3 Top Management QA Audits | 6 | Audit team joining from other product |
| 4 TQC Promotion through TPM | 2 | division. |
| 5 Company Wide QA Information System | 4 | All of claims and quality incident |
| 6 QA Campaigns and Educations | 6 | to QA head quarter. |

TABLE **4**  Modelling Applications of differential equations

| 1-dimension<br>(Axial symmetric) | 2-dimensions<br>(Axial symmetric) | 3-dimensions |
|---|---|---|
| | Aluminum Hydrate<br>Thermo-reaction<br>(Exact) | |
| Furnace wall life<br>(Exact) | | |
| | Aluminum billet<br>Thermo-homogenizing<br>(Exact) | |
| Heatflux sensing<br>response(Exact) | Calibration<br>heatflux sensing<br>(FEM-thermal) | Thermal response<br>soil plantation<br>(FEM-thermal) | Fish package mold<br>design(FEM-stress) |
| Graphitizing<br>furnace insulation<br>(Galerkin) | Ring furnace<br>thermo ditribution<br>(FEM-themal) | Carbon particle<br>formation & sintering<br>(FEM-thermal) | Plastic blow molding<br>design(FEM-stress) |
| Plastic molder<br>cooling(Exact) | FeSi furnace<br>design(FEM-thermal) | | |
| Flon reaction<br>dynamics(RungeKutta)<br>thermo-reaction<br>front (Exact) | Aluminum cell anode<br>thermo ditribution<br>(FEM-thermo-electro) | | High pressured water<br>electrolysis<br>(FEM-stress) |
| Shaft kiln<br>thermal ditribution<br>(RungeKutta) | Ultra high pressed<br>crystall.mold<br>(FEM-stress) | | Magnetic shield<br>Alminum cell<br>(FEM-magnetic) |
| Ethyene cracking<br>(Gear) | Distillation tower<br>(FEM-stress) | Electrode joint<br>design<br>(FEM-electro, | Forklift pallet<br>design<br>(FEM-stress) |
| Fibre spinning<br>(RungeKutta) | Semi-conductor<br>thermal design<br>(FEM-thermal-stress) | thermal,stress)<br>Chocoralsky<br>christallation | Plastic automobile<br>parts design |
| | Optical lens molding<br>process(FEM-thermo-stress) | (Fluid dynamics,<br>FEM-thermal,stress) | Aluminum casting<br>design(Solidification) |
| | Electrolysis current<br>design(FEM-potential) | MOCVD<br>(Fluid dynamics,<br>FEM-thermal,stress) | Plastic food can<br>design(FEM-shock) |

TABLE **5**  Modelling Applications of Algebraic equations, Stastical analyis and Time series analysis

| Algebraic<br>modelings | Statistcal (Multivariate)<br>modelings | Time (Spectral)<br>modelings |
|---|---|---|
| Chlorination<br>of propyrene | NaCl electrolysis<br>reaction yield<br>(PCA,Multi-regressions) | Off-shore structure<br>design<br>(sea wave spectra and |
| Radical polymerizations<br>and molecular ditributions | Quality evaluations<br>GaP process<br>(PCA,Multi-regress) | structural object interactions) |
| Venture plannings<br>(LP) | Engineers,their profiles<br>(PCA,KAI,Markov) | X ray deffraction,polymer crystallations<br>(Gauss-Chauchy,3 order splines) |
| Feed blendings<br>(LP) | | |
| Optical screening<br>steric compounds | Liquid chromat calibrations<br>(Log multi regress) | Polymer reactor dynamics<br>and factor cross-correlations<br>(Akaike's AIC criteria and AR modeling) |
| Plastic inflation<br>die design | Powder sieve size<br>design | Water power reservoir<br>predictions<br>(Akaike's AIC and AR modeling) |
| Optical lens ray tracing | Polyethylene grade design<br>(PCA,Multi-regress,PC multi regress) | |
| Process material and<br>energy balance simulations | Phenol resin production<br>quality analysis<br>(Multi-regress) | FeCr pellet process<br>quality evaluations<br>(Akaike's AIC and AR modelings) |
| Distillations simulations | Fertilizer pelletizing<br>(Multi-regress) | |
| | C1 project catalyst<br>evaluation<br>(PCA,Multi-regress) | Chocoralsky process<br>dynamic modelings<br>(Akaike's AIC ,etc) |
| | Protein process molecular<br>design(PCA,Multi-regress) | |

# QUALITY CONTROL AND QUALITY ASSURANCE — QC AND QA SYSTEMS IN JAPAN

## H. Makabe

*Tokyo Institute of Technology, Japan*

**Abstract**. Quality control and quality assurance have achieved important roles as management technology for developing and manufacturing product and system of good quality. Especially at present, high technologies are developing and complexed product and system are making use, and advanced quality control and quality assurance system is required. In this comment report, overviews on quality control and quality assurance system are given. The discussions are focused on quality system which was developed based on Japanese cultural climate and economical background.

**Keywords**. Statistical quality control (SQC); Total quality control (TQC); Quality assurance; QC circle; QC audit; Total productive maintenance (TPM); Policy deployment; Prevention by prediction (mizen-boshi); Seven steps of quality assurance; Quality function deployment (QFD)

## INTRODUCTION

In this report, in section 1, in order to grasp outline quality control, development of it is reviewed putting emphasis on Japanese type quality control (TQC), and in section 2, some features of quality control of Japanese type (TQC) are explained based on QC symposium discussions in 1987. In sections 3 and 4, some aspects on quality assurance which is based on TQC activities are given and seven steps in quality assurance system are explained.

## 1. The Development of Quality Control

In this section, we describe chronicle history of quality control which started in United States in the 1920's. This consideration is useful for reviewing the characteristics of quality control in Japan and for assessing future aspects on quality control activities in the world.

### (1) Enlightment of statistical quality control by W. A. Shewhart

Statistical quality control (SQC) was proposed by Shewhart in the 1920's. Shewhart's idea was to utilize mathematical statistics for controlling production process and improving product quality. The importance of SQC concept and methods was recognized, and ASME (The American Society of Mechanical Engineers) and other organizations began to prevail SQC in many factories in U.S.A. making good manuals on SQC.

Thus, in the 1930's SQC methods such as control chart and sampling inspection technique was spread graduately in many factories. In the next decade, U.S. Government introduced SQC for procurement management. In second world war, the government needed SQC to acquire many products of good quality at moderate cost.

### (2) Researches on statistical theory

In the same decade when SQC had spread, Statistical Research Group (SRG) was organized in some universities in U.S.A. and promoted to research statistical theory which had been the basis of SQC method developed and applied successfully later.

### (3) Quality control (SQC) in Japan

Statistical quality control which contributed for manufacturing many good products in U.S.A. was accepted with keen interests in Japan, and was systematically introduced early in the 1950's.

SQC thus prevailed in Japan was successfully applied for manufacturing good product and reducing percent of defectives. Through those experiences, the management and the engineers came to recognize the importance of SQC, and SQC of Japanese style was established.

### (4) Quality control (TQC)

after successful application of SQC in manufacturing department, concept of quality control was graduately adopted in company-wide system. In the company where SQC was active, the new concept came to be accepted in management stage not only in manufacturing and engineering. Thus quality control based on the new concept have called "total" quality control which is abridged "TQC".

Trade liberalization problem in Japan and modernization of Japanese industries accelerated the progress of TQC late in the 1960's. Organization office for QC circle was established in 1962, and QC circle activities was spread in many Japanese companies.

### (5) Quality control in the 1970's and 1980's

Quality control which is called TQC have developed and have become to be characterize in some aspects. In the paragraph, some main trends in TQC are summarized in the following three items and detailed characteristics will be explained in the next section.

(i)  Quality management and TQC
     TQC have recognized as a basis for quality
     management, where quality management means
     effective management regarding quality as
     important.

(ii) TQC extending in tertiary industry
     TQC activities have spread from the secon-
     dary to the tertiary industries. This
     phenomenon means TQC which was grown in
     the manufacturing industry can be applied
     to the others such as service industry
     (for example, electric power and transpor-
     tation) and construction engineering
     industry.

(iii) Needs for quality assurance
     As new technology is grown and system of
     high performance is constituted by new
     complexed component, the requirement for
     reliability and perfect quality is in-
     creased. For acquisition of the system
     and component, quality control need to
     involve in its TQC system new concept of
     quality assurance and quality assurance
     system.

2. The Characteristics of Quality Control in
Japan (TQC)

In this section, some features and characteristics of quality control developed and constructed
in Japan under the influence of Japanese own
social structure and cultural climate is discussed. For the discussions, it is convenient
way to take up ten characteristics of TQC which
was summarized and authorized in Quality Control
Symposium (QCS) held in 1987. Ten characteristic
item are given in the following table.

TABLE 1   Ten Characteristics

(1)  QC activities promoting under the leadership
     of top management and involving all
     organization and all people
(2)  Management which is thoroughly based on an
     idea and a concept of "quality first"
(3)  Policy deployment and its management
(4)  QC audit (diagnosis) and its utilization as
     management implementation
(5)  Quality assurance activity which is
     consistently implemented at all stages
     extending from product planning and
     development to sales and after-service stage
(6)  QC circle activities
(7)  QC training and education
(8)  Development and application of QC method and
     technology
(9)  Introduction of quality control in tertiary
     and other industries
(10) Nation-wide QC production system

Items (1) and (2) in the table concerns on the
quality control activities at company management
level. In fact, in some company group, TQC is
introduced not only as company-wide but also
group-wide activities under leadership of the top
management. Concerning items (3) and (4), the
management by vision is explained as typical case
study. In those cases, company has vision indicated by broad target and policy of long range
management plan as shown in Figure 1. The vision
then is deployed in details as mid-range and
short-range (one Year)planning. Results of the
activities implemented based on planning are
evaluated and analyzed in QC audit by top and
middle management for finding difficulty of plan,

and then corrective action is taken as counter-
measure.

longrange plan ------> "vision"
        ↓
midrange plan
        ↓
one year plan ------>  policy ---> { target
                      (hoshin)       policy

Fig. 1.  Vision and policy

Item (5) describes characteristics of quality
assurance developed in Japan as cross-functional
management oriented for the "quality first". The
detailed explanations is given in next section.

3. Quality Assurance and its Role in the System
of TQC

In the 1960's, there was some new trend in TQC
activities represented by quality assurance. In
early stage in development of SQC, quality improvement were focused on reduction of number of
defectives and defects, however, in this decade,
as new product planning and development became
important in quality management system, new concept which we are calling quality assurance came
into TQC activities.

The increase of new complexed products stimulated
the need to establish quality assurance system,
and in many company, quality assurance department
was organized as head quarter to promote cross-
functional management for quality and to install
quality dike for defects and failures.

By the movement of consumerism which originated
in consumer protection policy advocated by
Kennedy, requirement for safety and perfectionism
was increased, and early warning system was
adopted in development and design phases of
quality assurance system. In many Japanese companies such as automobile and electronic
component manufacturing units, notion and technique of reliability was introduced as an
important tool in quality assurance system. It
is natural that reliability management and methods are useful tool for preventing defects and
hazards by prevention through prediction ("mizen-
boshi"). Thus "Quality" in terminology "quality
assurance" can be regarded as constituted by two
notions as follows;

quality { quality (in narrow sense)
        { reliability

In the 1970's, as quality competition was coming
harder, another new notion which is call
"attractive quality" was introduced. Basically,
for assuring the quality it is necessary to
reduce member of defectives or defects. But at
the same time it is also important to satisfy the
customers' needs. This consideration suggests us
to consider customer satisfaction. For achieving
new role of quality assurance, new technology
such as market survey, product planning by quality function deployment and design review was
introduced in quality assurance system. Following table is given in order to summerize above
discussions in the simple form,

TABLE 2   Summary of Quality Assurance

(1) Quality control in Japan started from
    quality improvement in process of factory in
    1950's.  The main object of Japanese QC was
    to reduce percent of defects and amount of
    reworks.
(2) In above process of the improvement, we found
    out that improvement of quality is the
    shortest way to efficient productivity and
    reduction of cost.
(3) After succeeding in the improvement of
    quality and productivity, object of Japanese
    QC was focused on development of new products
    in the middle of 1960's.
(4) In same decade, for new movement such as
    consumerism, concerns for
    safety and reliability increased gradually in
    Japanese QC activities which is called"TQC".
    QA and "reliability" was adopted as one of
    important subjects of QC education in the
    early of 1970's.
(5) TQC is now extending to engineering
    construction group and to tertiary industry
    such as electric power company.  Applications
    of elementary statistical method such as
    Pareto analysis and fish-bone diagram are
    also popular.
(6) As a result of hard market share competition,
    concept of "attractive quality" is adopted as
    an important motion of quality.
    Quality Function Deployment (QFD) is used as
    a popular tool for QA.

4. Seven Steps of Quality Assurance

To describe detailed significance and activity of
quality assurance,  explanations of the following
steps (phases) are necessary:

    (i)    Market Research (Marking)
    (ii)   Product Planning
    (iii)  Research and Development, Design
    (iv)   Manufacturing Preparation (Production
           Engineering)
    (v)    Manufacturing (Mass Production)
    (vi)   Sales and After Sales Service
    (vii)  Vendor Relations

In the followings,  some of above item are
explained in detail.

(1) Market research

In this step,  collection and analysis of  market
information are performed.  Market research based
on the analysis concentrates on
    (a) needs of customers for the product in
        the market
    (b) understanding the usage and
        environmental condition for the product
        in the market

(2) Product planning

The  result of market research must be  reflected
to  this  phase work.   There are  two  important
quality   assurance   activities  which  are   as
follows;
    (a) needs and specification
    (b) early detection of failure problems
In (a), we must deploy the need of customers into
specification of quality planning,  and if necessary we can use quality function deployment (QFD)
for assuring quality to meet customer needs.  The
result can be evaluated by customer satisfaction.
In (b),  defect and quality failure must be  predicted  and  prevented at early stage  in  early
warning   system,   and   we  can  use   popular

reliability technology such as FMEA and FTA.

(3) Research and development; design

The  results of product planning step  activities
are  described in planning document as  detailed
product plan specifications.  In this step,  the
specification must be  materialized in  design
paper and in proto type model.  For  completing
activities,  QC methods such as design  review,
engineering model experiments,  FMEA and FTA, and
design  of experiment are utilized  for  assuring
product quality.

(4) Manufacturing preparation (Production
engineering)

Important factors concerning quality are referred
to as 4M (machine, man, method, material), but in
this stage assurance of 3M (material excepted) is
taken up.
For  manufacturing product of good  quality,
quality machine (equipment) must be assured  at
first.  The  quality of machine is evaluated  by
process capability Cp which is required  to  be
above 1.33.  In order to assure other characteristics of machine including process  capability,
quality  and process function deployment  ("koho-
tenkai")  are utilized,  and in many cases  total
productive  maintenance (TPM) is also popular  in
Japanese industry.

(5) Manufacturing

After manufacturing preparation and once manufacturing starts, quality data is collected in early
warning  manufacturing system and is examined  if
process is in stable state till the stability  is
confirmed,  and  SQC method such as control chart
and prevention recurrence ("saihatsu-boshi") system are used for controlling the process.

In summary,  TABLE 3 (QA system) is given to show
are case study.

TABLE 3(i)   QUALITY ASSURANCE SYSTEM (SEVEN STEPS)

(Q1) Marketing                          user's needs
                                        object of product use
                                        circumstance

                                                    QFD
(Q2) Product Planning                   user's needs ——→ product specifications
                                        circumstance ——→ product defects

(Q3) Development and Design             DR
                                        FMEA/FTA
                                        design of experiments

(Q4) Pre-production Demonstration       process capability
     Production Engineering             QA by process equipment
                                        prevention by prediction (mizen boshi)

(Q5) Production                         prevention recurrences (saihatsu boshi)
                                        process control (control chart)
                                        fool proof (FP) system
                                        training and education (workers)
                                        QC circle activities

(Q6) Sales and After Sales Service      statistical analysis
                                                    (analysis of market share)
                                        recommendation of products use
                                        field data analysis

(Q7) Vendor Relations
     Procurement Management

TABLE 3(ii) QUALITY ASSURANCE SYSTEM (A CASE STUDY)

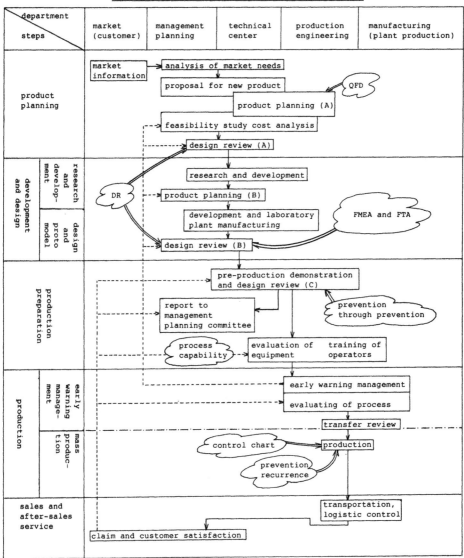

# USE OF SPC AND RGA FOR ON-LINE MONITORING OF CHANGING PROCESS CONDITIONS

## N. Jensen

*The Process Design, Dynamics and Control Group, Instituttet for Kemiteknik,*
*Technical University of Denmark, DK-2800 Lyngby, Denmark*


The Process Design, Dynamics and Control Group
Instituttet for Kemiteknik, Technical University of Denmark, DK-2800 Lyngby, Denmark

*Abstract:* The paper introduces and describes a new tool called the relative standard deviation (RSD) for on-line monitoring of changing process conditions. The industrial control engineer spends a significant amount of time troubleshooting control loops due to changing operations or process conditions. The technique introduced and described here can help the control engineer by alerting him to process changes, which may requirer re-tuning of control loops.

The relative standard deviation is based on the combined use of statistical process control (SPC) tools and the relative gain idea, which for years has been applied to control loop configuration. It is proposed to extract the process information from control charts based on relative standard deviations. The selection of sample frequencies and group sizes are discussed. The paper shows how the relative standard deviations are derived.

Examples of its use to identify changes in the parameters of a simple first order system are given. Finally results of applying the relative standard deviation array to experimental data from a pilot-plant distillation column with heat pump are shown and discussed.

## 1   Introduction

Many process control computers installed in chemical plants around the world often come with standard history collection packages. Standard history collection packages usually gives the history data for all points in the control system, whether actual process inputs or calculated variables. On some systems one get spot values at one or two intervals in addition to for example six minute, hourly, daily, weekly and monthly averages.

Most history data are collected, but never used. The operators, process engineers and control engineers only monitors a few key variables out the many thousand on typical process control computer. Only when the cause of a major upset is searched for is all available data used. The cognitive abilities of most humans limit the amount of information, which can be gliemed by scanning plots or worse tables of history data for one or a few inputs. Thus the history data constitutes a huge potential source of information about the process. The problem seem to be how to condense the data and extract the information. It is with this extraction and condensation statistical process control (SPC) can be of help.

In this paper SPC techniques and relative gain array (RGA) concepts are used to develop a new tool called the relative standard deviations (RSD) for on-line monitoring of changing process conditions. While ordinary SPC techniques may be used to monitor process stability, they do not readily pin-point control loops in need of attention. As an example is in figure 1 shown averages and standard deviations for key process variables in the distillation column pilot plant at Instituttet for Kemiteknik. In the fig-

ure the averages and standard deviations are for a 12 hour period of a $2\frac{1}{2}$ day experiment. From the figure the stability of the operation over that period can be determined. If the calculation period is reduced to for example 3 hours and a rooling calculation is performed, then such a display could be a useful aid for the operator or process engineer in detecting process upsets. The control engineer could use the display for assessing where feedback or feedforward control is currently inadequate, and new control strategies need to be implemented.

The relative standard deviation (RSD) is introduced in section 3, but first some background infor-

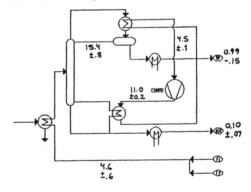

Figure 1: Distillation Process with Simple SPC Data

mation on SPC techniques in the process industries and RGA concepts are reviewed. The paper finishes with two examples on the use of the RSDs.

# 2   SPC and RGA Background

SPC techniques has now been used in the process industries for a number of years, however, the literature on on-line use of SPC is very limited. The evidence suggest successful and profitable applications do exist (Guillory(1988), Hovermale (1988) and Pfeifer (1988)). In the following subsection some ideas on the use of simple SPC techniques in on-line monitoring of continuous processes and their control is discussed.

## 2.1   SPC Techniques

The SPC tools in on-line use in the process industries are CuSum charts, moving average control charts and almost normal control charts. An excellent introduction based on work at DuPont is given by Hess (1989).

Process diagrams, such as in figure 1 with averages and standard deviations of key process variables, are very useful for communicating information about process stability to management and others. The diagram can be improved by using indicators (a star) for values, that are out of control based on the last 30 days of operation.

An almost normal control chart of setpoint minus process value is an excellent tool for monitoring control loop performance. This is similar to the technique Bothe (1988) advocates for short production runs and it makes the monitoring of control loops independent of production levels. Almost normal indicates, that instead of range is the estimated standard deviation of the group used. This requires a somewhat larger group size, which is not a problem if the calculations are done automatically by the process control computer. The group refers to a series of from 9 to 15 equidistant datapoints sampled at from one quarter to one third the dominant time constant of the system. In a plant, such as our distillation column with heat pump, the sample time criteria may have to be violated in order to have all control loops monitored the same way. If the estimated group standard deviation is divided with the difference between the upper control limit and the lower control limit dimensionless numbers for comparing the performance of different control loops, e.g. a temperature control loop and a concentration control loop, is obtained.

Users of normal or almost normal control charts in the process industries one should be aware, that the sampling is rational and not random (McGue and Ermer (1988)). This means many of the rules for interpreting control charts, such as seven in a row on the same side of the mean, may not apply. However, out of control situations based on the control limits can still be detected.

Moving average charts and CuSum charts can both be used to alert the operator to an undesirable situation. When CuSum charts are used on infrequent laboratory tests, then one must recognize, that two or more processes are involved: the process producing the material and the process(es) measuring its property(ies). In order for the results of the

laboratory tests to provide reliable feedback signals the measurement process(es) must be in good statistical control at all times.

The univariate process capability index defined by

$$\frac{allowable \quad process \quad spread}{actual \quad process \quad spread}$$

really is not applicable to chemical processes, which are mostly multivariable and nonlinear with considerable interaction. Recently Chan et al. (1989) defined a multivariate measure of process capability, which is applicable to multivariable chemical processes such as distillation and polymerization.

Figure 2: System P with m-inputs and n-outputs

## 2.2   RGA Techniques

The relative gain array (RGA) concept was introduced by Bristol (1966) as the ratio of the open loop gain to the gain, when all other outputs were perfectly controlled. Bristol showed that the Hadamard product could be used to evaluate the relative gain array based on only the open loop gains

$$\Lambda = K_P \circ [K^{-1}]^t \qquad (1)$$

The RGA has found widespread use in screening control system configurations for example for distillation columns (Shinskey (1984)).

Since the RGA is a function of the process gains and is dimensionless it would be a good candidate for monitoring for process changes, which could require control loop changes, e.g. re-tuning. It is, however, difficult to determine the RGA experimentally with sufficient accuracy and ease to make it useful for this purpose. This is where SPC techniques may help as outlined in the following section.

# 3   Relative Standard Deviations

Consider the system P in figure 2 with m-inputs $u_1, \ldots, u_m$ and n-outputs $y_1, \ldots, y_n$. If we introduce a step change in the i'th input $u_i$ and record the changes in all n outputs we can determine the n-gains $k_{ij}, j = 1, \ldots, n$. If we introduce simultaneous step changes in $u_i$ and $u_l$ ($i \neq l$), and again record the changes in all outputs, we have for the j'th output

$$\Delta y_j = k_{ij}\Delta u_i + k_{lj}\Delta u_l$$

Hence we can no longer determine all the individual gains from the final steady state changes.

If a known sinusoidal change, $\sin(\omega t)$, is introduced in the i'th input $u_i$ and the changes in all n outputs are recorded the n gains $k_{ij}, j = 1, \ldots, n$. can of-course again be determined. Then introduce the known sinusoidal simultaneously at two inputs $u_i$

and $u_l$ $(i \neq l)$, and record the changes in all outputs we have for the j'th output

$$\Delta y_j(t) = k_{ij} \sin(\omega t + \phi_{ij}) + k_{lj} \sin(\omega t + \phi_{lj})$$

Hence we can no longer determine all the individual gains simply from the changes in the n-outputs. If we record the changes for four different times within the period of the sine, we can however determine both the gains and the phase lags.

The above analysis shows, that by taking advantage of the time varying input, we can get more information about the system, that by using simple step changes at one point in time. This is nothing new, it is the fundamental principle of all closed loop estimation schemes.

If in place of four spot data sets we sample the system at four periods, which are not multiples of each other, and use SPC to calculate the standard deviations of each data series and of the input sampled the same way four equations are obtained from which the two gains and two phases could be estimated.

During normal operation a process is subject to random input fluctuations. These will give rise to output fluctuations with a different variance. The magnitude of these fluctuations can be determined using SPC techniques, e.g. on-line control charts. Then the process gains can be estimated using SPC techniques. Let $\sigma_{y_j}$ be the standard deviation of the j'th output and let $\sigma_{u_i}$ be the standard deviation of the i'th input, then for a SIMO system one has the *relative standard deviation*:

$$k_{ij} = \frac{\sigma_{y_j}}{\sigma_{u_i}} \tag{2}$$

and the relative standard deviation array $K = [k_{ij}]$. In general the relative standard deviation array will be singular since the process inputs normally will not excite the outputs sufficiently and since inputs and outputs may be correlated due to feedback control loops. The square of the relative standard deviation will be proportional to the closed loop gains. For a SISO system the following relationship holds

$$k \approx \frac{k_C k_P}{1 + k_C k_P} \tag{3}$$

and hence

$$k_P \approx \frac{k}{1 - k} \frac{1}{k_C} \tag{4}$$

where $k_C$ is the proportional gain of the regulator, $k_P$ is the open loop process gain and $k$ is the closed

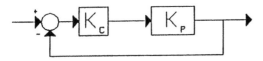

Figure 3: A MIMO system with a controller

loop relative standard deviation. For a MIMO system one gets an array of standard deviation ratios, which may have information about the system gains.

The system is of course equipment plus all closed control loops. For the system in figure 3 the relationsship is:

$$K \approx (I + K_C K_P)^{-1} K_C K_P \tag{5}$$

hence

$$K_P \approx K_C^{-1} K (I - K)^{-1} \tag{6}$$

where $K$ is the matrix of ratioes of closed loop standard deviations. The many SISO loops in industry may be monitored for changes in performance using the relative standard deviation as defined in equation 2.

I recommend automatic control charts of the RSD for each control loop, and the reporting of out of control loops on a daily basis.

It is of course trivial to extend the definition to disturbances. Such relative disturbance standard deviations could be useful in monitoring when changes in feed-forward control loops were necessary.

# 4   Examples of Use of RSD

The following subsections give to examples of the use of the RSD array to extract information from apparently useless noisy data. The first example uses simulated data from a distillation column model. The second example uses experimental data from a pilot plant distillation column at Instituttet for Kemiteknik at the Technical University of Denmark.

## 4.1   Distillation Column Model

Waller et al. (1988a) used the following linear model structure to model a pilot plant distillation column separating a binary mixture:

$$y = Gu + G_F d \tag{7}$$

with $y = (T_4, T_{14})^t$, $d = (F, x_F)^t$ and different input sets $u$. The temperature on tray 4, $T_4$, is used as an indicator of top product composition, and similarly is the temperature on tray 14, $T_{14}$, used as indicator of changes in bottom product composition. The disturbances are feed flow, $F$, and feed composition, $x_F$. For $u = (\frac{D}{L+D}, \frac{V}{B})^t$ the following reconciled, see Häggblom and Waller (1988), transfer function matrices are found (Waller et al. (1988b)):

$$G = \begin{pmatrix} \frac{6.4e^{-0.5s}}{15.0s+1} & \frac{1.0e^{-0.5s}}{25.0s+1} \\ \frac{23.0e^{-1.3s}}{26.0s+1} & \frac{34.0e^{-0.5s}}{15.0s+1} \end{pmatrix} \tag{8}$$

$$G_F = \begin{pmatrix} \frac{0.001e^{-1.0s}}{6.0s+1} & \frac{-0.029e^{-7.5s}}{18.0s+1} \\ \frac{-0.05e^{-1.0s}}{7.5s+1} & \frac{-0.89e^{-1.0s}}{7.5s+1} \end{pmatrix} \tag{9}$$

Simulations with a sample time of 2 minutes and a group size of 15 gives with random gaussian noise input the following RSD array:

$$\begin{array}{cc} 0.8460 & 0.6185 \\ 7.6458 & 5.0927 \end{array}$$

For changes in for example $k_{22}$ or $k_{11}$ the corresponding row in the RSD array changes by a proportional amount. The simulations also revealed, that time constant and gain changes cannot be distinquished, but as expected a time constant change has the opposite effect of a gain change. The simulations also showed, that changes in the parameters of the off-diagonal transferfunction elements appears as proportionately smaller changes in all the elements of the RSD array.

## 4.2  Pilot Plant Distillation Column with Heat Pump

Two experiments were performed on the pilot plant shown in figure 1. The equipment is described in Hallager et al. (1986). Each involved multivariable identification of the plant. The measured variables in both experiments were the low pressure (P8) in the heat pump circuit, which is closely related to the column pressure, the high pressure (P10) in the heat pump and the estimated concentrations of methanol on trays 1, 10 and 19 (X1, X10 and X19) (trays numbered from bottom to top). The estimates were based on simultaneous temperature and pressure measurements. The manipulated variables were feed concentration (XF) and feed flow (FF), reflux flow (RF), cooling in the heat pump circuit (P10S) and pressure loss (DP) in the heat pump circuit. The

| Experiment 26.April 1988 | | | |
|---|---|---|---|
| P8 | 4.14± 0.14 | XF | 0.50± 0.03 |
| P10 | 11.04± 0.22 | FF | 4.59± 0.36 |
| X1 | 0.10± 0.02 | RF | 15.81± 0.38 |
| X10 | 0.51± 0.08 | P10S | 11.03± 0.22 |
| X19 | 0.97± 0.07 | DP | 0.52± 0.11 |

| Experiment 5.May 1988 | | | |
|---|---|---|---|
| P8 | 4.47± 0.11 | XF | 0.50± 0.04 |
| P10 | 11.01± 0.24 | FF | 4.57± 0.59 |
| X1 | 0.10± 0.07 | RF | 15.46± 0.75 |
| X10 | 0.51± 0.13 | P10S | 11.01± 0.25 |
| X19 | 0.99± 0.15 | DP | 1.00± 0.13 |

Table 1: Average experimental operating conditions for two experiments given as mean values ± the standard deviations.

average operating conditions for the two experiments are summarized in table 1. All pressures are in Bar, flows in litre/min. and concentrations in molefraction methanol. The other component in the binary system studied is isopropanol.

The operating conditions in table 1 for the two experiments appear quite similar. The relative standard deviation arrays for the two experiments based on 12 hours of data are shown in table 2. Since the dominating time constants of the pilot plant distillation column is about 7 and 180 minutes respectively a sample period of 2 minutes and a group size of 15 was used in the RSD calculation. This follows the guideline of a sample period equal to $\frac{1}{3} - \frac{1}{4}$ the fastest time constant, and a group size time sample time equal to 3-4 times the dominating time constant.

About half of the values in the RSD array change by 25% or more. Thus one would expect significantly different steady state gains at the two operating conditions in tabel 1. The steady state gains estimated using a multivariable identification procedure and shown in tabel 3 confirms this.

## 5  Conclusion

This paper suggest the use of automated control charts of relative standard deviation to monitor the performance of control loops. The standard deviations of groups of sampled data are used. A sampling period of about one quarter the dominating time constant is suggested. The group size should be choosen, so

| RSD Exp. 26.April 1988 | | | | |
|---|---|---|---|---|
| 3.13 | 0.24 | 0.45 | 0.52 | 0.99 |
| 5.86 | 0.46 | 0.81 | 0.93 | 1.91 |
| 0.50 | 0.04 | 0.06 | 0.10 | 0.12 |
| 2.37 | 0.18 | 0.30 | 0.46 | 0.60 |
| 1.05 | 0.07 | 0.18 | 0.18 | 0.32 |

| RSD Exp. 5.May 1988 | | | | |
|---|---|---|---|---|
| 2.24 | 0.16 | 0.14 | 0.38 | 0.72 |
| 5.67 | 0.41 | 0.34 | 0.95 | 1.81 |
| 0.68 | 0.05 | 0.03 | 0.09 | 0.20 |
| 2.16 | 0.15 | 0.12 | 0.33 | 0.66 |
| 1.20 | 0.08 | 0.06 | 0.14 | 0.34 |

Table 2: Relative standard deviation (RSD) arrays for two experiments. A column corresponds to one measure variable and a row to one manipulated variable.

| Gain Exp. 26.April 1988 | | | | |
|---|---|---|---|---|
| -0.0064 | 0.0080 | -0.0325 | 0.9361 | -0.1365 |
| -0.0142 | 0.4755 | 0.0800 | 0.6176 | -0.3749 |
| -0.0656 | 2.0542 | 0.3225 | -1.7775 | -0.2922 |
| -0.0089 | 1.0562 | 0.3235 | -1.5780 | 0.0580 |
| -0.0093 | 0.7728 | 0.2534 | -1.0179 | 0.0196 |

| Gain Exp. 5.May 1988 | | | | |
|---|---|---|---|---|
| 0.0220 | 0.0478 | 0.0608 | 0.9402 | 0.1612 |
| 0.0410 | 0.0675 | 0.0837 | 0.9403 | -0.1979 |
| 0.2420 | 0.4081 | 0.5071 | -2.3378 | 0.3851 |
| 0.4153 | 0.4772 | 0.6797 | -3.2412 | 0.5305 |
| 0.4221 | 0.4540 | 0.7708 | -3.4593 | 0.4987 |

Table 3: Steady state gains for the two different operating points.

the time covered by one group of data is 3-4 times the dominating time constant.

Simulations of a simple distillation column model and experimental data from a pilot plant distillation column confirm, that relative standard deviations can be used to monitor for process changes, which may requirer control loop tuning.

# References

[1] Bristol, E.H. (1966): "On a new measure of interaction for multivariable process control", IEEE Trans. Automatic Control, **AC-11**, p.133-134.

[2] Bothe, D.R. (1988): "SPC for short production runs", Quality, December, p.58–59.

[3] Chan, L.C.; Cheng, S.W.; Spiring, F.A. (1989): "A Multivariate Measure of Process Capability", To appear in Int. J. Model. Simu.

[4] Guillory, A.L. (1988): "Statistical process control in a paper mill", Chem. Eng. Progr., April, p.52–57.

[5] Hallager, L.; Toftegård, B., Clement, K., Jørgensen, S.B. (1986): "A distillation plant with an indirect heat pump for experimental studies of operation form, dynamics and control", IFAC Symposium on Dynamics and Control of Chemical Reactors and Distillation Columns, Bournemouth, England.

[6] Hess, J.L. (1989): "Managing quality", Chemtech, July, p.412–416.

[7] Hovermale, R.A. (1988): "Quality management system in high-performance plastic films", Chem. Eng. Progr., April, p.36–44.

[8] Häggblom, K.E.; Waller, K.V. (1988): "Transformations and consistency relations of distillation control structures", A.I.Ch.E. Journal, **34**(10), p.1634–1648.

[9] McGue, F.; Ermer, D.S. (1988): "Rational samples – not random samples", Quality, December, p.30–34.

[10] Pfeifer, C.G. (1988): "SPC in the process industries", Quality, December, p.38–40.

[11] Shinskey, F.G. (1984): "Distillation Control", 2nd Edition, McGraw-Hill Book Company, New York.

[12] Waller, K.V.; Finnerman, D.H.; Sandelin, P.M.; Häggblom, K.E.; Gustafsson, S.E. (1988a): "An experimental comparison of four control structures for two-point control of distillation", Ind. Eng. Chem. Res., **27**(4), p.624–630.

[13] Waller, K.E.; Häggblom, K.E.; Sandelin, P.M.; Finnerman, D.H. (1988b): "Disturbance sensitivity of distillation control structures", A.I.Ch.E. Journal, **34**(5), p.853–856.

Copyright © IFAC Production Control in the
Process Industry, Osaka, Japan 1989

# APPLICATION OF NONLINEAR STATE ESTIMATOR IN QUALITY CONTROL OF A POLYESTERIFICATION

## Haitian Pan, Shuqing Wang and Jicheng Wang

*Institute of Industrial Process Control, Zhejiang University, Hangzhou, PRC*

**Abstract.** The application of nonlinear state estimation in quality control of in industrial polyethylene-terephthalate (PET) condensation reactor is discussed in this paper. Besed on a simplified model of the reactor, a nonlinear state estimation algorthm is developed by using extended kalman filter technique. The estimator predicts the unmeasured state variable-conversion in the reactor from on-line turbidity reaction temperature measurement. Then the intrinsic viscosity of polymer, which is quality index of PET, is calculated according to a found viscosity equation. Moreover, it is diseussed how an adaptive state estimation scheme is used to inprove the convergence behaviour and the acuracy of the estimator. The simulation results show that the estimator can reconstruct the unmeasured state variable and the estimated results and the real reacting process have good agreement. The predicted value of the viseosity can meet the needs of production quality control for PET.

**Keywords.** polymerization reactor, Kalman Filter, Nonlinear state estimation, state variable, Dynamic model, simulation.

## INTRODUCTION

Polymerization process is very complex and strongly nonlinear. If the optimization of polymer quality is performed by means of automatic control, the process information discribing reaction state must be obtained or the dynamics of a process must be known. In practical reactor operation, the key states are the variables concerning with produced polymer propertyies, such as molicular weight and its distribution, particle size distribution, polymer rheology and conversion. However, most of the variables can't be measured by on-line, owing to the lack of reliable and robust on-line polymer characterization instrumentation. The key problem is how to get the states relating to polymer properties in polymer quality control, and it becomes an area of interest in the polymerization reaction engineering.

A powerful tool to overcome the difficulty is the state estimation techniques. By utilizing the online state information obtained, the variables unmeasureed are real-time estimated. A number of researchers have applied the technique at various polymerization reactors. Jo and Bankoff (1976) applied extended Kalman Filter (EKF) to a vinyl acetate free radical polymerization carried out in an experiment CSTR. Monomer conversion and the weight-average molecular weight of the polumer are estimated from the refractive index and temperature, schuler and papadopulau (1986) studied a decoupled estimator for the real-time estimation of the chain length distribution and conversion in a batch polystyrene reacter, and demonstrated the estimation algorithm in the experiment. Ellid and coworkers (1988) beveloped a two time seale filter to estimate monomer and initiator conversions as well as the polymer molecular weight distribution from on-line measureed, of temperature, monomer conversion, and gel permeation chromatoraphy in a batch methyl methacrylate (MMA) polymerization reactor. In addition, the state estimation techniqcces have had better application in other processes.

In this paper, we study the product quality control problem in an industrial polyethylene-terephthalate

(PET) condensation reactor. The conversion is the important state variable of polycondensation reaction. It shows directly the polymerization degree of PET melt. If the information about the comversion can be obtained, the average molecular weight or the viseosity of produced PET will be controlled on the desired qualty index. But there are no online measurement devises that are used to follow the tracks of conversion. In the real production, the viscosity of PET melt is used as an indirect index. Based on the analysis of polumerization kinetics, a simplified dynamic model of PET condensation reactor which predicts the temperature and conversion, is found.. And the relation of PET viscosity with conversion is employed to build a viscosity equation. Based on Kalman Filtering technique, a nonlinear state estimation algorithm is developed by using above reactor model. The estimator predicts the conversion from only temperature measurement, the viscosity of PET is calculated according to the viscosity equation. The predieted viscoisty value can be used to update optimal quality control strategy for the polycondensation reacter. At the same time, the influence of model error and meascuement noise is also filtered.

In the present work, we test the efficiency of the estimation algorithm and discucss the simulation results. The algorithm is numerical in the case of different reaction condition. It order to inprove the convergence performance of the state estimator, an adaptive state estimation scheme is designed. At last, the simulation results of estimateor are presented and compared with practical reaction prosess.

## MODEL DYNAMIC OF REACTOR

In the considered PET reactor, the whole reaction can be divided into two process, reversible polycondensation reaction and a series of degradation reaction. These degradation reaction is mainly thermal degradation at higher reaction temperature. Consider aboue two reaction, the whole reaction system can be written as follows,

$$2 \sim \langle o \rangle - COOC_2H_4OH \xrightarrow[k/K]{k}$$

(A)

$$\sim \langle o \rangle - COOC_2H_4OOC - \langle o \rangle \sim + HO-C_2H_4-OH$$

kd     (B)         (C)

$$CH_2 = CHOOC - \langle o \rangle \sim + \sim \langle o \rangle COOH$$

(D)

Where k is polycondensation rate constant, K is reaction balance constant, and kd is degradation rate constant.

According to above reaction equation, the polycondensation kinetic equation can be formulated.

$$\frac{dC_B}{dt} = k\, C_A^2 - \frac{k}{K} C_B C_C - k_d\, C_B \qquad (1)$$

Also, the energy balance gives

$$V \rho\, C_p \frac{dT}{dt} = F p C_p (T_f - T) + (-\Delta H) v k \exp(-E/RT) C_B$$
$$- hA\,(T-T_C) \qquad (2)$$

Through hypothesis and synthesis, Eqs.1 and 2 can be modified to

$$\frac{dx}{dt} = \alpha_1 \exp(\beta_1/T) \cdot [1-x+\tfrac{1}{2} D \cdot t]^2 - DX \qquad (3)$$

$$D = \alpha_2 \exp(\beta_1/T) \cdot p + \alpha_3 \exp(B_2/T) \qquad (4)$$

$$\frac{dT}{dt} = \gamma_1 (T_f - T) + \gamma_2 \exp(B_1/T)(1-x)$$
$$- \gamma_3 (T - T_C) \qquad (5)$$

Where x is monomer conversion ( $x = \frac{C_{Ao} - C_A}{C_{Ao}}$ ) , p

and Tc, the vacuum in the reactor and the temperature in the jacket, are controled. The parameturs $\alpha_i$ (i = 1 - 3), $\beta_i$ (i = 1 -2), and $\gamma_i$ (i = 1 -3) are all model parameters: These will be identified by using experimental data in the plant. So Eqs.3, 4 and 5 form the fundmental reactor model. It describes two states, conversion and temperature, in the reactor.

In the practical operation, temperature in the reactor is an on-line measutement variable. We get output equation of reactor model

$$y = T \qquad (6)$$

On the other hand, the intrinsic viscosity of PET melt and conversion have following relation-ship

$$IV = \xi_1 [a(\frac{1}{1-x}) + b] \xi_2 \qquad (7)$$

where a is the molecular weight of structure unit, b is the molecular weight of end group. $\xi_1$ and $\xi_2$ are coefficients. Eq.7 is called viscosity equation.

Figs. 1 and 2 are simulation results in the case of different reaction conditions. They conform with actual results. As the results of simulation study, we found that the reaction is mainly polycondensation of chain propagation at the temperature T = 280°C and the resident time τ = 200min and the degradation reaction can be neglected. So the above

model can be further simplified into following form.

state equations

$$\frac{dx}{dt} = \alpha_1 \exp(\beta/T) [(1-x)^2 - \alpha_2\, px] \qquad (9)$$

$$\frac{dT}{dt} = \gamma_1 (T_f - T) + \gamma_2 \exp(\beta/T)(1-x) - \gamma_3(T-T_C)$$

output equation

$$y = T \qquad (10)$$

Obviously, this model structure is simpler than the former. It is suited to real-time ealculation by computer.

## NON-LINEAR STATE ESTIMATOR

The reactor model investigated can be described by the nonlinear stochastic differential equation

$$\dot{x}(t) = F[x(t),\, u(t),\, t] + \omega(t)] \qquad (11)$$

$$y(t) = H[x(t),\, t] + V(t) \qquad (12)$$

The vector F is a nonlinear function corresponding to the determinstic model Eqs 8 and 9. $\omega(t)$ is model error vector, V(t) is measurement noise vector. They are all zero mean gaussian noise and have following statistic property.

$$E[\omega^T(t)\omega(t)] = \theta(t)\, \delta(t-\tau)$$

$$E[V^T(t)V(t)] = R(t)\, \delta(t- )$$

$$E[\omega^T(t)V(t)] = 0$$

where Q and R are respectively model and measurement error covariance matrixs.

In fact, the sampled temperature signal is time-discrete. Eq.12 should be a discrete form

$$y(k) = H[x(k),\, k] + V(k) \qquad (13)$$

In accordance with above reactor model, the only temperature measurment is enough to be used in the estimation of another unmeasurement state —— conversion, because the reaction system is complete observable.

Based on the extended Kalman Filter, Consider the continrune state equations and discrete output equation, the nonlinear state estimator has continuous-discrete structure.

During sampling time $t_K$ to $t_{K+1}$, the state is predicted by integrating Eq.11.

$$\dot{\hat{x}}(t_{K+1}/t_K) = F[\hat{x}(t_{K+1}/t_K),\, u(t_{K+1}),\, t_{K+1}] \qquad (14)$$

using $\hat{x}(t_K/t_K) = \hat{x}(t_K)$ as the initial state

The prediction error covariance matrix

$$\dot{p}(t_{K+1}/t_K) = Ap + pA^T + Q ,$$

$$p(t_K) = p(t_K/t_K) \qquad (15)$$

At samping time $t_{K+1}$, an new measurement $y(t_{K+1})$ is obtained, the above state prediction value $\hat{x}(t_{K+1}/t_K)$ is updated

$$\hat{x}(t_{K+1}/t_{K+1}) = \hat{x}(t_{K+1}/t_K) + k(t_{K+1}) \cdot$$
$$\cdot [y(t_{K+1}) - H[\hat{x}(t_{K+1}/t_K),\, t_{K+1})]] \qquad (16)$$

where $\hat{x}(t_{K+1}/t_{K+1})$ is the estimation value of x from measurement $y(t_{K+1})$, $K(t_{K+1})$ is so called filtering gain

$$K(t_{K+1}) = p(t_{K+1}/t_K)C^T(t_{K+1}) \cdot$$

$$[C(t_{K+1})P(t_{K+1}/t_K)C^T(t_{K+1})+R]^{-1} \quad (17)$$

The entire estimation error covaniance is

$$P(t_{K+1}/t_{K+1}) = p(t_{K+1}/t_K) - K(t_{K+1})$$

$$\cdot C^T(t_{K+1}) \ P(t_{K+1}/t_K) \quad (18)$$

In the above equations, $A(t_{K+1})$ and $C(t_{K+1})$ are the linearizing result of the model about the current estimation $\hat{x}(t_{K+1}/t_{K+1})$. i.e.

$$A(t_{K+1}) = \frac{\partial F}{\partial x}\Big|_{\hat{x}(t_{K+1}/t_K)},$$

$$C(t_{K+1}) = \frac{\partial H}{\partial x}\Big|_{\hat{x}(t_{K+1}/t_K)} \quad (19)$$

The estimation algorithm consists of two parts. At first, a set of differential equations are soled by Runge-Kutta method, then the final estimation at the time $t_{K+1}$ is found by iteration. The structure of the state estimator is shown in Fig.3.

The specific estimation algorithm is only algebra iterating calculation and avoide matrix operation. It is suited to real-time estimation computation on-line. After the conversion state is obtained from the estimator, the viscosity of PET is correspondingly infered by Eq.7. This estimation value of viscosity can be used to guide reactor operation and control the quality of produced PET melt.

## ADAPTIVE ESTIMATOR SCHEME

In order to inprove estimator behaviour and increase estimation accuracy, a factor S is drawn into the developed algorithm. Eq.15. is rewritten into following form.

$$\dot{P}^*(t) = AP^* + P^*A^* + Q + S(t) \ P^*(t) \quad (20)$$

Because of $S > 0$, camparing Eq.20. with Eq.15, we have

$$P^*(t_{K+1}/t_K) > P(t_{K+1}/t_K)$$

$$K^*(t_{K+1}) > K(t_{K+1})$$

Then Eq.16. is rewritten

$$x^*(t_{K+1}/t_{K+1}) = [I - K^*(t_{K+1})C(t_{K+1})]$$

$$\cdot x^*(t_{K+1}/t_K)+K^*(t_{K+1})y(t_{K+1})$$

$$(21)$$

Therefore it can be seen that the effect of current measurement $y(t_{K+1})$ on estimation is strengthened, but the influence of historical dada is weakened.

How to select S is important. If s is selected as a constant, a steady-state error may be caused. Here, we use a heuristic reasoning method to determine S,

At $t_K$, Definiting:

Estimation error    $E(t_K) = \hat{x}(t_K) - y(t_K)$    (22)

Estimation error rate

$$EC(t_K) = \hat{x}(t_K) - \hat{x}(t_{K-1}) - y(t_K) + y(t_{K-1})$$

$$(23)$$

Through the simulation investigation of nonlinear state setimator, we can summerize a series of fuzzy statements, such as

IF $E(t_K)$ is large and $EC(t_K)$ is large then S is large.

According to these statements, we may define a fuzzy reasoning relation R, After each estimation, $E(t_K)$ and $EC(t_K)$ is calculated, thus S, which is induced by the product of E and EC, is given by the compositional rule of inference, that is,

$$S(t_K) = [\ E(t_K) \times EC(t_K)] \cdot R \quad (24)$$

In the application, the fuzzy redation R is calculated off-line in advance and stored in the computer. It is obvious that the scheme increase the adaptire function and Robustness of the former estimation algorithm.

## SIMULATION AND RESULTS

The performance of developed nonlinear state estimator is tested by numerical simulation in the computer. The whole simulation program includes following contents,

· Dynamic simulation of PET reactor

· Prediction of reaction temperature and conversion

· Producing sampled temperature messurement signal

· The state estimation uplate and noise filtering process

· Calculating the intrinasic viscosity of PET melt in the polycondensation reactor.

When the state estimation is applied, the certain perior information about the reaction process must be known, to determine the initial values of model error covariance Q, measunement noise covariance R and estimation error covariance P. In the actual problem, the assumption is made by the experiments in the reactor system.

The estimation error at t=0 is assumed about $\pm 5\%$ for the actual value of temperature and couversion. The nondiagonal elements of $P_0$ is equal to zero

$$P_0 = \begin{bmatrix} 0.25 \times 10^{-2} & 0 \\ 0 & 0.25 \times 10^{-2} \end{bmatrix}$$

The stochastic noise influence on the operation state is mainly the change of resident time caused by the stirring speed and the rate of input. Consider the model error owing to simplization and assumption, model error covariance Q is given by $Q=0.01 \ P_0$.

Sensor failure and signal processing lead to ineasure ment noise. Generlly, the measurement error covariance R is determined in between 0.01 to 0.001. It can meet the needs of real measurement error, here R = 0.005.

Figs 4 and 5 show the simulation result of nonlinear state estimator. The conversion x is estimated only from temperature measure ment with the estimation algorithm. The sampling time interval is 5 min. The viscosity of PET is calculated finally. From the comparison of estimation results and actalal measurement, it can be seen the estimated viscosity IV and temperature T converge respectively to the measurement data of the plant, in spite of the choice of defferent initial states. It shows that the estimator has better stable behaviour and can be adaptable to the object of the PET reactor.

The matrix Q must be chosen to be representative of modelling error. The good choice of a Q matrix mesults in a stable estimator. Without firm knowledge of the process noise term, it is advisable to determine Q using smaller positive quantities on the principal diagonal. But the matrix Q is selected too small so that the estimation is insensitive to the most recent measurement and the Riccati equation reaches a very small solution p; It will make that the convergence behaviour of the estimator is poor, especially the nonlinear system like the polycondensation reactor. The desingned adptive estimator scheme reduce the use of the matrix Q. The faster convergence of adaptive estimation compared with the general state estimator is obtained as showing in Fig.6. The performance of the estimator is improved grceatly.

The estimator is tested also in the case of various reaction contitions. The simulation shows that satisfactory estimation results can be applied in the real PET polycodensation process, because the results have agreement with the actual measurement even when the dynamic model is a simplified one.

## CONCLUSION

A nonlinear state estimator has been developed for reconstructing the unmeasured state —— conversion in a PET reactor from only temperatare measurement The viscosity of PET melt, which is the quality index of PET product, is infered by utilizing the found relation-ship of viscosity and conversion. The simulation shows that the estimation results are in godd agreement with the real measure ment. The estimation algorithm has better stability. The adaptire estimator scheme improve the convergence performance and estimation acuracy of the estimator. and play an important part in optimal estimation. Because the total computing time is negligible compared to the sampling intervol, the estimation algorithm can be used for the real reactor process. The nonlinear state estimator is the base to realizing optimal control of PET quality. It can be used in a two-level computer control system for the polycondensation reactor. The estimation algorithm will be further demonstrated in the practical ptant.

## REFERENCES

Ellis, M.F., Taylar, T.W.and Jensen, K.F. (1988). Estimation of the Molecular weight Distribution in Batch Polymerization. AIChE J., 34(8) , 1341-1353

Gelb, A., ed., (1974). Applied Optimal Estimation , MIT Press, Boston

Geedwin, G.G.(1984) Adaptive Filtering, Predrction and Control Prentice-Hall

$J_O$, J.H., et.al. (1976). Digital Monitoring and Estimation of polymerization Reactor. AIChE J., 22(21), 361-369

Pan Yuqiang, et. al (1986). Semi-empirical Formula of PET Reaction Kinetics J. CHem. Ind. Eng. (China), 1, 95-101 (in chinese)

Ray, W.H, (1981) Advanced Process Control, McGraw-Hill Inc

Schuler, H. and S. Papadopoulou. (1986). Real-time Estimation of the chain-length Distribuation in a Polymerization Reactor: II. Comparison of Estimated and Measured Distribution Function, Chem. Eng. Sci., 41, 2681-2683

Zadkh, L.A. (1965) Fuzzysets Inform. cntrol., 338-353

Acknowledgement – this work was sppported partly by National Natural Science Foundation of China.

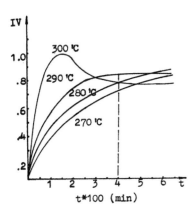

Fig.1. The effect of temperature on
the polycondensation reaction

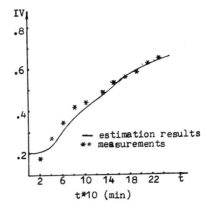

Fig.4. Estimation of the intrinsic
viscosity from the estimation
of the conversion

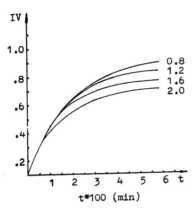

Fig.2. The effect of vacuum on the
polycondensation reaction

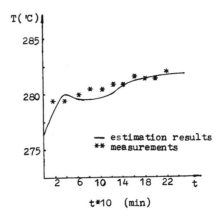

Fig.5. Estimation of the reaction
temperature

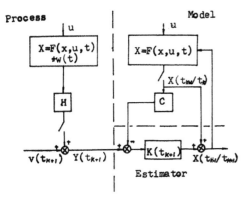

Fig.3. The structure of the state estimator

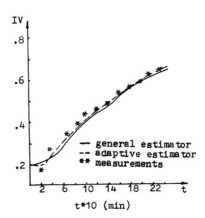

Fig.6. Conparison of simulation results
of adaptive estimator with
gerieral estimator

# STABLE PRODUCTION OF THE OPTICAL MEMORY DISK BY AN ITERATIVE LEARNING CONTROL SYSTEM

## Y. Hanakuma,* K. Yoshinaga,* K. Kojima* and E. Nakanishi**

*Technical & Engineering Department, Idemitsu Petrochemical Co. Ltd.,
1-1 Anesaki-kaigan, Ichihara, Chiba 299-01, Japan
**Department of Chemical Engineering, Kobe University, Nada-ku, Kobe 657, Japan

Abstract. This paper presents a new stable production system of polycarbonate optical memory disk by an new iterative learning control scheme. It has been confirmed that much more superior performances in reference tracking, reducing molding operations and keeping the disk weight uniformly can be attained by using an new iterative learning control scheme as compared to the conventional PID control system. The operation system employing this control scheme is today perfectly running and contributing the stable production of polycarbonate optical memory disks with high quality.

Keywords. Injection molding machine; iterative learning control; optical memory disk; production control.

## INTRODUCTION

A new stable production system of polycarbonate optical memory disk ( OMD ) has been currently developed by Idemitsu Petrochemical Company. Highly accurate control of not only both dimension and sharpeness but also birefringence distribution are required for the commercial use of OMD. In order to satisfactorily maintain these performances in the continuous production of OMD, it is of primary importance how to keep each disk weight uniformly among lots. However, the stable production of OMD to keep its weight uniformly has not been successful in the conventional PID control system which was utilized so far in the mold opening processes.

Recently we have improved the mold opening process by employing an new iterative learning control system and have succeeded in achieving stable production of OMD with uniform weight. The improved opening process is operated by such cascade control system that the primary feedback controller which was of conventional PID mode is replaced by an new iterative learning controller which gives the hypothetical reference pattern to the secondary controller at each molding operation. In this control system, the error which is the difference between reference pattern and controlled variable at previous molding operation is adjusted by error modifying function in order to obtain the desired controllability. By using this control scheme,the difference between reference pattern and controlled variable has successfully been minimized to almost zero within a couple of repeated molding operations.

It has been confirmed that much more superior performances in reference tracking, reducing operations and keeping the disk weight uniformly can be attained by an new iterative learning control scheme as compared to the conventional PID control system. The operation system employing this new control system is today perfectly running and contributing the stable production of OMD with high quality which are supplied to 70 companies in Japan.

An outline of production system by using an new iterative learning control system is presented in this paper together with demonstration of the results of its application.

## CONTROL STRUCTURE OF INJECTION MOLDING MACHINE

### OVERVIEW OF INJECTION MOLDING MACHINE

The structure of injection molding machine is shown in Fig. 1. The resins from a hopper are melted, mixed and conveyed forward through heating cylinder and injection screw. Afterward, they are filled in mold cavity by advancing of injection screw with high speed. On the other side, the clamping force is given to the mold for keeping OMD more stable quality, such as well-coining capability of formatting and well-uniforming of optics.

The control architectures of automatizing the production include following techniques:

* Control system of injection molding pressure
* Control system of mold clamping force
* Temperature control system of heating cylinder and mold
* Control system of injection screw speed revolution

## CONTROL METHOD OF INJECTION AND COMPRESSION PROCESS

The main control system of targeted precision injection molding machine for production of OMD is divided into two schemes as injection and compression process control schemes (Fig. 2). The injection process control system has reference tracking control scheme which controls injection molding pressure of cavity for tracking the reference pressure pattern. This control scheme makes it possible to control directly fluid dynamics of resins in cavity.

On the other side, the compression process control system has mold opening control scheme which is that after the mold opening distance was kept constant by setting mold clamping force in lower level, the mold clamping force is given to injection mold. This control scheme makes occasionally mold opening distance unstable by the unbalanced force between injection pressure and mold clamping force. But the mechanical structure of mold is compacted.

However, these control systems had serious problems such as the fluctuations of weight, thickness, birefringence and face swing in a continuous production of OMD.

## DESIGN OF ITERATIVE LEARNING CONTROL SYSTEM

### BASIC PROCEDURE OF ITERATIVE LEARNING CONTROL

The iterative learning control system has been applied to the technical fields, such as robot manipulator and batch reactor ( Kawamura, et al., 1985; Kawamura, et al., 1986; Katoh, et al., 1988 ). In this control scheme, the manipulated variable and error pattern which is the difference between reference pattern and controlled variable at previous operation are memorized. Then, the manipulated variable pattern at present operation is adjusted by error modifying function in order to obtain the desired controllability. By using this control scheme , the difference between reference pattern and controlled variable has successfully been minimized to almost zero within a couple of repeated operation.

### DESIGN OF NEW ITERATIVE LEARNING CONTROL

Based on abovementioned iterative learning control scheme, we have developed an new iterative learning control scheme which gives the hypothetical reference to the usual feedback controller. Figure 3 shows the schematic diagram of new iterative learning control scheme, where $G_c(s)$ is transfer function of controller and $G_p(s)$ is transfer function of process. This scheme is the equivalent method as the total transfer function of control loop is equal to 1.

The control target was mold opening distance in mold opening process which is equipped with the PID controller of mold clamping pressure. Figure 4 shows the block diagram of new iterative learning control system for mold opening process. This control algorithm is shown in Fig. 5. This control algorithm includes the following practical techniques in reducing molding operations for desired reference tracking performance:

(1) The process dynamics is approximated with the ARX model by 1st operation data of reference pattern and controlled variable. The hypothetical reference pattern to mold clamping pressure controller at 2nd molding operation is possible to calculate by this ARX model. The recursive substitution of the reference pattern into the controlled variable of this ARX model leads to the hypothetical reference pattern.

(2) The optimal error modifying function of this control scheme can be designed by using this ARX model.

The design procedure of this control system is described as below. At the 1st molding operation, this control system gives reference pattern $\Pi$ to mold clamping pressure controller, and the molding operation is executed. The process dynamics of mold balanced force model (Fig. 6) is approximated with the simple model of Eq. (1). Identification data makes use of mold opening distance $y$, mold clamping pressure $F_1$ and cavity pressure $F_2$.

$$y^{(1)}(k+1)=ay^{(1)}(k)+bF_1^{(1)}(k)+cF_2^{(1)}(k)(1)$$

These model parameters a, b and c in Eq. (1) are identified by the Recursive Maximum Likelihood method. Thus, the model parameters are given by Eq. (2). Figure 7 shows the result of model identification.

$$y^{(1)}(k+1)=0.98y^{(1)}(k)+0.069F_1^{(1)}(k)$$
$$-0.07F_2^{(1)}(k) \qquad (2)$$

The hypothetical reference pattern $\Pi^{(2)}$ at the 2nd molding operation is calculated by Eq. (3). The recursive substitution of the reference pattern $\Pi$ into the mold opening distance $y$ leads to the hypothetical reference pattern $\Pi^{(2)}$.

$$\Pi^{(2)}(k)=[ \ \Pi(k+1)-0.68 \ \Pi(k)$$
$$+0.07F_2^{(1)}(k)]/0.069 \qquad (3)$$

Further, the design of the error modifying function in a new iterative learning control algorithm is very important for reducing molding operations. This error modifying function $G_E(z)$ is expressed by Eq.(4).

$$G_E(z)=\sum_{i=0}^{1} \gamma_i z^i \qquad (4)$$

$$| \ 1-G_E(e^{j\omega t})G(e^{j\omega t}) \ | \ < \ 1 \quad ( \ 0<\omega t<1 \ ) \qquad (5)$$

Where,

$\gamma_i$     is error modifying coefficients

$Z=e^{j\omega t}$     is sift operator

$G(z)$     is total transfer function of feedback control loop

and $\gamma_i$ are determined in such a way to satisfy Eq.(5). Namely, $\gamma_i$ are calculated by the optimal converged condition $G_c(z)G(z)=1$. Thus, this procedure for determination of $\gamma_i$ gives $\gamma_0=10$, $\gamma_1=2.5$.

## APPLICATION TO INDUSTRIAL USE

This iterative learning control system was applied to mold opening process of industrial use. The weight performance of OMD accomplished by the new control scheme presented in this study is compared to those of the conventional PID control system ( Fig. 8-(a) ). The performance of conventional PID control system showed the great weight fluctuation of OMD while that of the new iterative control system exhibited the small weight fluctuation.

In order to confirm the control performance, this control system was examined by using the artificial disturbance tests in the following three cases. Figure 8-(b) shows the results of tests where the control performance in test cases made based on the evaluation of "IAE ( Integral of Absolute value of Error" between the reference pattern and controlled variable.

<Test-1> $2kg/cm^2$ setting up of injection molding pressure

<Test-2> $4kg/cm^2$ setting up of injection molding pressure

<Test-3> $8kg/cm^2$ setting down of injection molding pressure

In the above three cases, test-3 is the rare case which will really not happen.

The results of test-1 and test-2 have confirmed that good performance in converging weight fluctuation of OMD within a couple of repeated molding operations can been attained by this control system.

Futhermore, the result of test-3 has shown that good performance in converging weight fluctuation of OMD within 10 repeated molding operations can be attained by this system.

Figure 9 shows the result of good control performance at a continuous molding operation. Compared to the conventional PID control system, the difference of weight fluctuations among lots has been decreased to about 30% by this control system. Figure 10 shows the reference tracking performance. By using this iterative learning control system, the difference between reference pattern and controlled variable has successfully been minimized to almost zero.

## CONCLUSION

We have developed a new production system of OMD by employing a new iterative learning control scheme which gives the hypothetical reference pattern at each molding operation to mold clamping pressure PID controller in mold opening process, and have succeeded in achieving stable production of OMD with uniform weight. Compared to the conventional control system at a continuous operations, the improvement of reference tracking performance has been confirmed to keep OMD weight uniformly and OMD birefringence distribution lower. The operation system employing this control system is today perfectly running and contributing the stable production of OMD, the saving labor and the improvement of product yield rate.

## REFERENCES

Katoh, N., K. Nakao and M. Hanawa(1988). Learning Control of a Batch Reactor, Proc. of PSE '88, Sydney, pp.206.

Kawamura, S., F. Miyazaki and S. Arimoto (1985). Hybrid Position Force Control of Manipulators Based on Learning Method, Proc. of '85 ICAR,Tokyo.

Kawamura, S., F. Miyazaki and S. Arimoto (1986). A Learning Control Method for Dynamic Systems, Keisoku Jido Seigyo-Gakkai Ronbunshu, 22, (4), pp. 56.

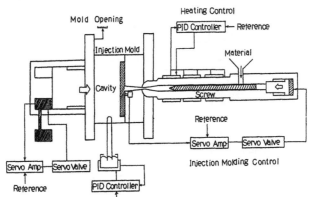

Fig. 1   Injection molding machine

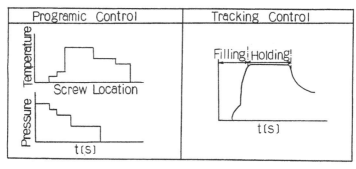

Fig. 2   Injection and compression
         process control schemes

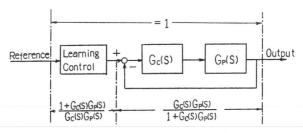

Fig. 3   Schematic diagram of new
         iterative learning control

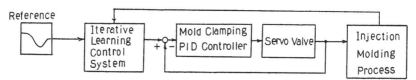

Fig. 4   New iterative learning
         control structure

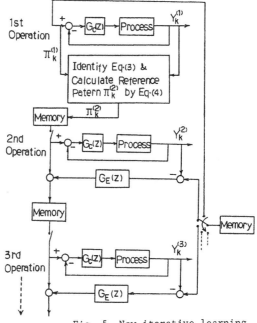

Fig. 5   New iterative learning
         control algorithm

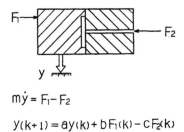

$$m\dot{y} = F_1 - F_2$$

$$y(k+1) = ay(k) + bF_1(k) - CF_2(k)$$

Fig. 6   Model structure of mold
         opening mechanism

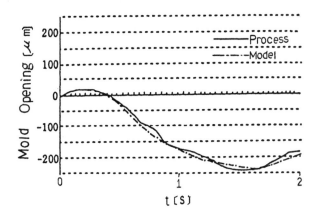

Fig. 7   Result of model identification

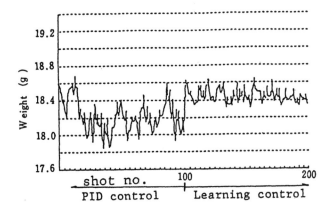

Fig. 8-(a)   Control responses by conventional
             PID controller and new iterative
             learning control

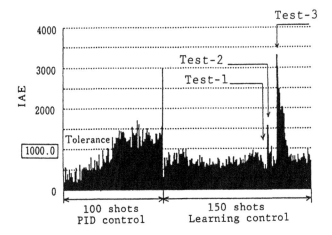

Fig. 8-(b)   Control responses by using the
             artificial disturbance tests

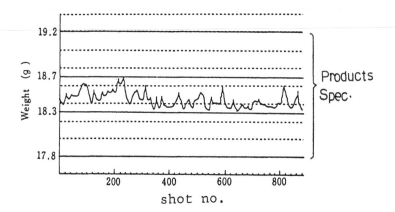

Fig. 9   Control performance in weight
         fluctuation at a continuous operation

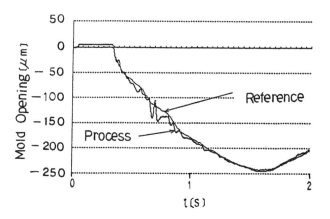

Fig. 10   Tracking performance

# AN INFORMATION SYSTEM FOR PRODUCTION MANAGEMENT IN THE CHEMICAL INDUSTRY

**J. König and W. Stockhausen**

*Department of Process Control, IN-PLT-ST, Bayer AG, Leverkusen, FGR*

### Abstract

Modern production control is one of the most important and challenging tasks in worldwide operating companies. Fully integrated, hierarchically structured information systems handling parts of different administrative, logistic and technical data bases are needed. Layout, modularization and standardization of these systems in production control is an important step towards the economical development and efficient introduction of these complex information systems. Especially important is the exact definition of all interfaces to the neighbouring process control and corporate management systems. Such a system has been developed by Bayer in the past five years; it is called Production Information System (ISP). While parts of this system are still under development, other parts are already running in the production departments of several business groups.

### Keywords

automation, chemical industry, computer control, control systems, information science, production control, software development

### Information systems in large companies

Large, worldwide operating companies are increasingly exposed to growing international competition for various reasons:

Products and markets change rapidly and diversification increases. The availibility of resources, i.e. raw-materials, energy, equipment, manpower and capital, poses practically no limit. In addition, production surpluses exist whilst the markets saturate. Increased efforts in research and development, as well as in production and marketing are needed to meet these problems. Improved feedback between research, development and the market and faster conversion of developments into production are important for shorter innovation cycles of the products. In the production process itself more and more powerful process control systems increase yields, product quality and flexibility of equipment and also ensure constant quality. Furthermore, efficient disposition and minimization of quantities of intermediate and final products on hand should be achieved.

Thus, in addition to capital and manpower, information and its processing become increasingly important as production resources. Consequently most companies have built powerful information systems. However, usually the systems are designed either for the specific needs of specific management

functions or the needs of individual business groups. Figure 1 shows typical data bases, normally not compatible and not integrated, for various functions in a company. Especially companies with complex product and production structures lack data models. This hinders the design of functionally structured and integrated information management systems that

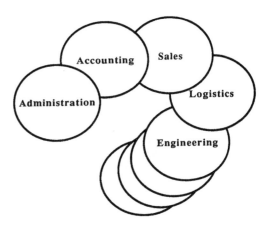

Fig. 1

118                       J. König and W. Stockhausen

organize the vertical and horizontal data flow
between and in the different information
management levels in a company. Only systems
integrating all levels of the information hierarchy
will solve the above mentioned deficiencies.

## Production Information System (ISP)

Figure 2 shows the four level information hierarchy
of computer integrated production. Bayer's
Production Information System (ISP) belongs to the
production management level.

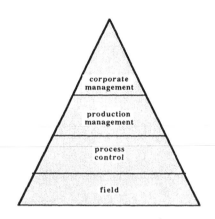

**Fig. 2**

The corporate management level above deals with
corporate-wide tasks such as controlling, logistics
and planning. Below the production management
level, the process control handles computer
operations interacting with the sensors and actors in
the field level.

Most important for the implementation of ISP is the
proper functional layout. ISP constitutes a
standardized basic system for the production
departments of all business groups of the company.
Its layout is such that most needs are covered by the
basic system and only minor extensions are needed
for specific functions in the individual production
departments. As ISP-modules have to run on various
computer systems, the user interface is the same on
different systems to allow for easy switching
between computers. The modularization of the
system permits stepwise introduction of the whole
system and also facilitates the move from central to
decentralized computer systems.

As shown in Fig. 3, functions of the ISP are grouped
into six modules:

• Inventory Management (IM),

• Inventory Disposition (ID),

• Warehouse Management (WM),

• Production Planning (PP),

• Demand Disposition (DD), and

• Production Control (PC).

These modules operate on a common data base and
communicate with each other either directly or via
the data base.

The first three modules are needed when orders can
be satisfied by quantities on hand. The latter three
modules are activated when orders cannot be
satisfied by quantities on hand and thus production
becomes necessary.

The module Inventory Management maintains and
updates inventory and furnishes inventory
information. The module receives the sales orders
from the corporate control level and informs
Production Planning about the current inventory. If
the quantity on hand is sufficient to meet current
sales orders, control is passed to Inventory
Disposition. This module reserves and allocates
material and passes requirements to Demand
Disposition to trigger production if needed.
Inventory Disposition also receives the material
requests as needed by the production. The next step
is then an order to Warehouse Management, where
the receiving and distributing of material is
controlled.

At first production requests by either the marketing
or the plant side are passed to the module Production
Planning. Requirements for final and intermediate
products are planned for short range (one month) and
medium range (one year) taking into account
potentially available resources. Economical batch

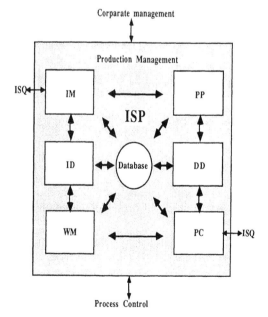

**Fig. 3**

sizes are determined by comparing and optimizing production and warehouse costs. Requirements and schedules for raw-materials are planned for all production steps and demands are then forwarded to Demand Disposition. There the requirements are merged with the requirements from Inventory Disposition and modified using up-to-date inventory information. The module then prepares the request to Production Control and furnishes the purchasing department with purchasing orders for external suppliers. The production request passes to the module Production Control which will be described in more detail.

## Production Control (PC)

The focal point of the module Production Control, shown in Fig. 4, is the administration of a master information data base. Most important is the knowledge about the production process, its possible

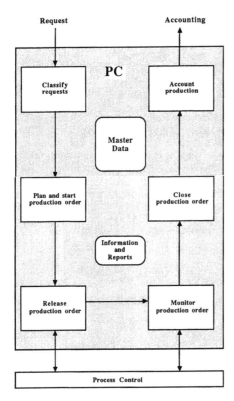

Fig. 4

variants and input raw-materials. This information is maintained in a recipe administration system.

The basic recipe for a product (Fig. 5) contains one or more recipes for the plant units, which may run either sequentially or in parallel. A unit recipe is related to a type of plant unit and is subdivided into phases containing the desired values of process and product parameters. Different raw-material qualities

and different production processes lead to variants of the basic recipe for the same product. It is important to mention that basic recipes are maintained in a 'neutral' way: quantities are normalized and the unit recipes are related only to a type of plant unit, not to a specific one. The specific data are added at production time and generate together with the basic recipe the so called control recipe. The control recipe then exists only temporarily during the production process.

Further master data maintained in Production Control are: material data, shop calendars, technical and structural data of plants and plant units.

As mentioned before, Production Control is entered from Demand Disposition by a production request for a certain quantity of a certain product. The first step is the classification of the request, estimation of raw material, and inspection of raw-material quality. With that information at hand one defines the basic recipe and its proper variant, which implies the definition of the desired product quality. Requests to the module Inventory Disposition then reserve the appropriate amount of raw material at the required quality level. The material is allocated at a later time.

In the next block of functions one determines and allocates plant units of the proper type demanded by the recipe. Then the request is fit into the current plant unit schedule either manually or automatically. This, however, is still a preliminary schedule as changes arising from complications and delays in the then running processes may shift this schedule continually. Now, knowing the recipe and a preliminary production schedule, the production order may be started.

The started production order is released in the next block of functions. The earlier reserved raw material is now allocated.

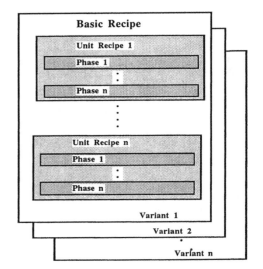

Fig. 5

Figure 6 shows how the basic recipe together with the desired quantity, the specific plant units, the order number and the nominal schedules generates the control recipe. In the case of computer-aided process control, this control recipe serves as the base for the generation of the control structures for the process control system. If there is no underlying computer system, working papers for the operating personnel are prepared from the control recipe.

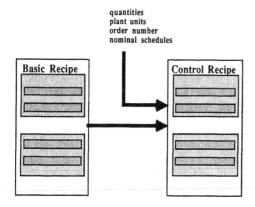

**Fig. 6**

The module Warehouse Management is informed about material that must be issued. If needed, the Quality Information System (ISQ), a sister system to ISP, gets orders to perform quality tests during production at this time to enable the laboratory to schedule these tests properly.

The fourth block of functions (Fig. 4) collects and analyses data from the running production order and corrects the quality of the product if needed. From the control recipe the process control gets the planned interruptions for quality checks. The process control level sends a message via Production Control to ISQ and receives the test result back the same way. The status of the production order is recorded continually via messages from the process control level. The real time lapses are compared to the planned times. Deviations influence pending production orders; corrections to the planned production orders due to the modified plant schedules are handled by Production Control automatically. Also production data of more general types are recorded and the actual material expenditures are measured. The communication with ISQ ensures that final product quality meets the specified quality. After the production order comes to an end the next fuction block is activated.

Here yields and expenditures are recorded for Inventory Management. Final test reports to ensure compliance with the quality desired by the customer are issued from ISQ and Demand Disposition is

informed about completion of the production order.

Final steps in Production Control are the accounting of yields, expenditures and production costs. This information is then passed to the central accounting systems.

## Implementation

The different levels in the information hierarchy (see Fig. 2) have different requirements on data processing. As described the data base for computer integrated production is a common one, but may be kept distributed. The functions in the different levels, operating on this common database, may run on various computers. In view of today´s hardware possibilities the four levels of the hierarchy suggest four levels of data processing: host-computer, midrange computers, process control systems, and controllers. ISP has functions of more administrative and others of more operative character. The latter modules with data links to the process control level preferably run on dedicated midrange computers. Also the relevant part of the data base resides there to minimize data flow between different information levels. This goes well along with the requirement of production departments that for reasons of data security production related data be kept on local systems. Especially important is this requirement for the basic recipes which contain the complete information about the processes. Thus, preferably the modules Production Control and Warehouse Management run locally. Inventory Management and Inventory Disposition are candidates to run in a shared mode, partly on the host and partly locally.

The ISP modules are written in COBOL. Usually the more administrative modules run on a central IBM 3090 host computer and keep data in an IMS-DL/I database. More operative modules run on dedicated local computers running UNIX and a relational data

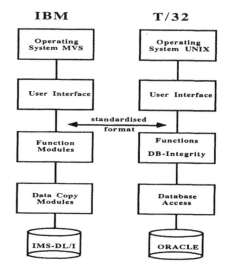

**Fig. 7**

base management system.

A local version of the module Production Control has been implemented on a fault-tolerant Nixdorf T/32 M400 computer system running UNIX. The module operates on the relational data base DDB/4, which will be replaced by ORACLE shortly, and use SQL as the data manipulation language. The software development follows X/OPEN guidelines for COBOL and SQL with additional restrictions to adhere to company-wide software development rules.

Figure 7 compares the basic building blocks of the system on the host and the local computer. The user interface uses the the same form layouts as the host system and the data format between functions and user interface is standardized.

## Acknowledgement

We greatly appreciate the numerous productive discussions with our colleagues from the central computing and corporate planning divisions. The further development and completion of the entire system ISP will be strongly dependent on continued good cooperation.

# PRODUCTION INTEGRATION SYSTEM IN
# A CHEMICAL MANUFACTURING COMPLEX

## T. Ono and T. Ogawa

*Kao Corporation, Wakayama Techno Manufacturing Complex, 1334 Minato,
Wakayama-shi 640, Japan*

OUTLINE. In order to ensure effective production activity in a chemical manufacturing
complex, it is important to achieve a smooth flow of goods in the up-stream, mid-stream
and down-stream flows in a complex. For this purpose, it is essential to integrate information systems for production control -- which gather total information real-time
and/or in time series related to production such as information on market demand for
products, delivery of materials to the complex, manufacture within the complex, goods
distribution and delivery of products to sales centers -- with personnel who are concerned with taking responses to unexpected situations, reviewing the present state of
affairs and creating a new production system. Positive introduction of the functions
of systems, control and information into production activity appears to be the key technology for survival in this era in which production is being switched to short-run operations and competition is sharp in cost-performance.

The contents of the key technology are:
1. Further introduction of engineering methods into production activity (Promotion of
conventional automation and approach to unmanned factory operation)
2. Establishment of a man-machine system specially designed to enable operators to give
full play to their human functions.
3. Establishment of an information system which will enable quick responses to changes
in the situation or will promote evolution of enterprise activities.
(1) A certain class of problems arising in the actual world are thoroughly formulated,
analyzed and computerized. However, at present, this work is done manually, therefor a
more effective method is desired.
(2) At the same time, there is another class of problems which are not easily definable
nor are answerable. For problem-solving in an enterprise, it is necessary to establish
a system of effectively processing information necessary for speedy decision-making.

Our company has made various attempts contend with this problem. Of our attempts, we
will discuss here the PIS (production integration system) plan which is now under way,
with respect to its content and background.

### PIS PLAN

#### (1) TCR activities

In our company, TCR (total creative revolution)
activities are under way as an established innovative movement embracing the entire company.

The TCR activities seek to achieve the following.
* Activities to develop our company's own unique
new technology/system.
* System to ensure a smooth flow of information
and goods
* Integration of sales/marketing -- research/pro-
duction -- goods distribution/flow of commercial
activities
* Organizational and personnel transformation
(sharing of information, multi-functional spe-
cialization)
* Labor productivity (for unmanned operation)
* Space Productivity (high-speed, high-density operation, just-in-time parts delivery, tank-less,
pipe-less operation, etc.)
* Resources Productivity (energy and natural resources conservation, marine transportation, etc.)

The above mean that it is necessary to basically
review the past work systems and to create a new
man-machine system based on new ideas and technology.

#### (2) Characteristics of production system

Our company domestically manufactures about 300
different kinds of household goods such as deter-
gents, hair care products and cosmetics, and about
1,500 basic industrial materials and intermediate
products including foodstuffs, pharmaceuticals,
steel products, and construction materials at
eight factories. The production systems are
characterized as follows.

1) Production is vertically integrated from materials to final products (from upstream to downstream). (Some go through as many as 15 processes,
and each process is operated on a short-run basis.)
2) Natural materials are for upstream use, and
conditions for upstream production are not fixed
because of differences of materials in chemical
composition according to different supply sources.
3) Production is generally geared to that of consumer goods, so that it is an important task to
minimize distribution costs from the points of
production to the points of sale.
4) The average lifecycle of products is about two
years, so that changes in production methods and
manufacture of new chemicals are frequent.

#### (3) Formulation of PIS Plan

Our company's Wakayama Factory accounts for 45% of

123

its total production and is a major production base with about 160 plants. The PIS plan started in December 1986 as a project at Wakayama Techno Manufacturing Complex and Knowledge and Intelligence Science Research Institute dealing with the development of "non-goods", Wakayama Laboratories engaged in the development of "goods." The plan was started in March 1987 for completion by March 1990. The PIS plan is defined as a "systems-oriented structure" for high level management of total production activities covering receipt of materials, manufacture and delivery to sales centers according to market needs, through utilization of information networks and computer technology (software and hardware). It consists of the following three mutually related levels.
Level 3. High level process operation (unmanned approach)
Level 2. High level process management (increased production efficiency)
Level 1. Management of production activities at many factories (integration)
The development schedules are also in the above-mentioned order, though these levels are partially overlapped with each other.

## (4) Progress of PIS Plan

4)-1 High level process operation (unmanned approach)  The PIS plan is aimed at smooth management of a broad area of production activities, and at obtaining the best answer to a given object and under given conditions. Establishment of basic conditions is important for the system which should be flexible and robust to the total functions and external changes. The basic object in a production system is a plant, so that it may be operated freely with a production control system incorporated in it. For this purpose, so that the 160 plants scattered in a compound of 420,000 square meters may be operated from one place (instead of 40 places as was in the past), the plan is required to do the following tasks.
* To review plant hardware from its foundations to control systems and strengthen it.
* To concentrate operation consoles so that a few operators may perform monitoring and remote-control operation.
* To install sensor-based data bases in each plant for the purpose of failure prediction from detailed information on the behavior of the plant and also for the purpose of development and management of high level control systems.
* To install an optical LAN(local area network) with a total length of 20 kilometers to facilitate the sharing of various kinds of information resources within the factory.
The scale of control systems incorporated into the plants inside the factory are given in TABLE 1.

TABLE 1   Scale of Plant Control Systems at Wakayama Factory

| | |
|---|---|
| No. of plants | 168 |
| Control computers *1 | 140 sets |
| Analog measuring points | 10,000 points |
| Control loops *2 | 3,000 loops |
| ON-OFF valves incorporated in control systems | 16,000 |
| Electric motors incorporated in control systems | 12,000 |

*1: PLCs (10 sets per line) contained in the filling lines (41 lines) are excluded.
*2: Field loops are excluded.

The task of Level 3 is to "integrate operations" and "share information". Furthermore, it is intended to "lay a foundation" and "create circumstances for the tasks of Levels 2 and 1. In other words, R&D personnel, engineers and operators are to be integrated through information and/or communication.

4)-2 High level process management (optimum production)  It is desirable that the efficiency of the production system can be examined at any time and the improvements can always be upgraded on the basis of real-time and broad-based information on the management of the above-mentioned unmanned operation-oriented plants and flows of raw materials and products.

The theme of this item is to perform the following tasks on the basis of real-time information on the operation of plants obtained as described in 4)-1, and day-to-day information on sales through utilization of signal processing, control theories, OR and AI, fuzzy inference and other engineering methods and also of information resources such as multi-media data bases.
* Upgrading of plant operation control
* Upgrading of plant failure diagnosis and maintenance technology
* Development of a real-time information system and sensors concerning plant quality
* Development of an analysis system for detection of bottlenecks in the production or flow of goods
* Production scheduling incorporating total inventory and inventory minimizing in the factory
At present, in the development work on these tasks (in future, development and research work will directly lead to production), some practical systems have been developed by utilizing the circumstances prepared by Level 3, that is, information gathering and high level processing of information. For instance,
1. Advanced control of distillation plants where changes in the composition of materials are large, through a model prediction control method
2. A practical on-line and real-time plant failure prediction system based of sensor-based plant data through utilization of arithmetic models, AI, fuzzy inference, etc.
3. Development of physical analysis methods of OHV, SV, IV, water content, other fats and oils to replace the conventional chemical analysis methods and an on-line sensor system for visual inspection of production lines by means of an image analysis method
4. Management of a production scheduling system (time axis development) considering inventory minimization in multi-stage and multi-material processes
The tasks of Level 2 concern the development of element technologies in the PIS plan, and are intended for the establishment of individual PIS functions for human "hunch" and "skill" through engineering methods, and is also intended for an organic integration of individual functions into an independent unit of functions.

4)-3 Management of production activities at many factories (integration)  Our company is engaged in the production of consumer goods, and for the sake of the need to reduce distribution costs, factories engaged in end-use products are located near large domestic consuming areas. As our company seeks to realize a vertical integration of production, it is desirable that production centers of raw materials and distribution centers scattered throughout the country are managed synchronously as if they were a single entity. Therefore, the tasks of this level are
* Development of factories described in 4)-1 (unmanned operation oriented) throughout the country
* Sharing of on-line, real-time factory data through a WAN (wide area network) connecting these factories

* Utilization of software developed in 4)-2 (optimum production) by many factories, and real-time utilization of high performance information processing resources by many factories
* Planning for distribution of productions among the factories, including production centers and goods distribution centers) (on a month-by-month basis, and on a shorter interval basis)
The tasks of Level 1 are intended for a comprehensive organization through cooperation among and integration of the functional units related to production.


### CONCLUSION

The PIS plan, viewed as an information system,
1. is superior to any conventional approaches to information gathering, processing and control.
2. In other words, it is a step ahead of office automation, factory automation and other systems designed for automation of repetitive operations which lend themselves easily to automation, and it functions as a good partner to man in the conduct of non-repetitive operations which can be done only manually.
3. Mechanization of human functions in an enterprise is an everlasting task (beginning with the replacement of hand and foot functions with piping, pumps and valves) and at present the most advanced mechanization reaches a part of brain functions. Automation of non-repetitive operations is considered to be a major theme in future. Constant studies, development and application of the mechanization of non-repetitive operations are considered to lead directly to production. The following can be said about the implementation of the PIS plan.
1) An enterprise is based on the coexistence of man and machine, and its organization is maintained by communication. So that an organization may function effectively, the existence of information (facts and their meanings) shared in common is essential.
2) While the importance of a systems-oriented approach or an integrated study of objects will be increasingly stressed in future, such an approach should presuppose integration with methods devised by existing engineering technologies.
3) The evaluation yardstick of the systems of this kind has been mainly their effect on labor saving, but commitment to labor costs which account for only several percent of the production costs is impractical.

Such systems are infrastructures which are effective for the planning of a strategy related to all the costs involved in production, and the utilization of such systems has a direct bearing on the improvement of human functions and enterprise capacity. We can say that circumstances have been created in our company for a constant and efficient development of these activities. The sophistication of the content of the PIS plan is our endless task.

# CIM IN PROCESS INDUSTRY

## Y. Yoshitani

*Department of Mechanical Engineering, Nagaoka University of Technology, Japan*

## Abstract

Success of CIM naturally depends on good design of the system.
However, it is a huge system and time consuming project for the
corporate. It require long rang sight for building up CIM. Paper
review several problems involved in building CIM system.

## Keywords

automation, process industry, control system

## Problems of CIM in Process Industry

Economy of the country depends on her
strength of process industry. However,
the process industry has not so good due
to the market stagnation since the oil
crisis. In order to overcome the poor
image of process industry, many companies
started various restructuring programs.
Among those programs, the CIM is supposed
to be a sound project for the
reconstruction of the corparate. CIM
will considered to promote better
communication between various levels.
However, it is a large system and
take time for building up the system.
Therefore, it require long rang view
for building up the CIM.
There are several problems in building up
the CIM in process industry.

1. Good organizer who knew entire
   management problems and production
   process are required. Also the good
   corporation of various professionals
   are very important.

2. Benefits of the CIM will be poor, if it
   applied to the present production
   system. It is important to improve
   the production processes in parallel.

3. Both the innovated production system
   and the CIM must be operable and
   maintainable by their employees.

4. Progress of technology and market are
   fast. Therefore, it is important to
   up-date both system and production
   process.

It will be easy to draw the system
structure of CIM. However, it is
highly complex system. There are so many
problems involved in both design and
operation of the system.
Through the studies of system, many new
ideas for the improvements will discover.
Those are one of the big by-products
of CIM.

## Environmential Problem

Environmential problems are very tough problem for the process industries. It will need high investiments without much return. However, if you think it is vital investiment for the existance of the industry, it will be payoff.

Japan is highly populated country and after the mess of the Minamata disaster, the industrialists start aware of the needs of improvement of environmental problems. However, it takes time to improve the environmental situations.

One of the big trouble of introducing control system in the process industries process industries are their poor enviroment. It very hard to maintain delicate instrument and computer system in such environment. By the improvement of the environment of the plant it become possible to introduce CIM, since the decade. One reason is the improvement of damage coused by dast fall and corrosion from poor atmospher.

Atmospheric pollustion mainly contribut energy consumption. Better control of energy supply system and consuming system will promot less energy consmmption and also contribute environential protection. Advancement of power electronics also help promoting optimal operation of fans and pumps, which reduce loss of valves and unnecessary flow.

If you can enable to produce your products with high raw materal field and less energy consumption, you can reduce your cost and also the emissions.

## Innovation of Production System

Mass production is a most economical production system even today. That the reason why the industrialists stick on the market share. However, products variety will be inerease due to both the development of technology and demands of market. Therefore, the process industry inevitably force to establish flexible production system. Technically speaking, it will be easier to produce in mass, sinceithas less changes of operation. Continous production process is not flexible process in terms of small lot production. Possibility of continous flexible production system depends on flexible information system because of frequent changes of set up of individual processing machines.

Manufacturing industries had been very much behind of automation because they dependent on variety of machining and handling. By the development of Robot and Machining center, they enable to introduce CIM. In another words, they enable to introduce CIM by changing their production process.

It will be difficult to gain the full

production process, especially in long term basis. Therefore it will be vital for the CIM project to have parallel improvements of production process.

## System must be operable and mainteinable

Environment of the industries will be change year by year. Those changes will be faster. Every system has to be operable and mainteinable at the stage of instellation. However, you need up dating by their employeer in order to meet various changes. It is important to have good operaters who knew essential knowhow of operations which are responsible for him. Also the same things could be say for the maintenance men. It is very hard to design CIM system like the NASA projects due to economical reason. If the system will design complet fale safe design, it would be hard to introduce to the industries today.

Design of the system, therefore, depends on the capablilties of operaters and maintenance crew. Therefore, the system must be understandable, or it will be understandable up to their reach.

During the cource of design, vender of the CIM system have not much knowlage of production system. Process industries does not want to show in detail of their process knowhow, especially the control knowhow. Degree of dependance of system design depends on how many design engineers you will have in your own.

## For the Future of CIM

Our design of CIM are based on our experience and various theories avariable today. As I mentioned previously, the CIM systems are huge system and it is very hard to protect failures like you see in various happenings occured recently. Also we could not design what we wanted because of the lack of essential informations of the production process itself. Compaired of the progress of computer science, progress of senser and production processes are relatively slow. Research investiment for those are very poor compaired of computer science. I feel the needs of different style of research laboratory which will concern about process information including specific sensers for individual processings.

## Conclusions

Progress of technology has been accerated year by year. Also the change of society and internatial affair-has been very fast. Design of CIM for the process industries will requested more flexibilities. I find a lack of good planner or organizer of CIM due to the lack of experience in process industry. My conclusions for the CIM system are

1. Corperation of various professional engineer are requested for the better design.
2. Innovation of production system and sub-system must be followed for the adoption of CIM.
3. Design must be operable and maintainable within their employees.
4. Up-dating of the system must be done by their own employees.
5. Therefore, the training and education of employees will be essential for the sound development of the CIM.

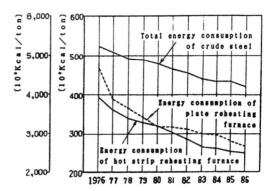

Fig. 1 Energy consumption per ton of steel in Japan

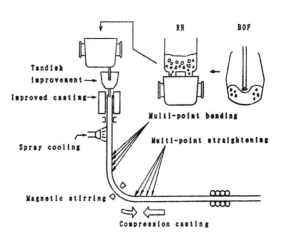

Fig. 2 Technology to produce non-defect semi-finished steel

| | Required Time | Temp. of Slab | Schedule Free | Reliability of Caster and Mill |
|---|---|---|---|---|
| CCR | several days | ○ | △ | △ |
| HCR | 20 hr | 400-700 | ○ | ○ |
| DHCR | 2 hr | 700-1000 | ○ | ○ |
| DR | less then hr | 1100 | ○ | ○ |

CCR  : Casting— Cooling ——————————Rolling
HCR  : Casting— Half Cooling —— Reheating—Rolling
DHCR : Casting— Holding furnace ——————Rolling
DR   : Casting——————————————Rolling

Fig. 3 Step towards Direct Rolling

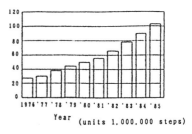

(a) Number of program

(units 1,000,000 steps)

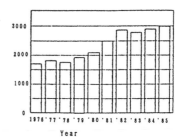

Year

(b) Number of employees in systems div.

Fig. 5 Trends of number of program and number of employees in systems div.

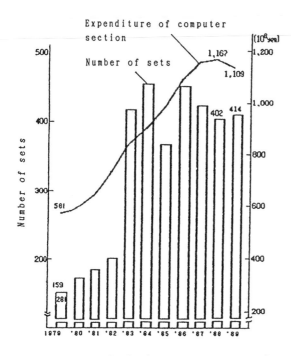

Fig. 4 Number of Business computer and expense of computer section in Japanese steel industry

# A NEW GENERATION CELL COMPUTER/ CONTROLLER IN THE CIM ERA

**K. Nezu**

*Deputy Division Manager, FA Sales Division, Yokogawa Electric Corporation, Tokyo, Japan*

**Abstract.** With the rapid expansion in the introduction of computer integrated manufacturing (CIM), the allotment of cell level functions and the placing of cell level systems become important. This paper describes the trend and the future image of the functions of cell level systems, that is, the cell computer/controller as system components of flexible manufacturing cells (FMC).

**Keywords.** CIM, FMC, Cell Computer/controller, Distributed System

## 1. Foreword

Manufacturing system development is advancing from flexible manufacturing systems (FMS) to factory automation (FA), and further, toward computer integrated manufacturing in the 1990's (Fig. 1).

Corresponding to such advancements in manufacturing systems, the investment tends to be concentrated not only in manufacturing machinery, but also in computers and communication technology.

That is, the investment is increasingly in software and system integration rather than in hardware. This tendency is also quite apparent when looking into the cell level (Fig. 2)[*1].

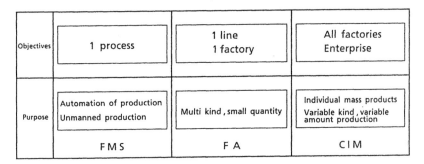

| Objectives | 1 process | 1 line<br>1 factory | All factories<br>Enterprise |
|---|---|---|---|
| Purpose | Automation of production<br>Unmanned production | Multi kind , small quantity | Individual mass products<br>Variable kind , variable amount production |
| | F M S | F A | C I M |

Fig.1    The Trend of Manufacturing System

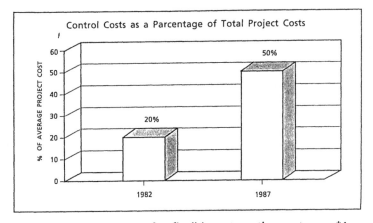

Fig.2    Control Costs for flexible automation systems    *1

Enterprises which plan to introduce CIM are increasing rapidly as they recognize CIM as the key to survival. They recognize the importance of building a production system capable of multi – product, variable lot production based on management strategy. This system will assess the market needs and demands precisely and convey the order information to the development and production departments on an on – line basis. The trends in the cell computer／controller as system components in levels 1 through 3 of the International Standard Organization (ISO) FA standard model (Fig. 3) are described in the following.

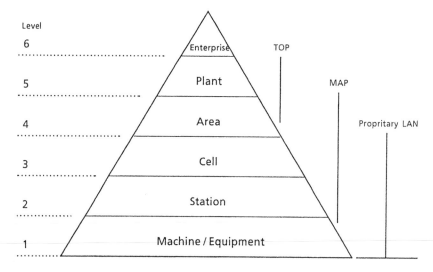

Fig.3    FA System Standard Model    (ISO)

## 2. Main Points of CIM Expansion

With the expansion from FA to CIM, the following points are observed regarding levels 1, 2, and 3.

   (1) Built – in intelligence in machinery／equipment

   (2) Automated／unmanned operation from the machinery／equipment through to the cell level

   (3) Flexible production planning and management

   (4) Emphasis on economy (spreading to small and medium sized enterprises).

With the functional improvement in machinery／equipment as a single unit, the development of compound functions and systematization will further increase. Thus, from the viewpoint of emphasizing simplified and small – scaled management and economy, it becomes increasingly urgent to move from FMS to FMC. As a result, improvements in flexibility and ease of integration are assured by fully implementing automation and unmanned operation down to the cell level  (Fig. 4)[*1].

## 3. Trends in the Cell Computer／controller

Figure 5[*2] shows the product segment trends in manufacturing computers and controllers as recently published by Dataquest. The figures show that control and management in levels 1 through 3 are increasingly tending toward distribution and therefore, the function of cell computers／controllers used in this area is changing. From the author's experience, the following trends can also be recognized.

   (1) Real – time, simultaneous and parallel operation management of various types of machinery／equipment

   (2) Expansion of cell work sphere
       This is in addition to simple data access and monitoring, and includes real – time scheduling, inter – cell management, and material flow tracking.

   (3) Conformance with MAP／TOP

   (4) Meeting JIT operational accuracy improvements

   (5) Increase in the manhours required for software development and system integration

In regard to item (3), it is difficult to achieve standardization using only MAP or TOP as shown in Fig. 6. LAN is progressing to joint and hierarchical use such as, in case of MAP, using both Full MAP and Mini MAP properly sometimes together with TOP. Thus, the cell computer／controller is forced to conform with these conditions.

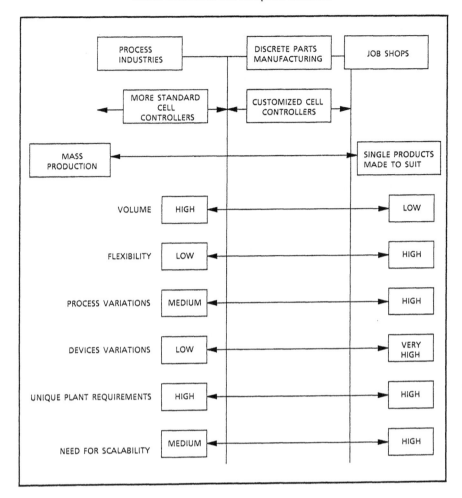

Fig.4    The need for customized cell controllers    * 1

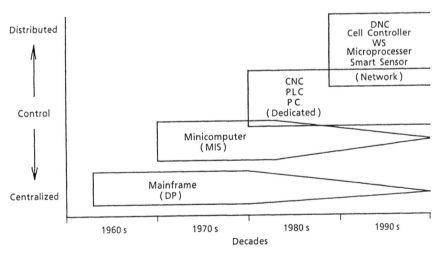

Fig.5    Manufacturing Computers and Controllers
Product Segment Trends    *2

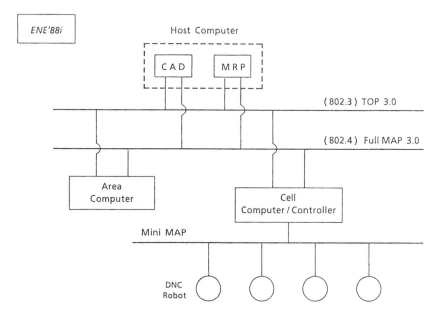

<u>Fig.6    MAP / TOP  Networking</u>

## 4. Functions of Next — Generation Cell Computers ∕ Controllers

Considering the above trends, it is thought that the next generation of cell computers/controllers will not be able to meet the needs of CIM if the following functions are not provided.

(1)  Unit system management within a distributed system

(2)  Improvements in flexibility

  − 1  Adaptability and connectability with a wide variety of machinery ∕ equipment

  − 2  High speed

  − 3  System expandability

  − 4  Software transportability

  − 5  Customization tool

(3)  Conformance with MAP∕TOP

(4)  Improvements in the man − machine interface function

(5)  Development of a cell control∕management language for factory engineers and oper· ators

(6)  Improved system reliability through the use of redundant systems

(7)  Improvement in tooling functions for real − time scheduling,production information display, inter − cell management and real − time material flow tracking

## 5. Conclusion

In order to build a flexible production management system, it is important to consider how the entire production line should be divided and automation and unmanned operation achieved. The key to the solution is in automation and unmanned operational systems built centered around the cell. Flexibility as referred to here means the ability to respond with quick production line changes as well as with diversified goods. Therefore, the substance of the usual saying, 'Distribute the control and centralize the management' is once again being advanced.

That is, control and management within the cell are distributed and simplified, while the information and data for total cell management are centralized. It is considered important that the system be expanded to the level of the enterprise through integration after these principles are achieved.

Although it may at first appear that CIM will result in large systems, it is thought that the probability of success will be higher if compound systems integrating comparatively small scale systems which can be managed by one person are developed. The function of cell computers ∕ controllers has been described from this viewpoint. The number of people who endeavor to introduce CIM is increasing in many industries, and it is hoped that this paper will be of help to them.

(References)

1.  David J, Larin : Cell Control, Manufacturing Engineering, Vol. 102, No. 1, January, 1989 ∗1

2.  Dataquest, April, 1989 ∗2

# OPERABILITY, CONTROLLABILITY AND OBSERVABILITY OF CHEMICAL PLANTS — A PRACTICAL INDUSTRIAL POINT OF VIEW

## O. A. Asbjornsen

*Systems Research Center and Department of Chemical Engineering,
University of Maryland, College Park, MD 20742, USA*

**Abstract.** A remarkable ignorance of the underlying process principles has been observed in the traditional treatment of operability, controllability and observability of chemical plants. Usually, the point of departure has been linear state-space models of a general form, as used in control engineering. However, if the plant production is approached from a systems engineering point of view, and the operational requirements are properly formulated. The point of departure now becomes the process technology itself.

But here there are fundamental properties and phenomena associated with the production, that yields a much better and more appropriate insight into the operability, controllability and observability of the plant. A class of those phenomena are the conservation principles that are always linear, and the products of intensive and extensive properties. The conservation principles are always constrained, and that brings the process constraints into focus at the early stage of operability analysis.

This paper shows how a systematic top-down model development in process operability analysis yields better understanding of the operational requirements and their realizability from a process operation point of view. Interestingly enough, the system engineering approach makes it possible to generalize to many similar production processes.

**Keywords.** Production Control; Modelling; Operability; Controllability; Observability

## INTRODUCTION

The operation of chemical production processes is governed by a few very fundamental principles that are specific to the physics and chemistry of the production. Those principles make it possible to model the entire production complex in a top-down hierarchical manner (Meyssami and Asbjornsen, 1989) according to the basic principles of system engineering (Asbjornsen, 1988). Those concepts and principles for a chemical production complex are mainly:

*The concept of extensive and intensive properties.*
*The classical conservation principle of matter and energy.*
*The conversion of matter and energy from one form to another.*
*The stoichiometry and kinetics of chemical reactions.*
*The transport phenomena of matter, momentum, and energy.*
*Thermodynamic equations for intensive properties.*

A mathematical formulation of these principles serves as a model for the operability, controllability, and observability of the production. The nature and characteristics of process principles make it possible to apply a top-down approach to the modeling also, where the same set of principles are applied at each level of the hierarchy of details, it is only the boundaries and the interior that change as one increases the level of details and the degree of decomposition of the production system.

It also becomes apparent, that the structure of the relationships between the basic principles themselves is of a hierarchical nature. For example, some of the terms in the conservation principle are derived from the conversion of matter or energy from one form to another, and these conversions may be derived from the stoichiometry and kinetics of chemical reactions, which again are derived from thermodynamic state equations describing the relationships between the intensive properties. Other terms are derived from transport phenomena, which also may be derived from thermodynamic state equations and intensive properties. This hierarchical structure forms a basis for the modeling of the production process, and those models are the ones that should be used for the evaluation of the operability, controllability, and observability of the process and its dynamics.

## THE CONCEPT OF EXTENSIVE AND INTENSIVE PROPERTIES.

Chemical production processes are modeled by two main classes of variables, the extensive variables which are proportional to the extension of the production mass or volume, and the intensive which are given as point variables. The extensive variables may be categorized as rates of flow, rates of conversion, and rates of accumulation. The rates of flow may again be sub-divided into convective flows and conductive flows.

### Convective extensive flows.

All constituent convective flows, grouped into a vector $f_{i,conv.}$ and carried by a total mass flow $w_i$, are then given by the product of the total extensive flow, on a mass, molar, or volume basis, and the intensive constituent property vector $c_i$:

$$f_{i,conv.} = w_i c_i \qquad (1)$$

In order to control the process, only total extensive convective flow variables are available for manipulation. But this implies that the extensive flows of all the constituent variables of the total flow will change in the same proportion.

The convective flows are specified as a product of a total extensive flow, e.g. a mass flow, and an intensive property characterizing the constituents of matter, energy, or cash. The cash flow is attached to the constituents by a unit price, which may be a complicated function of the concentration of the constituents of the total flow. Ignore the cash flow for simplicity, and consider the physical constituents only.

The product term in eqn (1) is responsible for most process interactions, and also the clue to decoupling. Consider a production complex with $n$ manipulable input streams. If $n$ is also equal to the number of constituents, then a desired and decoupled vector of input convective flow rates $f$ of the constituents may be obtained by a coordinated manipulation of all the $n$ input flows $w$ by a simple calculation:

$$w = C^{-1}f \qquad (2)$$

which is recognized as the basis for ratio control, and where the columns in the matrix $C$ are the constituent concentrations of the input streams. This control principle decouples the constituent control loops, but is sensitive to errors in the input constituent concentration measurements. Let that error be $\Delta C = C - C_o$. Then the corresponding error in the vector $f_{conv.}$, for a given set of calculated $w$, are:

$$\Delta f_{conv.} = C_o \Delta w =$$
$$= - \Delta C C_o^{-1}[I - \Delta C C_o^{-1} + (\Delta C C_o^{-1})^2 - ....]f_{conv.} \approx$$
$$\approx - \Delta C C_o^{-1} f_{conv.} \qquad (3)$$

This shows that the errors in the convective feed flows are proportional to the error matrix in the inlet constituent concentrations. This emphasizes the need for an observation of these concentrations, which may not be feasible, economic, or practical. In that case, the feed forward control errors will show up in the levels of accumulation. A direct feedback adjustment the the calculated flow rates may then be based on the desired, $l_{set}$, and measured or estimated levels, $l$ of accumulation of the constituents in the production system. Applied to the calculated flow rates, this control algorithm for this correction may be:

$$\Delta w = k\{(l_{set} - l) + \int(l_{set} - l)dt/\tau_i\} \qquad (4)$$

according to standard P-I control (Asbjornsen, 1986). Notice that by nature of accumulation, the l values are the integral of the imbalances in the actual extensive constituent flows $f_{conv.}$.

Conductive extensive transports.

The rates of interface transfer between phases in direct or indirect contact, are conductive extensive transports. They require a proper definition of a phase boundary, which is sometimes very hard to characterize and specify in gas-liquid and liquid-liquid systems in direct contact (e.g. total gas bubble or liquid droplet surface area). The contact area and transfer coefficient are not observable individually, but as a product.

All boundary transfer processes are typically defined by a transfer area, a transfer coefficient, and a potential difference. The potential difference is usually measured as a deviation in the bulk of the two phases from equilibrium. The transport flux is another intensive variable, per unit of transfer area. Every transfer process has an equilibrium where the driving force is zero and the net transfer flux is zero. Let the phase equilibrium relation between a property vector $y_e$ in one phase and a property vector $x_e$ in the other phase be:

$$g(y_e, x_e) = y_e - K(x_e)x_e = 0 \qquad (5)$$

This equation belongs to a set of physical property and composition dependent constitutive relations, and hence belongs to the class of constitutive thermodynamic relations. In the actual case of mass heat transfer process networks, $y$ and $x$ are not in equilibrium, but constitute a driving force for mass or heat transfer:

$$\Delta y = y_e - K(x_e)x_e \qquad (6)$$

Plotting this potential throughout the network for the constituents in question reveals the bottleneck for the transfer operation in the form of a pinch point for heat transfer (Linhoff and Hindmarsh, 1983), where $K = I$, or mass transfer (El-Halwagi and Manousiouthakis, 1989), at which the control and operability of the process is most heavily constrained, and at which the transfer operation is least effective and most expensive.

Usually, the equilibrium constant matrix $K$ is diagonal, and the diagonal elements are only functions of pressure and temperature. The deviation from equilibrium between the two phases constitutes the driving force for the conductive transport, which may then be expressed in scalar terms for constituent no. $i$: The extensive conductive transport is proportional to the transport conductivity $h_i$, the equilibrium potential difference $y_i - K_i x_i$, and the contact area $A$, integrated over the entire area:

$$_A\int h_i g_i(y_i, x_i)dA = _A\int h_i(y_i - K_i x_i)dA = f_{i,cond.} \qquad (7)$$

The equivalence to eqn (1) for convective flow is obtained by considering the conductive transport of all constituents in vector form. However, the transport conductivities now constitutes a matrix $H$, which is normally taken to be diagonal. From a point of view of irreversible thermodynamics, that assumption may not be applicable. For simplicity, the general conductive transport across a phase interface is written:

$$_A\int H(y - Kx)dA = f_{cond.} \qquad (8)$$

These transports are most certainly not measurable by standard instrumentation. In order to get an estimate of the conductive transport terms, one is forced to calculate from the conservation principle around the interface. Furthermore, the conductive transports are not directly controllable, but they may be affected indirectly. The transport conductivity may be increased by increased turbulence, the driving force may be increased by increased deviation from equilibrium, and the contact area may be increased by increased level of contacting. However, most conductive transports may be considered not manipulable and not measurable by direct means.

At phase equilibrium the constant $K = y_{eq.}/x_{eq.}$ is the ratio between the corresponding properties in the two phases. As such, there is a distinct similarity between equilibrium for interface transfer fluxes and equilibrium for chemical reaction or production and consumption fluxes. For both rates, equilibrium is established when the rates in each direction are equal. For temperature potentials measured in Kelvin and pressure potentials measured in absolute pressure units, the value of the equilibrium constant $K$ is $1$.

A very special case occurs in all those processes where the assumption of stage equilibrium is valid, because the mass transfer is finite even if the driving potential $(y_i - Kx_i)$ is zero. This is explained by the fact that the transfer coefficient goes to infinity:

$$f_{i,cond.} = lim\{h_i g_i(y_i, x_i)\} = lim\{h_i(y_i - Kx_i)\} \qquad (9)$$
$$h_i \to \infty, g_i \to 0 \qquad h_i \to \infty, y_i \to Kx_i$$

This assumption is frequently found in separation processes.

Extensive interior rates of conversion and of accumulation

The most important interior extensive rates of conversion in chemical engineering are the rate of accumulation, and the rate of conversion (production or consumption). If the interior has several phases, it is usually required to partition the interior into new sub-systems, one for each phase. The physical interfaces now become new boundaries and the interface transfers become a subset of boundary rates. For a given phase volume however, the distributed rate of accumulation and the rate of conversion are often given as a volume integral of a volume intensive rate, for example for the rate of chemical reactions:

$$f_{reac.} = MN_V\int(r_f - r_b.)dV \qquad (10)$$

where the forward and backward rates, $r_f$ and $r_b$ respectively, are intensive variables, the rate of forward and backward chemical reactions in moles per unit of time and volume, distributed over the interior volume, and $N$ is the matrix of stoichiometric coefficients, which are negative for consumption and positive for production. The diagonal matrix $M$ contains the moleweights along the diagonal, converting moles to equivalent mass units.

Similarly, for the rate of accumulation of the constituents, one finds that this is the only interior rate described by a time differential equation, all others are algebraic. Hence, the accumulation dynamics are derived from the expression:

$$f_{acc.} = d[_V\int c\rho dV]/dt \qquad (11)$$

where $c$ is the constituent concentration per unit of mass, distributed over the interior volume of the system, and $\rho$ is the intensive mass concentration per unit volume, also distributed over the interior volume of the system.

The interior conversion rates are not possible to manipulate directly, although they may be changed indirectly by temperature, pressure and concentrations. The accumulation rate is manipulated by the boundary variables only. The interior rates of conversion and accumulation are directly measurable by internal measurements. In all other cases they must be estimated from conservation equations around the boundaries of the system, where measurements may be more directly available. The fact that the rates of conversion and the rate of accumulation are distributed throughout the interior volume, makes them even less observable without fairly extensive interior measurements. As a general rule, the rates of accumulation are more accessible than the rates of conversion.

An important observation from the analysis of principles above, is that all rate processes, with the exception of the rate of accumulation are algebraic equations. This means that rates of accumulation are solely responsible for the process dynamics. The slower the rates of accumulation are, the slower the process dynamics are:

*The accumulation rates are solely responsible for the process dynamics, and a rough estimate of the time scale of accumulation dynamics is obtained by the inverse space velocity (time constant), mass hold-up divided by the mass throughput in the production complex, or energy hold-up divided by energy throughput.*

## THE CLASSICAL CONSERVATION PRINCIPLE OF MATTER AND ENERGY.

The concepts of intensive and extensive variables are used in the conservation principles, where only extensive variables are supposed to enter the conservation equations. These equations are basically linear in the extensive rates of the constituent variables, which is very useful in modeling, process diagnosis, control, etc.

The conservation principle is the most fundamental one in chemical engineering, complying with the classical laws of conservation of energy and mass. The application of this principle requires two equally fundamental specifications, namely the boundaries of the system considered and the definition of the variables to be conserved. At this point, a careful definition of the variables would avoid obvious misconceptions that for example, viscosity, or pH are being conserved, and less obvious, but actually quite common, misconceptions that for example momentum, enthalpy or entropy are being conserved (Mäkilä and Waller, 1980). The basic variables to be conserved are energy and mass, but energy may be converted to different forms, and matter may be transformed by chemical reactions and take different molecular forms.

In mathematical terms, the conservation principle for a set of constituent variables, for example energy or specific chemical components, may be expressed as a balance *always* of extensive rates (Asbjornsen et al., 1989, Meyssami and Asbjornsen, 1989), around a given boundary, and grouped into vectors $f_{in,j}$ for inputs (see fig. 1 below), $f_{out,j}$ for outputs, $f_{reac.}$ for chemical reactions, and $f_{acc.}$ for accumulation inside the boundary. Assume that there are $n_{in}$ input and $n_{out}$ output rates (convective and conductive), the production and consumption rates of the reactions are automatically taken care of by eqn. (10). Then the conservation principle for all the process fluid constituents is simply:

$$\sum_{j=1}^{n_{in}} f_{in,j} - \sum_{j=1}^{n_{out}} f_{out,j} + f_{reac.} - f_{acc.} = 0 \qquad (12)$$

Eqn (11) serves the purpose of a general linear model for the production complex, and should be the starting point for the building of ARMA models of the form:

$$\sum_{j=1}^{n_{in}} f_{in,j} - \sum_{j=1}^{n_{out}} f_{out,j} + f_{reac.} - r_{acc.} = \varepsilon \qquad (13)$$

where the right hand side $\varepsilon$ is the innovation process from the conservation principle. This general principle is generic, and may be formulated verbally:

*All conservation balances are linear, and the terms belong to a class comprising extensive rates of transportation, conversion, and accumulation. Further model break-down requires models for the rate terms. The deviation from zero in the conservation balances is the innovation process for model identification purpose.*

This observation may also be formulated verbally as a constraint or generic characteristic of the conservation principle:

*All terms in the conservation principle belong to one of the two classes, interior rates or boundary rates. The boundary rates again belong to the class of convective or conductive transport rates, while the interior rates belong to one of the two classes of conversion or accumulation. Interior rates are the total effect of volume intensive rates distributed over the interior volume, and they are therefore affected by interior transports.*

The conservation principle applies to all levels of process decomposition, and is the most important tool for the top-down hierarchical analysis and modeling of a production complex. The guidelines for process decomposition are the definitions of the boundaries for the sub-processes, as indicated by the decomposition of a distillation process shown in fig. 3-5 below. However, it should be borne in mind, that the finer details of the system interiors becomes less and less observable from the boundaries where the observations are normally done, the higher the degree of decomposition is.

## THE CONVERSION OF MATTER AND ENERGY FROM ONE FORM TO ANOTHER.

The conservation of total mass and total energy is invariant to any conversion processes taking place inside the production complex, and the production rates are not observable in the total rate balance for mass:

$$\sum_{j=1}^{n_{in}} w_{in,j} - \sum_{j=1}^{n_{out}} w_{out,j} - d[\int_V \rho dV]/dt = 0 \qquad (14)$$

and similarly for the total energy.

### The stoichiometry of chemical reactions

An important feature chemical reactions have in common with the general conservation of mass above, is the general conservation of atoms (Avogadro's law) and electric charge (the electro-neutrality principle). This leads to the fact that the "concentration" of atomic elements and that the sum of the concentration of positive and negative electric charges are invariant to the chemical reactions. The reaction invariance for the "concentration" of atoms is expressed by the Avogadro's law of stoichiometry, which may be characterized by the orthogonality of the atomic structure matrix $A$ and the stoichiometric matrix $N$, and written (Fjeld et al. 1974):

| | |
|---|---|
| Atomic symbol vector: | $a$ |
| Molecular structure vector: | $m = Aa$    (15) |
| Reaction mechanism formulae: | $N^T m \leftrightarrow 0$ |

The atomic elements are preserved during the chemical reactions, leading to the general and fundamental formulation of Avogadro's law, as the orthogonal relationship of the matrices $A$ and $N^T$ (Fjeld et al., 1974, Asbjornsen et al., 1989):

$$N^T A = O \qquad (16)$$

The interpretation of the atomic matrix $A$ is based on the molecular formulae, which may be understood as a product of a symbol vector, for example $[A,B,C,D,E]$ and an integer number vector, for example $[a,b,c,d,e]$, as illustrated by the general molecular formulae below:

$$[A,B,C,D,E]*[a,b,c,d,e] \Rightarrow Aa + Bb + Cc + Dd + Ee \Rightarrow$$
$$\Rightarrow A_a B_b C_c D_d E_e$$
$$(17)$$
$$[A,B,C,D,E]*[a,0,c,0,e] \Rightarrow Aa + B0 + Cc + D0 + Ee \Rightarrow$$
$$\Rightarrow A_a C_c E_e$$

This concept is extended to the conservation principle, where the reaction rates are substituted by the definition in eqn. (10). Multiplying by $A^T M^{-1}$ makes the reaction rates disappear by the orthogonality, and the result is a set of reaction invariant conservation equation of atoms:

$$A^T M^{-1} [_{j=1}\Sigma^{nin} f_{in,j} - _{j=1}\Sigma^{nout} f_{out,j} - f_{acc.} ] = 0 \qquad (18)$$

Similarly, the reaction variant space is reduced to a dimension equivalent to the number of reactions, by multiplying by the matrix $(N^T N)^{-1} N^T M^{-1}$:

$$(N^T N)^{-1} N^T M^{-1} [_{j=1}\Sigma^{nin} f_{in,j} - _{j=1}\Sigma^{nout} f_{out,j} - f_{acc.} ] +$$
$$+ _V\int (r_f - r_b) dV = 0 \qquad (19)$$

which leads to the fundamental conclusion on process observability and controllability:

*The chemical conversions are not visible in the reaction invariant subspace, but the accumulation dynamics are, with the reactions in situ. Hence it is possible to estimate the process dynamics independent of the reactions, but still with the reactions in place. Once the accumulation dynamics are estimated, the reaction rates may be estimated in a reduced space with the accumulation dynamics in place.*

### The kinetics of chemical reactions

The rules for the modeling of the reaction rates have a close resemblance to the transfer rates of heat and mass, in that they also have an equilibrium where there are no net consumption or production, and the net reaction rate is zero. As in eqn (10) it is common to identify a forwards reaction rate $r_{i,f}$, (from left to right) and a backwards reaction rate $r_{i,b}$, (from right to left). For reaction number $i$, this leads to:

$$r_i = r_{i,f} - r_{i,b} = r_{i,f}(1 - r_{i,b}/r_{i,f}) = r_{i,f}(K_e - r_{i,b}/r_{i,f}) \qquad (20)$$

where $K_e$ is an equivalent equilibrium constant equal to unity, i.e. the forward rate and backward rates are equal. This shows that the reaction rate may be equilibrium controlled, $r_{i,f}$ and $r_{i,b}$ close together, or kinetics controlled, $r_{i,f}$ and $r_{i,b}$ wide apart.

The forward and backward reaction rates are functions of pressure, temperature, and composition:

*Because the chemical reaction rates are functions of concentrations, pressure, and temperature, the reaction rates are controllable by the reaction invariants, but not vise versa, in a single sided relationship.*

The intricate relationships between reaction rates and constituent concentrations are hardly visible from the production process boundaries, except for dominant relations like the temperature effects. Therefore, one may apply several common assumptions and approximations in order to elucidate the forward and backward rates and their relations to the intensive properties. For the sake of simplicity, consider homogeneous fluid reactions where the kinetics follow those of elementary reactions. Then a concentration function and a temperature function are required for the reaction rates:

$$ln(r_{i,f}) = ln(k_{i,f}) - \Sigma v_{i,j} ln(c_j); \; v_{i,j} < 0$$
$$ln(r_{i,b}) = ln(k_{i,b}) + \Sigma v_{i,j} ln(c_j); \; v_{i,j} > 0 \qquad (21)$$

Here, the exponents in the concentration functions are the stoichiometric coefficients. The rule for modeling the reaction

rate constant is based on the Arrhenius' model, which may read for the forwards reaction:

$$ln(k_{i,f}) = ln(k_{i,f}') - E_{i,f}/(RT)$$
$$ln(k_{i,b}) = ln(k_{i,b}') - E_{i,b}/(RT) \qquad (22)$$

Formulae for chemical kinetics may also be considered a set of physical property and composition dependent constitutive relations, belonging to the family of such relations. The observation in eqns. (21) and (22) may be formulated verbally:

*The rate of reaction is proportional to the deviation from equilibrium in the forward and backward rates. At the equilibrium point, the ratio between forward rate is unity, which is equivalent to the equilibrium constant. When the reaction rate is kinetically controlled, it changes dramatically with temperature, but also with composition of the reactants (forward rate) or of the products (backward rates). When the reaction rate is equilibrium controlled, its dependence on temperature is usually weaker than when it is kinetically controlled.*

### The energy aspects of chemical reactions

There is an energy conversion involved in all chemical reactions, where the exothermic are of special interest because of the positive feed back from heat of reaction to the reaction rate. The energy conversion at constant pressure follows the Avogadro's law of stoichiometry, but with an adiabatic temperature rise as a stoichiometric number for the energy conversion, or the temperature change:

Intensive heat released per unit volume by reaction no. j:

$$q_j = (-\Delta H)_j (r_{i,f} - r_{i,b}) \qquad (23)$$

The equivalent total adiabatic temperature increase caused by all reactions is:

$$\Delta T = (-\Delta H^T)(r_{forw.} - r_{backw..})/(c_p \rho) =$$
$$= (\Delta T_{ad}^T)(r_{forw.} - r_{backw..}) \qquad (24)$$

### Stability of chemical reactions

The positive feed back in exothermic reactions is well known to introduce stability problems. A rough estimate of the upper bounds for latent instability, may be observed in the reaction mechanisms themselves. For that purpose, the sensitivities in the reaction rates are used, as derived from a linearization of the reaction rates in the Jacobian matrix $J$:

$$J = \{\partial(r_f - r_b)/\partial c\}N + \{\partial(r_f - r_b)/\partial T\}\{\Delta T_{ad}^T\} \qquad (25)$$

The observation of latent instabilities of a set of reactions is provided by the eigenvalues of the Jacobian matrix $J$. If all eigenvalues are in the left half plane, the reactions are definitely stable, because the exchange and accumulation of heat will both contribute to increased stability. By scanning through a possible range of reaction rates, one may identify areas of potential instabilities or run-away conditions, which will make the reactor inoperable without stabilizing control actions.

### THE SENSITIVITIES OF CONDUCTIVE TRANSPORT PROCESSES.

The conductive transport processes are controllable only indirectly through the rates of the two total fluid flows in direct or indirect contact. The sensitivities are proportional to the deviation from equilibrium between the phases, and the cost of operation of the conductive transport equipment increases inversely proportional to the deviation from equilibrium. The phenomenon behind this is found in the logarithmic nature of the describing equations for the conductive transport. To illustrate, assume proportionality between $x$ and $y$ along the interface of contact, i.e. constant flow rates $w_x$ and $w_y$, counter current operation and a constant, diagonal nature of the equilibrium relations. Then each constituent will be transported in a decoupled manner:

$$w_x dx_i = h_i(y_i - K_i x_i)dA = df_{i,conv}.$$ (26)

$$w_y dy_i = h_i(y_i - K_i x_i)dA = df_{i,conv}.$$ (27)

giving an exponential relationship between the deviation from equilibrium and the dimensionless distance $\xi$ into the contacting equipment:

$$\Delta y_i(\xi) = (y_i - K_i x_i) = \Delta y_i(0)exp(h_i A(1/w_y - K_i/w_x)\xi)$$ (28)

Depending upon the slope of the equilibrium line, $K_i$, and the operating line of the equipment, $w_x/w_y$, the deviation from equilibrium either decays as one moves along $\xi$ into the equipment, $w_x/w_y < K_i$, or expands, $w_x/w_y > K_i$. The sensitivities of the equilibrium deviation to flow rates, assuming constant transfer coefficient, is seen to be proportional to $\Delta y_i$ itself:

$$\partial \Delta y_i / \partial w_x = \Delta y_i h_i A K_i \xi / w_x^2$$ (29)

$$\partial \Delta y_i / \partial w_y = - \Delta y_i h_i A \xi / w_y^2$$ (30)

The cost $C$ of obtaining a given ratio between the inlet and outlet deviations from equilibrium grows as a logarithmic function, assuming cost is proportional to the contact area (proportionality factor $k_c$):

$$C = k_c ln[\Delta y_i(1)/\Delta y_i(0)]/[h_i(1/w_y - K_i/w_x)]$$ (31)

The formulations of the basic principles above lead to a systematic top-down analysis of operability, controllability, and observability based on a sensitivity analysis of the basic equations, starting with the plant wide operation. As the requirements to details of the interior processes increase, the process is decomposed into logical areas and functional operations, still applying the same principles of boundary processes and interior processes. As a consequence, the model equations become more involved, but one should keep in mind that the observability of the finer details of a model decreases with model complexity.

## A CHEMICAL PRODUCTION COMPLEX

A chemical production complex is characterized by a boundary through which there are flows of matter, energy, and cash. Those flows are either of the convective or conductive types, usually convective, as indicated in fig. 1. Inside the production complex is where the conversion processes for matter and energy take place, and there is where accumulations of matter and energy contribute to the over-all dynamic behavior of the production complex, as seen from the outside.

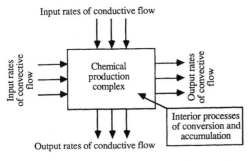

Fig. 1. A chemical production complex.

The simplest assumption for assessing the level of constituents inside the complex, both for conversion and for accumulation, would be to take a simple volume average assessment. Obviously, this is insufficient for design and operation purposes, and more elaborate decomposition of the process interior will be required. The general approach to this

decomposition is illustrated schematically in fig. 2 (Meyssami and Asbjornsen, 1989, Asbjornsen et al. 1989).

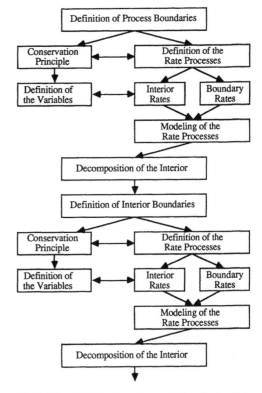

Fig. 2. Hierarchical top-down decomposition of a chemical complex.

The definition of the boundaries of the total production complex identifies the potential variables for control and operation of the whole plant. The conservation principle may be used to control inventories of the various constituents, by requiring that the rate of accumulation should be zero, and that the rate of production should be constant. For that purpose, eqn. (12) is searched for potential control variables that make this happen, exactly, or in a least square sense. The following optimal control algorithm may then be designed for the mass flows selected:

*Maximize the net value of the production with due regards to the constraints imposed by the maximum and minimum level of accumulation, with respect to the selected flow variables, subject to the constraint imposed by the total mass balance, the production invariant balances, the limitations in the control variables, and eventual constraints on the intensive variables.*

Let the mass flow rates be $w_{in,j}$ and $w_{out,j}$ and let the deviation in the levels of intensive variables in those streams be $c_{in,j}$ and $c_{out,j}$. Let the combined production and consumption levels by reactions be $f_p$ and $l_{max}$ and $l_{min}$ be the maximum and minimum constraining levels of accumulation, respectively. Multiplying the rates with a cost factor matrix for each constituent (negative for expenses and positive for revenues) leads to a practical objective function for optimal control. This control algorithm may be written:

$$Max\{C[\sum_{j=1}^{n_{in}} w_{in,j}c_{in,j} - \sum_{j=1}^{n_{out}} w_{out,j}c_{out,j} - f_{acc}.]\}$$ (32)

with respect to the selected control variables among $w_{in,j}$ and $w_{out,j}$, and subject to the constraints:

On the accumulation of constituents:

$$l_{min} < \int f_{acc}.dt < l_{max}$$ (33)

On the span of variation of control variables:

$$0 < w < w_{max} \qquad (34)$$

On the total accumulation of mass inside the system:

$$W_{min} < \int [ \sum_{j=1}^{n_{in}} w_{in,j} - \sum_{j=1}^{n_{out}} w_{out,j} ] dt < W_{max} \quad (35)$$

At this point, it becomes obvious that the most difficult task is to identify the interior variables that leads to the outlet variables, to the production variables, and to the accumulation variables. This may identify what the observations should preferably be, and how one might assess the interior processes of conversion and accumulation. This again requires the process to be decomposed, whereby new interior boundaries are defined, leading to new interior control variables which may be identified at the interior boundaries. This procedure repeats itself as a top-down modeling and analysis process according to the basic principles of system engineering (Asbjornsen, 1988), as indicated in fig. 2.

Inside the production complex there are total rates of accumulation, production and separation. The principle of conservation applies to the rates of flow, the rates of conversion, the rates of accumulation, and the rates of phase transfer. The complex is only controllable through the extensive total flows at the boundaries of the complex or through interior total extensive flows at the interior boundaries. The variation of those interior flows are constrained by the conservation principle applied to the total system boundaries. Hence, the steady state conservation of mass and energy for the total production complex is invariant to any interior flow manipulations.

As an example is shown a distillation process for binary separation in fig. 3, where the top-down analysis of operability, controllability, and observability may be obtained by a hierarchical modeling of the process. The crucial phenomenon here is the phase conversion by conductive transport for separation and not so much the conversion by chemical reactions.

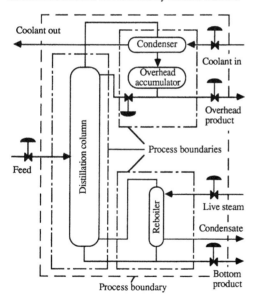

Fig. 3. Interior and boundaries for a distillation column.

First the total process boundaries identifies the boundary convective rates as potential disturbances or control variables. Then the process is broken up into functional units, where interior boundaries are identified, as shown in fig. 3. At those boundaries, new internal convective control variables are identified. It is then recognized that these controls are only possible because the system has internal accumulations at the top and the bottom of the distillation column, and that they must satisfy the constraints on accumulation.

Finally, the phase separation stages are decomposed int the two phases, usually considered to be in completely mixed equilibrium. If completely mixed equilibrium conditions can not be assumed, the stage may be further decomposed into mixing compartments with phase contact, as shown schematically in fig. 4. However, the practical rationale for this sophistication is usually very weak.

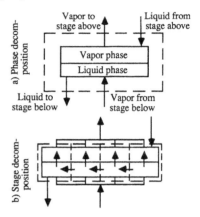

Fig. 4. Distillation stage decomposition - a) Phase decomposition, and b) Stage decomposition.

## CONCLUSION

A careful examination of the hierarchical structure of process knowledge leads to a top-down analysis of operability, controllability, and observability of a chemical production complex. At each level of the top-down process are assumptions and approximations made, in order to simplify the analysis. Further details in the models for the explanation of the production process are required as soon as higher degree of precision in the analysis of operability, controllability, and observability is necessary. This top-down and bottom-up iterative systems engineering approach to the analysis of the process dynamics, economics and plant wide operation serves a very valuable purpose in that the same procedures are used in the top level analysis as in the minor details lower down in the tree of functional analysis.

## REFERENCES

Asbjornsen, O.A., B. Meyssami and C. Sørlie (1989): "Structuring the Knowledge for Process Modeling from First Principles." *Proceedings, IAKE '89 - Knowledge Engineering Today's Marketplace,* University of Maryland, College Park, 1.

Asbjornsen, O. A. (1988): "A Systems Engineering Approach to Process Design and Operation." *PSE '88, Third International Symposium on Process Systems Engineering, Proceedings,* 347, Sydney, Australia, September 1988.

El-Halwagi, M.M. and V. Manousiouthakis (1989): "Synthesis of Mass Exchange Networks." *AIChE Jl.,* **35**, 1233.

Fjeld, M., O. A. Asbjornsen and K. J. Åström (1974): "Reaction Invariants and Their Importance in the Analysis of Eigenvectors, State Observability and Controllability of the Continuous Stirred Tank Reactor.", *Chemical Engineering Science,* **29**, p. 1917-1926.

Linhoff, B, and E. Hindmarsh (1983): "The Pinch Design Method for Heat Exchanger Networks." *Chemical Engineering Science,* 38(5), 745.

Mäkilä, P. M. and K. V. Waller (1981): "Energy balance in modeling gas-phase reactor dynamics." *Chemical Engineering Science,* **36**, 643.

Meyssami, B. and O. A. Asbjornsen (1989): "Process Modeling from First Principles - Method and Automation.", *Proceedings, Summer Computer Simulation Conference,* Austin, Texas, July, 1989, 292.

# PRODUCTION CONTROL IN
# PETROCHEMICAL COMPLEX

## M. Ogawa and G. Emoto

*Engineering Department, Mizushima Plant, Mitsubishi Kasei Corporation,
Kurashiki, Okayama, Japan*

ABSTRACT. In Mitsubishi Kasei Mizushima plant , we have constructed Total
Information Control System , which central computer , process computers and
distributed control systems are organically connected from 1983 , and now
it is almost completed. This Total Information Control System shows it's
power in every aspects of production control by hierachical partial
charged rule which makes the most of their feature. Especially in the
application of the process computer , we have developed many original
systems on it's flexible programming environments

KEYWORDS. Chemical Industry; Petrochemical Complex; Production Control;
Process Control; Optimization; Fault diagnosis.

## INTRODUCTION

The Mitsubishi Kasei Mizushima plant is a
petrochemical complex which consists of
many petrochemical processing units. We
strive for international competitiveness by
reacting quickly to changes in the economy
and energy and raw materials markets.
With the recent progress in computer technology , instrumentation and communications
we have been able to construct a total information control system that can deal with
delicate production management operations.

The Mizushima plant total information control system is a three stage hierachical
structure. The three stages are:
(1) The Distributed Control System (DCS)
(2) The process computer.
(3) The central computer.
The system is completely networked and controls from production to shipment.

This paper discribes production control and
control data processing in the petrochemical
complex, especially focusing on the system
applied process computer which we are currently using.

## SYSTEM STRUCTURE AND FUNCTION

### Structure of the Total Information Control
System

The structure of the total information control system in the Mizushima plant is shown
in Fig. 1. This three stage hierachical system connected by a network. Moving down the
hierarchy , the connection becomes stronger.
That is, the lower the stage in the hierar-
hy , the higher the frequency of information
processing and exchange.

The process computer is divided into a single-user type for mainstay processing units
and a multi-user type for downstream processing units. At the Mizushima plant , the
multi-user type process computer has a front end processor per 2~3 processing units
in order to keep real-time processing for
those units.

The multi-user process computer is connected to the front end processor by way of the
process computer's Local Area Network (LAN),
so it is very easy to reconstruct the system or increase the number of front end
processors. Information is also communicated through the single-user type process
computer by way of the process computer's
LAN. The process computer also communicates
with the central computer , mainly delivering production management information.

We began construction of the Mizushima
total information control system in 1983
and have now completed the central computer
network. However, the other two levels , the
process computer and the DCS , are still
under construction. They are now completed
70% and 55% respectively.

## Hierarchical Function of The Total Information Control System

In constructing the total information control system, we succeeded in making the system flexible and extensive by making clear the functions of each stage, taking into account each computer's features and performance. The hierachical distribution of each computer's function is shown in Table 1. A summary of Table 1 follows.

### (1) Plant Central Computer
The plant central computer functions as a large scale on-line computer to totally manage production, delivery and transportation, quality control, and budget and settlement of accounts.

### (2) DCS
The DCS functions as the brain of the processing units operations and must be simple to operate and extremely reliable. System shutdown caused by hardware trouble or software error must not occur. Thus, it is important to make the system simple and strong.

### (3) Process Computer
The process computer covers the gap between the central computer and the DCS. Based on information from the processing unit's on-line data base, the process computer deals flexibly with process control and production management. It also maintains real-time response and reliability.

## FUNCTIONS OF THE PROCESS COMPUTER

The main functions of the process computer can be grouped into production management, operational support, maintenance of processing unit safety, and process analysis and advanced control.

### Four Main Functions of The Process Computer

As described above, the process computer has four main functions. Each function will be described here.

### (1) Production Management
Production management information such as product and consumption of material per processing unit is generated everyday and sent to the central computer in order to manage total production at the Mizushima plant. After being sent to the central computer, these production data are fedback to every processing unit and used for performance analysis and setting conditions for processing unit operation.

### (2) Operational Support
Due to the automation of processing units, as mentioned above, the number of unit operations has decreased recently. In addition to that, CRT operation and advanced control have made the job of processing unit operator quite difficult and complicated. Therefore, especially when processing unit operational conditions change or problems occur, the operational support of the process computer is vital to the operator.

### (3) Maintenance of Plant Safety
In processing unit operations, it is most important to prevent hazards such as explosions or fires. In order to maintain safety, we are developing a fault diagnosis system and a simulation of process dynamics system to be used when the processing unit is in an unstable state.

### (4) Process Analysis and Advanced Control
Although the attempt to reduce plant operational costs is mainly influenced by external factors such as the cost of oil and the exchange rate, process analysis and advanced control also have an effect on energy and raw materials savings. Regardless of the scale of this effect, we must continue to develop and refine process analysys and advanced control if we are to remain competitive.

## The Process Data Base (PDB)

The four functions described above are accomplished based on information from the on-line Process Data Base (PDB). The PDB must be able to access freely, by tag number, process data in the required form, volume, and response time.

The PV, SV, MV, control status, and Alarm status of a processing unit is accessed by tag no. and that information is stored in the historical data file. Those data are displayed on a CRT as a trend graph and data modification is possible.

The process computer we use at Mizushima plant is a HIDIC V90/x5 series. HIDACS is the software package for chemical plants which we use and it contains all of the functions as we mentioned above. We have also added other functions such as x-R chart and information communication between processing units.

PDB is generated from process data gatherd by communication with the DCS. This data communication and exchange is extremely important.

The data concerning the design and size of PDB per processing unit are shown in Table 2. Performance in communicating data between one processing unit and another is shown in Fig. 2.

## EXAMPLE OF ACTUAL APPLICATIONS

Next, we will describe actual applications of the functions of the process computer.

## Production Management

Our production management system performs the following four functions:

(1) Processing of data through the PDB.
(2) Gathering of source data from the PDB and storing in the data files.
(3) Calculation and printing out of daily and monthly reports.
(4) Exchange of information between the central computer and the process computer.

Quality information communication is very important. Process analysis data and production analysis data analyzed in the Analysis center is communicated by gas chromatography computer to the central computer. Also, this information is sent to the process computer to be used in the adjusting of operational conditions.

The production management system is made up of many program modules. The data flow and the order of execution of the program is very complicated. The total number of steps amount to 20K. Therefore, original software development for each processing unit would require enormous man power.

Hence, we use a relational data base which has original data from several processing units ( e. g. the design data of instrument for correction of measurement values). The main program is common to each processing unit. This approach helps us to realize a reduction in cost of the development process. This also keeps the program simple and makes data maintenance easier.

## Fault Diagnosis of Instrumentation Using Sign Directed Graph (SDG)

Using SDG, this system diagnoses the malfunction of sensors and control valves. SDG, which represents qualitative physical cause and effect between process state variables, is effective in chemical processing units.

For example, as shown in Fig. 3, SDG is used for fault analysis of instrumentation.
In a processing unit, there is continuous reaction and separation, when high pressure reaction liquid declines in pressure, the gas and liquid separate. During this time, we must be careful of the problem of the sudden blowing through of reaction product from a high pressure line to a low pressure line. The fault diagnosis system helps to detect and solve this problem and has the following three functions:

(1) Detection of change in state variable.
(2) Malfunctin identification using SDG.
(3) Alarm function and reporting of the condition of, cause of, and counter-measure to the malfunction.

Here SDG needs some explanation. Please see Fig. 4 and Table 3. The control system shown in Fig. 3 with the graphical notation of SV, PV and MV relations become Fig. 4 when represented in SDG. Limiting the cause of malfunctions to sensors and control valves, we study the condition of state changes for each cause and can obtain decision tables (see table 3). State changes are passed on and on according to the process flow. We represent the extent of state changes by assigning a "stage".

In this case there are twenty-four decision tables dealing with fault origin. If a state change is detected, the fault diagnosis system compares patterns of state changes to decision tables. If state change matches with decision table, the malfunction cause is the cause that is displayed in the decision table. According to the problem, the appropriate counter measures are also displayed in order to solve the problem and maintain safety in the processing unit.

## Ethylene Plant Production Control System

In a petrochemical comlex, the product of an ethylene processing unit is material for downstream processing units. It is important to maintain the required volume of product as planned beforehand. Howeveer, there are so many disturbances, such as change of feedstock, change of operational conditions in distillate section, and deactivation of hydrogenation reactor's catalyst, that actual production deviates from planned production. With manual control, this deviation is sometime very large.

So, we have developed a production volume control system which, by using the process computer, controls the volueme of ethylene and propylene by manipulating the cracking temperature and the feed volume of materials in order to maintain the required volume of ehtylene and propylene. The control system consists of feedback and feed-forward functions and compensates for change of feedstock and demand. The system is robust and simple. By adopting the operator's know-how, we are able to take processing unit constraint into consideration in manipulating cracking temperature, feed volume of material, and setpoint. The control system and it's effect is shown in Fig. 5 and 6.

## OUTLOOK FOR THE FUTURE

As mentioned, the Mizushima plant total information control system hardware is almost completed. However, much work still needs to be done with the software.

Our main areas of emphasis will be:

(1) Automated planning of production
    Schedule.
(2) Quality control.
(3) Automation of processing unit operation
    at unsteady states.
(4) Prediction of abnormal phenomena.

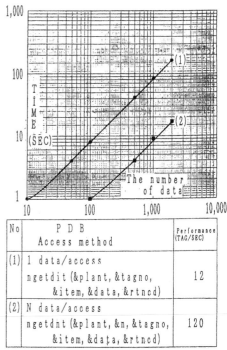

| No | P D B Access method | Performance (TAG/SEC) |
|----|---------------------|------------------------|
| (1) | 1 data/access <br> ngetdit (&plant, &tagno, &item, &data, &rtncd) | 12 |
| (2) | N data/access <br> ngetdnt (&plant, &n, &tagno, &item, &data, &rtncd) | 120 |

Fig. 2. Response of data communication between process computers

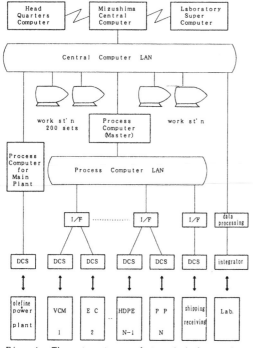

Fig. 1. The structure of total information control system

| DCS COMM. | | HISTORICAL | | | S P C | |
|-----------|------|-------|--------|------|-------|------|
| CYCLE (SEC) | TAG | CYCLE | PERIOD | TAG | CYCLE (SEC) | LOOP |
| 10 | 50 | 1 Min | 1 Hour | | | |
| 30 | 100 | 1 Hour | 31 Days | 1,000 | 15 | 20 |
| 60 | 500 | 1 Day | 3 Years | | | |
| ( SUM | 650 ) | | | | | |
| | 350 | | | | | |
| SUM | 1,000 | SUM | 1,000 | | SUM | 20 |

TABLE 2   The design and size of PDB

| | PRODUCTION MANAGEMENT | PROCES CONTROL | OPERATION MANAGEMENT | FEATURE |
|---|---|---|---|---|
| CENTRAL COMPUTER | Total Production Management <br><br> Total Production Planning <br><br> Total Cost Reductionn | | Offer of Balance Data ( MIzushima Plant ) ( For Headquarters ) | System down for maintenance is unavoidable <br><br> Difficulty in handling real-time process data |
| PROCESS COMPUTER | Data Communication between Central Computer | Process Analysis <br> Process Identification <br> Design of Control System <br> Process Optimization | Operation Management Data Communication to Central Computer <br> Realtime Data Communication between Plant <br> Quality Information Data Communication <br> Advanced Process Monitoring | Realtime Process Database <br> Facility in on-line development of process analysis and control <br> Facility in Set Point Control |
| | Gathering and Calculating of Production Data <br><br> Production Report ( Daily, Monthly ) | Fast Speed Control ( Disturbance Compensation) <br> Set Point Control ( To DCS ) | | |
| D C S | Data Communication to Process Computer | PID Control <br> Sequence Control <br> Advanced Control | Process Monitoring | Reliability and Operability is avobe everything <br> Simple function <br> Online maintenance is difficult |

TABLE 1   The hierachical distribution of each computer's function

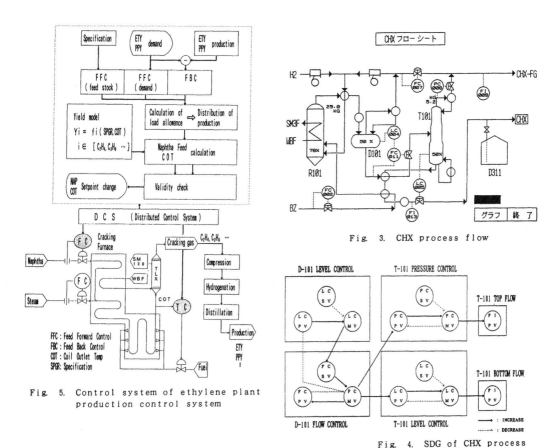

Fig. 5. Control system of ethylene plant production control system

Fig. 3. CHX process flow

Fig. 4. SDG of CHX process

Fig. 6. Effect of ethylene plant production control system

TABLE 3   Decision table

| ORIGIN | D-101 LC Lvel gage Fault Indicate | | | |
|---|---|---|---|---|
| STAGE | 1 | 2 | 3 | 4 |
| D-101 L C | SV | | | | |
| | PV | + | + | + | + |
| | MV | + | + | + | + |
| D-101 F C | SV | + | + | + | + |
| | PV | + | + | + | + |
| | MV | + | + | + | + |
| T-101 P C | SV | | | | |
| | PV | | | | + |
| | MV | | | + | + |
| OVER HEAD F I | PV | | | + | + |
| T-101 L C | SV | | | | |
| | PV | | | | + |
| | MV | | + | | + |
| BOTTOM F I | PV | | + | | + |

# DISCUSSION

## Session 3: Productive Control

*Chairmen*
**C. H. White**
*E. I. Du Pont, USA*
**E. Nakanishi**
*Kobe University, Japan*

In this session, the productive control in the
process industries is discussed mainly in the
aspect of Computer Integrated Manufacturing or
Computerized Production Integration. In the
below, attention is focused on the current CIM
trend in chemical industries in Japan.

### Current Situation in Chemical Industries in Japan

Today chemical industries in Japan are in trouble situations in both internal and external
phases. Internally, the chronic lack of labor
power is everlasting and the tendency of young
workers inflowing to the service industries is
making a serious damage to the manufacturing
industries such as chemical, steel and so on.
Externally, Japan is going to be challenged in
the cost competition of chemical products as
seen in the steel products in the past decade by
NIES where still the cheaper labor is available.
Furthermore, the cost for safety of plant operation is becoming another major burden on the
domestic chemical factories, most of which locate in densely populated urban areas.

### Toward to CIM in Chemical Industries

In the circumstances as mentioned above, the
chemical industries are required to change their
production system in views of manufacturing
products being prior to market need, practicing
plant operation with labor saving and safety.
Especially, the strong market need for speciality, manifoldness and high value-added of
products seems to motivate to promote the dras-
tical change of production system in chemical
industries. Most Japanese speakers in this
session consider that CIM is one of promising
solution to the reorganization of production

system in the chemical industries.

Originally CIM is the production concept created
in the machining factory where FA has been established based on FMS, while the modernization
in the process industry has been developed based
on PA. Is it reasonable to introduce CIM prac-
ticed successfully in the machining factory
directly to the process industry despite the
fact that there are fundamental differences
between both manufactures as shown in Table 1?

### Problems in practicing chemical CIM

Talbe 2 shows the comparison of production control as usual and in CIM. In order to practice
CIM in chemical industries, there are many problems to be cleared. In the first, chemical
plant is usually not only larger in scale but
also more complex in structure compared to machining factory. In the second, each plant is
almost independent in the machining factory,
while unit operations in the chemical process
are interconnected by complicated pipe line
systems.

Not to mention, the aim of CIM is the integration of production system. Taking account of
the facts indicated above, the integration of
production system may not be so easy in chemical
industries as in machining factories. In practicing CIM really fitted to chemical
industries, much remains to be done even in
chemical engineering aspects; realizing more
flexible production system aiming market need
priority for products, exploiting value-added
products with high quality, improving plant
operation system with high efficiency and full
safety, and so on.

Table 1    Comparison of process and machining factories

| description of factory | scale of factory | production output | manner of operation | variety of products |
|---|---|---|---|---|
| continuous plant | large | planned output | inflexible | few |
| batch plant | middle and/or small | planned output and/or order-received output | flexible | few and/or many |
| machine shop assembly line | small | order-received output | flexible | many |

Discussion

Table 2   Comparison of production control
as usual and in CIM

| production control | | as usual | in CIM |
|---|---|---|---|
| production management | production planning | factory initiative | sales initiative |
| | inventory management | baffer in factory | minimization |
| | quality management | factory initiative | customer's need initiative |
| operational management | scheduling | static | dynamic |
| | process control | conventional control | advanced control |
| | automation | process unit | total system |
| equipment management | maintenance | preventive | predictive |
| | planning | individual | integrative |

# AN OPTIMALIZED COMPUTER CONTROL SYSTEM OF POLYVINYL CHLORIDE POLYMERIZED (PVC) REACTOR

## Huan Chang Lu and Zhang Yu*

*Computer Station of Jiangsu Chemical Research Institute,
Center of Jiangsu Computer Application in Chemical Engineering, Nanjing,
Jiangsu, PRC*

Abstract. Basic principles and system constitution as well as functions
and industry effeciency of an optimal control system with a SDS-1 micro-
computer are presented, in which a PVC enamel reactor is controlled using
model reference adaptive prediction strategy. Interlock and start up between different stages of PVC process (including quantity measurement,
temperature elevation, temperature stabilization and pressure reduction)
are performed automatically by corresponding control programs.

The mathematical model is established by statistical regression and
heater equilibrium analyses with characteristic parameters of the process
and regulating parameters got through computer simulation, making a good
remedy to the affection on the control quality of the control system by
nonlinearity, unstability as well as time varying and dead time of the
object. Using the control system, temperature of the PVC reactor waves
within the interval of $\leqslant \pm 0.1\,°C$, while the production period of a single
reactor is reduced by more than 1 hour, which means a sharp saving of
energy and raw material.

The sequencial control and closed-loop control of the whole production
process which consists of stages including measurement, feed-in, cool-
stir, elevation of temperature, transition, temperature regulation and
reduction of pressure etc. Fig.1 is realized by the control system by
possesing distributed control and centralized monitoring. Additionally,
functions like CRT display monitoring, parameter modification, judging
and warning of the first failure, print of data and curves, process
display, manual and automatic activities and remote-controlled non-inter-
ference alternation etc. are installed.

The technical materials and constitutional devices of the system are
ready to supply to factories and corporations as a completed technology
if needed.

Keyword. Adaptive control; cascade control; iclentification; nonlineer
control systems; commumieations control applieations; computer control;
optimisation; programming control; time-varying systems.

## INTRODUCTION

The control quality of batch reactor relates to a variety of factors, for example, nonlinearity, unstability, time
variation property and large time delay
of the process. In the past, due to the
limitation in automatic instrumentation,
it is not very ideal to overcome the delay. With the wide use of computer,
modern control theory and numeral simulation technique offer the essential to realize optimal control on the complicated
processes.

The SDS-1 Computer system includes model
founding, numerical simulation, pressure
prediction and parameter identification.
According to each stage in polymerization
reaction, for example, meterage, temperature raising, transition, reaction and
pressure dropping ect., the different control programs are used to implement two-
level computer control, i.e. destributed
control and centralized monitor.

## EFFECTING FACTORS

The flow sheet of PVC reaction Process
with control loops is given in Fig.1.

There are many factors effecting on the
quality of resin, e.g. temperature, pressure, stirring rate, impurity in raw
material, innitiation, dispersing agent,
the proportion of vingl chloride to water
as well as the pressure and temperature
variation of cooling water or heating
steam. Some of them are required to be
controlled in proceding stages, or have
need of measurement accuracy, others may
be controllable or uncontrollable. The
temperature and pressure in reactor are

---

* XIAN JIAN NING

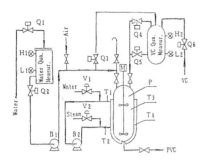

Q1 ~ Q6: Air starting control valve
V1, V2: Regulating valve
B1, B2: Pump          M: Motor
T1 ~ T4: Temperature  P: Reactor presure
H1, H2: High Alarm    L1,L2: Lower Alarm

Fig. 1. The flow sheet of PVC reaction
       process with control loops.

considered as the main factors effecting
on control quality of PVC based on process
analysis.

The rate of polymerization is increasing
at higher reaction temperature. If the
temperature can not be regulated properly,
the heat in reactor will be accumulated
much more. It will lead to the deline in
the degree of polymerization or "explosive
polymerization", so as to effect the quality of resin. But at lower temperature
the rate of polymerization is slow. It
will decrease the productivity of resin.

In the earlier and middle stage of reaction, consider the linear relationship
between pressure and temperature in the
reactor. The time delay of temperature
measurement is about $10 \sim 15$ min. But the
lag of pressure measurement does not
exist. In the later stage of polymerization, there is no above linear relationship.

## DESING OF THE CONTROL MODEL
## AND CONTROL STRATEGY

### Design of the Adaptive PVC Control System with Model Reference.

The adaptive system with model reference
was developed earlier and raised in 50's
used in servo mainly. The basic idea in
these systems to make the object output to
follow the output of parmetic model. some
time later the systems were also used in
on-line identification of process models.
The method utilized the step-by-step para-
metic optimization theory in earlier period it, however, could not guarratee the
stability of the system. The Popov's
theory of super-stability, therefore, has
being accepted since 70's.

Suporse that state space equation of the
identified process is as follows.

$$\dot{X}=Ap \cdot X+Bp \cdot U, \qquad X(o)=Xo , \qquad (1)$$

The state space equation of the simultaneous estimative model is

$$\dot{Y}=As(v,t)Y+Bs(v,t)U, \qquad Y(o)=Yo , \qquad (2)$$

The definition of the state widesense
error is

$$e=X-Y , \qquad (3)$$

According to supper-stability theory the
following idetification algorithm can be
derived.

$$v=D \cdot e ,$$

$$As(v,t)=\int_{o}^{t} FA \cdot V(GA \cdot Y)^{T} d\tau +As(o) \quad (4)$$

$$Bs(v,t)=\int_{o}^{t} FB \cdot V(GB \cdot U)^{T} d\tau +Bs(o) \quad (5)$$

Where FA, FB, GA, GB and D are the positive definite matrices of the corresponding dimensions respectively. And if the
selection of D makes the transfer function

$$H(S)=1/D(S1-Ap) , \qquad (6)$$

To keep this rigiously, it is neecessary
that for any initial conditions of X(o),
Y(o), As(o) and Bs(o) be piecewisely con-
tinuons, thus

$$\lim_{t \to \infty} e(t)=0 \quad (stability) ,$$

is guarrenteed moreever, when the following three conditions are satisfactroy:

· The object identified is completely controllable,
· The components of vector U are linear
  independent,
· The number of frequencies involved in
  vector U is noless than $(n+1)/2$,
one can expect:

$$\lim_{t \to \infty} As(v,t)=Ap ,$$

$$\lim_{t \to \infty} Bs(v,t)=Bp , \quad (convergence) .$$

The problem one has to solve for PVC pely-
mer-reactor is to identify the parameters
in first-order system. Assume that the
process identified is given by:

$$\dot{\theta}1=-(1/T1) \cdot \theta1+[1/T1 \cdot a] \cdot p' , \qquad (7)$$

The parameteric model is

$$\dot{\theta}s=-As \cdot \theta s+Ks \cdot p , \qquad (8)$$

The wisesense error is

$$e=\theta 1-\theta s , \qquad (9)$$

In formula (7) $p'=p$

When FA, FB, GA, GB and D all equal to 1,
then the identification algorithm can
perform as following,

$$As = \int_o^t (FA, e, \theta s) d\tau + As(o) , \quad (10)$$

$$Ks = \int_o^t (FB, e, P) d\tau + Bs(o) , \quad (11)$$

The reason is that the temperature control system with pressure pre-estimation (TCPPE) is implemented on a total distributed system.

It is difficult for a primary computer to identify two parameters of As and Ks. But it is known from simulation that effect on the control pricision is trivial even though As is not very acurate. Therefor in this method, only one parameter, Ks is identified, Then the reference model is

$$\dot{\theta} s = -(1/T1') \cdot \theta s + (K/T1')P , \quad (12)$$

The identification algorithm is

$$K = \int_o^t (Fb, e, p) d\tau + Ko , \quad (13)$$

During implementing, at first according to equations (10) and (11) make on-line identification of T1 and a by a singleloop controller when the regulator of the original loop is just used as a identifier, not as a controller then take the identified value of T1 as T1', take the reciprocal of the identified value of a as the initial value of K · 1/T1' is recommended as the value of Fb. The simulation results have shown that the dynamic characteristics of the adaptive system are quite encouraged.

## The Control Strategy for PVC Polymer-Reactor

The modelling of polymerization reactor is the key to realizing the optimal control of reactor. The pressure measurement response is much faster than that of the temperature measurement. So we use the method that the pressure signal is used to predict the temperature. In order to simplify the calculation and improve the accuracy, the approximate energy balance equations are used in the model, and the error of approximation is corrected by following equations.

$$\theta c = ap + \theta o , \quad (14)$$

$$\theta o = b \int (\theta 1 - \theta c) , \quad (15)$$

where, a and b are constants, $\theta o$ is the correcting value, $\theta c$ is the set point, $\theta 1$ is the measurement value, and p is the pressure measurement value.

The variation of temperature causes the charge of pressure in the reactor. If $\theta c \neq \theta 1$, the correcting calculation is taken place by the computer according to eqs. (14), (15). When $\theta c = \theta 1$, the system is in the steady state. The

composition of control system is showed in Fig.2.

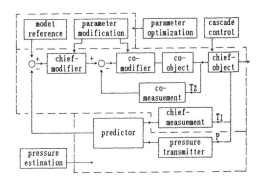

Fig. 2. The Composition diagram of Control system.

Based on the modern control theory, the whole process, realizes the computer control, including meterage, temperature raising, transition, polymerization and pressure dropping. The control programs are divided into four parts.

In the temperature raising, the control algorithms includes judgement, preset of initial value, and data tables.

In the transition, the temperature approching the actual value is predicted. Then the time for opening the steam valve or the cooling water valve is determined. The programs have structure combining logic judgement with time comparing.

The polymerization stage keeps longer time. There are many disturbance in this stage, and the parameter variation with time and delay are more serious. So the control programs have three functions.

· The cascade control algorithms of to the reaction temperature in reactor with jacket temperature.
· Predicting the reaction temperature according to the reaction pressure.
· A predictor is set up in main measurement loop to overcome the dynamic difference of system.

In the pressure drop stage, the relationship between pressure and temperature does not exsit. The programs shift to openloop control. The logic judgement combining temperature, pressure and time is used to determine whether the polymerization has finished.

### SIMULATION OF THE CONTROL SYSTEM OF PVC POLYMER-REACTOR

#### The General Algorithm of Numarical Simulation

As a matter of fact, an operational amplifier for simulation in an analog computer acts as integrator.

On other hand, numerical simulation has to solve the problem of the integration calculation. Therefor the numerical integration method has become the basic method used in numerical simulation of control system. The common numerical integration methods are

Euler method. Suppose equation

$$dy/dt = f(t,y) , \qquad (16)$$

$$y(t_{n+1}) \cong y_{n+1}$$
$$= y_n + f(t_n, y_n)[t_{n+1} - t_n] , \quad (17)$$

Where $h_n = t_{n+1} - t_n$ is the calculated stepsize.

Equation (17) does be the form of Euler integral equation Euler method can be, of course, written in form of definite integral and it may be calculated via rectangular formula.

Both differential equation and definite integral equation can be in implicit form, If one wants to write the differential equation in expliceit form, what he needs to do is just to solve constant coefficient differential equation and to proceed integration with the stepsize.

The character of Euler method is simple, but not accurate. Usually the shorter stepsize is used to reduce the error.

Trapezoidal method. This method is a modification of Euler rectangular method. The general equation is

$$y_{n+1} = 1/2 y_n \cdot h_n[f(t_n, y_n) + f(t_{n+1}, y_{n+1})] , \qquad (18)$$

Because there are two unknown variables $(t_{n+1}, y_{n+1})$ in prior in equation (18), iterative algorithm is the best way to solve it and in practice, one iteration is enough to get a good resolution.

Here $y_{n+1}$ in equation (18) is the corrected value of $y_{n+1}$. It is obvious that calculation in trapezoidal method is double comparasing with that of Euler method when the same step size are used.

Fourth-order Runge-Kutta method. It has been proved that the First-order Runge-Kutta method is eguevalant to Euler formula, the second-order Runge-Kutta to the trapezoidol method. Usually, in practical use, Runge-Kutta method implays the Fourth-order method, i.e.

$$y_1 = y_0 + h/6(k_1 + 2k_2 + 2k_3 + k_4) , \qquad (19)$$
$$\text{where} \quad k_1 = f(t_0, y_0) , \qquad (20)$$
$$k_2 = f(t_0 + h/2, y_0 + h/2 k_1) , \qquad (21)$$
$$k_3 = f(t_0 + h/2, y_1 + h/2 k_2) , \qquad (22)$$
$$k_4 = f(t_0 + h, y_0 + h k_3) , \qquad (23)$$

h is the stepsize and $k_n (n=1,4)$ are the coefficeents in the Taylor-series expansion.

Because Runge-Kutta mathod is more accurate, it is widely accepted in practice.

The stability and the trancation error are tightly relative to the stepsize in all three methods mentioned above, and the relation between them is very complicated. The trancation error is directly-proportional to squar of h in Euler method, to third or fourth power of h in trapezoidal method, and to fifth power of h in Runge-Kutta respectively.

The numerical simulation means program design, debugging and some other processes. And the mathematical abstracted from a chemical process is generally very rough, so that it is considered that the error of no more then 0.5% in numerical simulation is satisfied for engineering and the method of having fixed step size may be used for the purpose of simplification. Usually, a step size for Fourth-order Runge-Kutta method is recommended.

$$h <= 1/5 \omega_c (\text{or } t_c/10, t_n/40) , \qquad (24)$$

where  $\omega_c$: open-loop shear frequency.
       $t_c$: rise time under step rise.
       $t_n$: transient time under step rise

In the case of multiple system existing the fastest reaction system should be selected among $\omega_c$, $t_c$, $t_n$. And sampled-data control system, the step size should be taken to be equal to sampling period or less than the latter and the latter must be n times the value of the former, where n is an integer more than 1,

Numerical Simutation of Main Lag Bit

First-order lag bit

$$X/U = K(TS+1) , \qquad (25)$$

where U is input, and X is output

Remark it with differential equation

$$dX/dt = 1 , \qquad (26)$$

Suppose K=1, T=10, h=0.1 and solve the response characteristic under the step rise will be as follow.

$$dX/dt = 1/10(1 \cdot U - X) = 0.1(U-X) , \qquad (27)$$

Second-order lag bit

$$X/U = K(T_1S+1)(T_2S+1) , \qquad (28)$$

where U is input, and X is output.
set K=1 and by substitution it is got,

$$d\omega/dt = 1/T_1 \cdot T_2[U - (T_1 + T_2)\omega - X] , \quad (29)$$

$$dX/dt = \omega , \qquad (30)$$

Thus the Runge-Kutta can be used to solve equations (29) and (30) simultaneously to get the step response in the site.

Suppose T1=0, T2=1, h=0.1 and substituting them in the equations (29) and (30), we will get.

$$d\omega/dt = (1/T_1)(U - \omega) , \qquad (31)$$

$$d\omega/dt = (1/T_1)(\omega - X) , \qquad (32)$$

## Pure lag bit

$$X/U = Exp - \tau s \ , \tag{33}$$

where U is input and X is output.

set $\tau = nh$ (h is a positive integer).

The numerical simulation program of the pure lag bit can be solved n dimensions space.

## Bit of PID regulator

$$Y/X = (100/P)[1 + (1/T1S) + T2S] \ , \tag{34}$$

Where X is input, y is output, p is proportion ality, T1 is integral time and , T2 is derirative time.

PID is the first-order approximat regulation and does not requir excessive precision in simulation. Usually calculate the differential equation with a fixed stepsize.

$$yn = yn-1 + (100/p)[Xn-Xn-1 + (100/p)(h/T1)Xn + (100/p)(T2/h) \cdot Xn-2Xn-1+Xn-2], \tag{35}$$

Equation (35) is the calculation formula for ordinary numerical regulator and is widely used in DDC, Its integral items can be calculated by Euler formula and its differential items approximately calculated just as in the case of First-order lag bit. Sometimes, of course, they are handled with trapezoidal method for the purpose of less error.

## Numerical Simulation

Numerical simulation technique is a important method to build up computer control system for chemical industry. The experiment can be done on model, not on actual processes. It is difficult to get the model of polymerization reactor by experimental method on a actual reactor. In order to predict the performance of PVC reactor, the SDS-1 computer control system is used to simulate it in numeral before the system build up. Fig.3. shows that how to realize the simulation model based on Eqs (14), (15). In Fig.3(a).

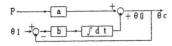

Fig. 3(a). The method that the pressure signal is used to predict, the temperature.

The temperature measurement $\theta 1$ is compared with the prediction $\theta c$, then the measured value is corrected. Because of the time-variation nature of the process, it causes the dynamic difference between $\theta 1$ and $\theta c$ so that $\theta c$ can not fully follow the changes of $\theta 1$.

In order to attain the aim of full follow, a delay unit is added in between $\theta c$ feedback and comparing unit, as show in

Fig. 3(b).

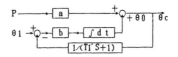

Fig. 3(b). Control Method of the pressure signal predict the temperature and time delay of the process.

Fig. 3(b) can meet the needs of follow. But stability can not assured. The reason is that it has something to do with model, identification. It is very difficult to find a appropriate Liapunov function. Popov's stability theory is used to design the system, as shown in Fig.4.

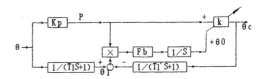

Fig. 4. The control of model reference adaptation and pressire predictive tempreture.

Fig.4 meets the needs of both compensation and stability.

The hardware and software of SDS-1 computer system can ensure that the control system has certain robustness under various disturbance.

## THE COMPOSITION OF SYSTEM AND CONTROL SOFTWARE

This computer control system divided into two levels. At second level, the computer is used as a monitor, and at first level the system has at most 50 PID control loops. The logic control is used to measure soft water and monomer. The adaptive prediction control strategy is applied to regulate the temperature and pressure in reactor. as show in Fig.5.

The Software Consists of Two Parts:

The software of computer in second level is programed by using assemble language into 120KB,and the whole program is divided into 25 modules, including main program, subroutines, communication, transformation and display subroutine etc., The progrem·is provided with menu form.

The computer software at first level has two parts.

The first one is the time sequence control. Several reactors share it to measure the soft water and monomer fed in the reactor.

The second is model reference adaptive prediction control part. It is used to control the polymerization reaction pro-

cess. Its function includes temperature
raising, cascade, prediction, identification, pattern switch reaction end judgement and heat or pressure alarm. as show
in Fig.6.

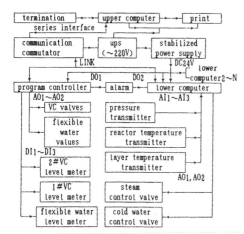

Fig. 5. Computer optimal Control system
for Polyvingl chloricle (PVC)
poly merized process.

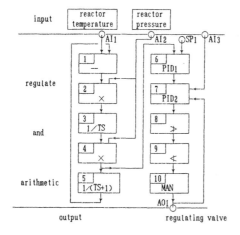

Fig. 6. The principle block diagram of two
level computer system.

Figure 6 illustrates the principle block
diagram of two level computer system, in
which the first to fifth module realize
the adaptive prediction control, the sixth
to tenth module implement optimal cascade
control.

## APPLICATION RESULT

The continue operation results for about
2000 hours in a plant have shown the SDS-1
computer system is successful and achieve
the goal of optimal control. Figures 7(a)
and (b) are the temperature record curves
of the manual manipulation and controlled

by SDS-1 computer system respectively as
show in

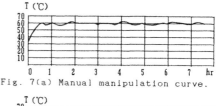

Fig. 7(a) Manual manipulation curve.

Fig. 7(b) Manual manipulation curve.

The new method is used to design the
system, control strategy, structure composition, software and the measure against
disturbance. The system gives the
following design requirements.

The variation of  temperature is control
within ±0.1 ℃. The production  period is
shortened over an hour. Singal reactor
can increase the productivity of resin 120
tons per year and save on water, electri-
city, steam and raw material. The capital
expenditure on the computer control system
can refund in half year.

The computer system has various functions,
e.g. display, print, alarm and manul-
atomatic transformance. The operation is
stable, reliable, robust, and very convenient. The control accuracy is high. The
temperature control problem of small  PVC
reactor with enamel jacket is solved quite
well, especially the control process is
nonlinear, unstable, time variation and
high time delay.

REFERENCE:

H.Amrehn, (1973).  Prozebrehner in Der
Chemie Computer. Proxis 8. 229-240.
H.Gran, (1974).  Application of a Degital
Computer in a PVC Manufacturing Plant.
SEE(7)7, 312.
Bertom, Corradi, Miani, (1974).  Digital
Control in an Electrochemical Process.
4TH IFAC/IFIP International Conferen-
ncen on Digital Computer Applications,
to Process Control 270, Part. 1.
A.E. Bryson, (1975).  《Applied Optimal
Control》.
H.Amrehn, (1977). Computer Control in the
Polymergation Industry. Automation
Vol. 13, 533-545.
L.D LanDau, (1979). Adaptive Control the
Model Reference Approach. Marcel
Desker. Inc.
Hoogendroorn, (1980).  Control of Polyme-
rization Processes, Procedinga of the
4TH IFAC Comference on PRP Automation,
ICC, Ghent. Belgiuns.
M.G.Singh,(1981). 《Decentratised Control》.

# DEVELOPMENT OF MODEL BASED
# PREDICTIVE MULTI-VARIABLE
# CONTROL SYSTEM

### H. Otani and S. Sugimoto

*Department of the 2nd System Development, Fujifacom Corporation, Tokyo, Japan*

Abstruct. Recently, in process control field, improvements on controllability
have been essential to produce high quality products and save energy. From this
point of view, multi-variable control, which improves the controllability of
interacting multi-variable processes, has attracted special interest.
We developed a new control system, that is called FPACS-MV(FACOM Process
Advanced Control System for Multi-Variable control). Multi-variable control
systems can be easily and efficiently established by using FPACS-MV at plant
operating sites. To confirm the efficiency, FPACS-MV were applied to a
distillation column, and good performances were confirmed.

Keywords. Model based control;predictive control;multi-variable control;
distillation column.

### INTRODUCTION

Recently, to cope with various customer's
needs and stiff competition among enterprises,
plant operators have been demanded
improvements on quality, saving energy and
frequent changes of operating conditions.
It is difficult to satisfy these demands by
only manual operations using usual process
control systems; intelligent and advanced
control systems have been required.
Under these circumstances, multi-variable
control, which achieves stable control for
interacting multi-variable processes, has been
attracted special interest. Decoupling
control with transfer function elements and
optimal regurator using modern control theory
have been tried for multi-variable processes,
but few actual processes have been applied.
One of the reasons is that those methods need
complicated parameter tuning.
From this point of view, we developed a new
control system, that is called FPACS-MV. A
multi-variable control system can be easily
and efficiently established by using FPACS-MV
at plant operating sites. The merits of FPACS-
MV are as follows.

- Robust control and simple control parameter
  tuning will be performed, since FPACS-MV
  introduces model predictive control, which
  has superiority on them.

- Multi-variable control systems will be
  smoothly structured, since FPACS-MV is
  composed of data collection, identification,
  simulation and on-line control functions.

- Multi-variable control systems in a facility
  will be continuously structured, since

plural multi-variable control
systems(maximum 10) are independently
structured by using FPACS-MV.
To confirm the efficiency, FPACS-MV were
applied to a distillation column in a oil
refining plant, and control experiments were
carried out.

### ALGORITHM OF MODEL
### PREDICTIVE CONTROL

FPACS-MV introduces model predictive control
with impulse response model(PCIM). To
simplify the explanation, SISO system will be
handled in this chapter. PCIM computes the
future controlled variable over a range of the
predictive horizon by convolution calculation
with impulse response models, and the future
manipulated variable as the future controlled
variable will approach the reference
trajectory, as shown in Fig.1. The outline of
PCIM is as follows.

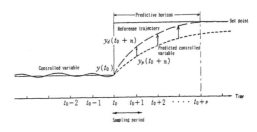

Fig.1 Algorithm of PCIM

## Reference Trajectory

PCIM introduces the 1st-order reference trajectory $y_d(t_0+n)$, which is computed with the present controlled variable $y(t_0)$ and the set point R at each sampling time.

$$y_d(t_0+n) = \alpha^{(n)} \cdot y(t_0) + (1-\alpha^{(n)}) \cdot R \qquad (1)$$

where $\alpha$ is the response coefficient of the reference trajectory. $\alpha$ is converted from time constant of the reference trajectory.

$$\alpha = e^{-\frac{\Delta t}{Tc}} \qquad (2)$$

where $\Delta t$ is sampling period.
Small Tc decreases the period that the controlled variable arrives to the set point, while large Tc increases the period. Then, the controllability can be easily adjusted by Tc.
Therefore, the stable control can be maintained by changing Tc even in the case that process response may change in on-line controlling.

## Predictive computation of the controlled variable

The controlled variable at n sampling future $y_m(t_0+n)$ is computed by convolution calculation with past manipulated variable u, disturbance variable d and corresponding impulse response models $H_u(i), H_d(i)$.

$$y_m(t_0+n) = \sum_{i=1}^{N} H_u(i) \cdot u(t_0+n-i) + \sum_{i=1}^{N} H_d(i) \cdot d(t_0+n-i) \qquad (3)$$

where N is the number of coefficients of impulse response models, $u(t_0+i)=u(t_0)$, $d(t_0+i)=d(t_0)$ ($i=1,2\cdots$).
To correct the prediction error due to unmeasured disturbances or model uncertainty, the predicted value $y_m(t_0+n)$ is corrected by the prediction error $\{y(t_0)-y_m(t_0)\}$.

$$y_p(t_0+n) = y_m(t_0+n) + \{y(t_0)-y_m(t_0)\} \qquad (4)$$

## Computation of the manipulated variable

The manipulated variable $u(t_0)$ is computed as the predicted controlled variable $y_p(t_0+1) \sim y_p(t_0+p)$ will be approached to the referense trajectory $y_d(t_0+1) \sim y_d(t_0+p)$. Namely, $u(t_0$ is computed as the objective function J will be minimized.

$$J = \sum_{i=1}^{p} \{y_p(t_0+i)-y_d(t_0+i)\}^2 \qquad (5)$$

where, p is the predictive horizon.
In the case of MIMO system, the objective function $J_m$ is as follows.

$$J_m = \sum_{j=1}^{L} Q_j [\sum_{i=1}^{P} \{y_{pj}(t_0+i)-y_{dj}(t_0+i)\}^2] \qquad (6)$$

where L is the number of controlled variables, and $Q_j$ is weighting factor of each controlled variable.
In general, the extension of the predictive horizon may improve robustness.

## STRUCTURING PROCEDURE OF MULTI-VARIABLE CONTROL SYSTEM

Plural process inputs effect plural process outputs on interacting multi-variable process as shown in Fig.2.

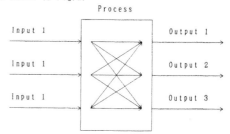

Fig.2  Interacting multi-variable process

FPACS-MV introduces PCIM for multi-variable control. Impulse response models are used for PCIM, then they are obtained with identification using time-series data. Furthermore, simulations are carried out for the parameter tuning and the stability confirmation.
This structuring procedure of multi-variable control is shown in Fig.3(a), and functions of FPACS-MV corresponding to the procedure is also shown in Fig.3(b). The outline is as follows.

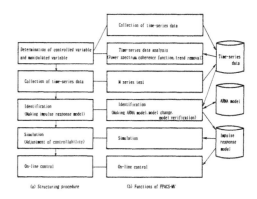

Fig.3 Structuring procedure and FPACS-MV

## Determination of controlled, manipulated, disturbance variables

Controlled, manipulated and disturbance variables are determined by understanding process dynamics using operating data or step response test data.
The function of time-series data analysis is efficient for this work.

## Collection of time-series data

In general, impulse response models of chemical processes are hard to be obtained from physical or chemical equations, since these are nonlinear and/or distributed systems.

Therefore, M series signals, which are artificial pseudo random binary signals, are introduced to process variables. Time-series data are collected, and impulse response models are identified as shown in Fig.4.

The proper waveform of M series signals is determined with approximate values of transfer functions, which are estimated from step response test data.

If the collected time-series data have a trend, a low frequecy noise, it is removed using approximation by a polinominal.

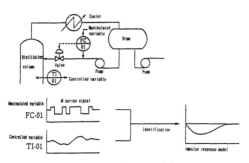

Fig.4 Identification method

## Identification

Two methods can be used to obtain impulse response models from time-series data. One is direct method, that coefficients of response are directly estimated by least square method, and another is indirect method, that they are obtained by changing from ARMA models, identified in advance.

Since the former method needs estimation of all coefficients of response, oscillating coefficients may be obtained as shown in Fig.5(a). Such a oscillating model is hard to be used for PCIM.

Since the latter method estimates a few parameters, smooth coefficients will be obtaind as shown in Fig.5(b).

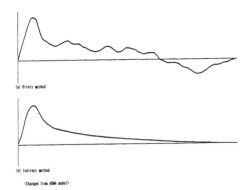

Fig.5 Impulse response model

ARMA model is that present output y(t) is expressed by linear combination with past values of input u and output y.

$$y(t) = \sum_{i=1}^{n} a_i \cdot y(t-i) + \sum_{i=1}^{n} b_i \cdot u(t-i) \qquad (7)$$

Where, n is degree of ARMA model.

Degree of ARMA model is determined using the results of the following model verification. Estimated models are verified by the coefficient of correlation and the error variance between the original data of controlled variables and the simulated data of controlled variables, which are computed with the estimated models and the original data of manipulated variables.

## Simulation and on-line control

To adjust the controllability, control simulations are performed on various values of control parameters. In addition, to confirm the robustness, the simulations are also performed with the differences between the models used in PCIM and the models for the subject process. The control parameters of PCIM are as follows.

· Time constant of reference trajectory.
· Predictive horizon.
· Weight of each controlled variable.

## MULTI-VARIABLE CONTROL FOR A DISTILLATION COLUMN

PCIM has superiority on robustness and simlicity for the adjustment of the controllability. PCIM were applied to a distillation column, a typical interacting multi-variable process.

## Distillation column

Distillation columns have been widely used in oil refining plants or chemical plants, and are essential processes for separations of products. The objectives of column control are to maintain the purity of products at the specified values.

Since distillation columns are interacting multi-variable and sometimes large time constant processes, it is difficult to prevent disturbances and adjust controlled variables to the set points with PID control. For this reason, multi-variable control has been required for the distillation column control.

The subject of this paper is a distillation column in a oil refining plant.(Fig.6) The column performs a binary separation of LPG and gasoline.

## MULTI-VARIABLE CONTROL SYSTE'

The objective of the multi-variable control is to keep a fraction of gasoline in LPG and that of LPG in gasoline at specified values under frequent flowrate and ingradient changes of feed oil.

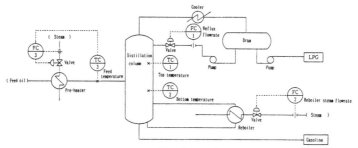

Fig.6 Distillation column

Since the top temperature and the bottom temperature are closely related to the purity of LPG and gasoline respectively, these temperatures were selected for the controlled variables. While, the manipulated variables were fixed to the reflux flowrate and the reboiler steam flowrate. The feed temperature, that had been one of the candidates for the manipulated variable, was not selected for the large time constant against the top temperature.
The multi-variable control system were fixed as shown in Fig.7.

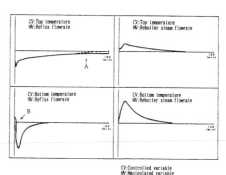

Fig.8 Impulse response model of distillation column

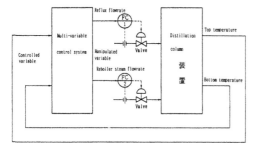

Fig.7 Multi-variable control system of distillation column

### Results of the experiments
(1)Identification
ARMA models of the column were computed from the time-series data, that had been collected while M series signal had been introduced to the manipulated variables.
The degree of ARMA model was fixed for 5 by model verifications, where the coefficients of correlation were more than 0.8 and the error variances were less than 0.1.These results indicate that the estimated model is sufficient for PCIM.
The coefficients of impulse response model were obtained by changing from ARMA model.(Fig.8) In the original impulse response model,there were two improper parts, A and B in Fig.8.
In the part of A, the coefficients of the response were discontinued at the maximum limit, which is due to the restriction of the computer hardware.
PCIM needs the impulse response model,which settles to zero. If there exists a stepwise response such as A, the algorithm of PCIM may disturbes the manipulated variables itself. Therefore, the part of A was modified to the broken line as shown in Fig.8.

In the part of B, there was a inverse response, which had never been observed in the data of step response tests.
PCIM with a inverse response model may compute the sudden change of manipulated variables in the case of short period of the predictive horizon.The part of B was also modified to the broken line
(2)Simulation
For the column control system, the following controllability had been required.
· The bottom temperature would be more precisely controlled than the top temperature.
· The sudden change of the manipulated variables would be minimized for avoiding the influences to the other processes.
To obtain the proper values of the control parameters, the set point change simulations were performed on various values of the parameters. The simulation results are shown in Table 1, where the controllability is evaluated with the following criteria.
· Settling time(ST):
   top temperature----------less than 60min
   bottom temperature-------less than 30min
· Maximum change of the manipulated variables (MCMV):
   reflux flowrate----------less than 4.0KL/H
   reboiler steam flowrate--less than 0.4 T/H

Table.1 Simulation results

| No. | PH (min) | WCV | | TCRT(sec) | | ST(min) | | MCMV | | Total evaluation |
|---|---|---|---|---|---|---|---|---|---|---|
| | | Top | Bottom | Top | Bottom | Top | Bottom | Reflux | Reboiler | |
| 1 | 10 | 1 | 1 | 800 | 800 | 53 ○ | 39 ○ | 4.4 × | 0.56 × | × |
| 2 | 10 | 1 | 1 | 1200 | 600 | 66 × | 33 ○ | 4.0 ○ | 0.50 × | × |
| 3 | 15 | 1 | 1 | 800 | 400 | 55 ○ | 26 ○ | 3.8 ○ | 0.44 × | × |
| 4 | 15 | 1 | 1 | 1200 | 400 | 69 × | 33 × | 3.2 ○ | 0.36 ○ | × |
| 5 | 20 | 1 | 1 | 800 | 400 | 59 ○ | 29 ○ | 3.3 ○ | 0.38 ○ | ○ |
| 6 | 20 | 1 | 1 | 1200 | 600 | 73 × | 36 × | 2.9 ○ | 0.30 ○ | × |
| 7 | 20 | 1 | 1 | 800 | 800 | 58 × | 38 × | 3.2 ○ | 0.34 ○ | × |
| 8 | 20 | 1 | 5 | 800 | 800 | 58 × | 38 × | 2.7 ○ | 0.25 ○ | × |
| 9 | 20 | 1 | 10 | 800 | 800 | 58 × | 42 × | 2.4 ○ | 0.23 ○ | × |

PH:Predictive horizon
WCV:Weight of controlled variable
TCRT:Time constant of reference trajectory
ST:Settling time
MCMV:Maximum change of manipulated variable

Long period of the predictive horizon(PH) decreases MCMV on the same time constant of the referense trajectory(TCRT). While, TCRT mainly affects ST, but weights of the controlled variables(WCV) hardly affects ST. The control parameters were fixed at NO.5 in Table 1. The simulation result of NO.5 is shown in Fig.9.

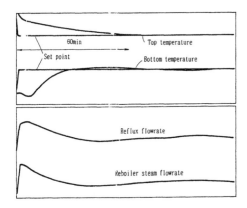

Fig.9 Result of simulation

(3)On-line control
On-line control experiments were carried out with the control parameters obtained at the simulations.
The result is shown in Fig.10(a), compared with the result of PID control, shown in Fig.10(b). The result of the multi-variable control shows the smooth and quick(about 30min) changes, while that of PID-control shows the oscillations.
The set point changes, which had been taken about 2 hours by manual operations, were reduced by the multi-variable control.

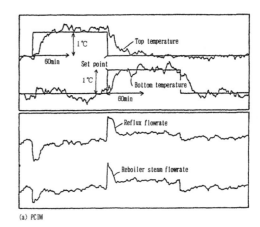

(a) PCIM

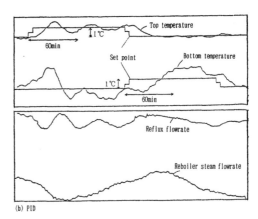

(b) PID

Fig.10 Results of on-line control

CONCLUSIONS

We developed a new multi-variable control system FPACS-MV, by which control systems can be easily and efficiently established at plant operating sites.
FPACS-MV were applied to the distillation column, and good performances were confirmed. For future development, we plan to make a overall multi-variable control system combined with a optimizing control system.

REFERENCES

Richalet, j., A. Rault, J. L. Testud, and J. Papon. (1978). Model predictive heuristic control-applications to industrial processes. Automatica, Vol. 14, pp 413-428.

Nishitani, H. (1986). Synthesis of process control systems-topics in chemical process control. System & control (in Japanese), Vol. 30, pp 16-25.

Mimachi, C., Kibashi, J., and Otani, H. (1989). Optimal control of distillation column by multi-variable control system. FUJITSU (in Japanese), Vol. 40,1, pp 48-54.

# CONTROL SYSTEM FOR LNG RECEIVING
# TERMINAL

## T. Fujii

*Department of Engineering, Osaka Gas Co. Ltd., Osaka, Japan*

**Abstract.** The operation control system of an LNG terminal must reduce production cost through labor saving and optimization, while providing the safe and uninterrupted gas service. To do this, a hierarchical computerized system based on the distributed computer control system has been implemented at the Osaka Gas LNG terminal. This system has the advantage of allowing the use of digital backup systems, large scale displays, and wireless debugging tools. Data communication between the process control system and the production management system is being introduced to enhance gas productivity.

**Keywords.** Direct digital control; Hierarchical systems; Computer architecture; Digital computer applications; Natural gas technology; Production control.

## INTRODUCTION

Osaka Gas Co., Ltd. supplies town gas to approximately 5,100,000, customers in 6 prefectures in the Kinki District of Japan. It is well known that the clean energy of LNG (Liquefied Natural Gas) has been widely used for electric power generation, industries, and air conditioning as well as town gas, occupying an important position as a substitute energy for petroleum. Osaka Gas has LNG terminals at Senboku Works I and II and Himeji Works to supply gas over a wide region. A large-scale distributed computer control system consisting of hierarchical systems is being introduced at these terminals, for greater reliability, labor saving, and efficient maintenance work. This paper describes our LNG terminal computer control system.

## DESCRIPTION OF AN LNG TERMINAL

Figure 1 shows an outline of the gas manufacturing processes at our Senboku Works. The natural gas is collected at the place of origin, liquefied at the extremely low temperature of -162°C, transported to the production plant by LNG tanker. The production plant receives and stores LNG, then regasfies it, adjusts its calorific value and odorizes it and distributes it. Since the processes at LNG plants are simpler than at general chemical plants, automation is comparatively easy. The major characteristics of LNG plants are:

· LNG at extremely low temperatures is handled.
· A large amount of LNG is received and stored safely.
· Safety is particularly necessary since the plant is the main source of town gas.
· Ways to effectively recover LNG cryogenic energy are pursued.

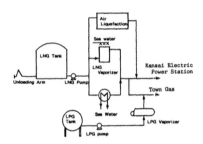

Fig. 1. Process flow.

## MAJOR FUNCTIONS REQUIRED
## FOR OPERATION CONTROL

The most advanced LNG terminal operation control system requires the following major functions.

### Supervisory Control System

· Daily report, emergency data logging, and other logging
· Metering of transactions with related companies
· Integrated management including other works
· Interfacing of telemeters, LA (Laboratory Automation) systems, and other systems

### Overall Plant Control System

· Optimal load distribution for each equipment
· Optimal control of the equipment according to the gas demand
· Judgement of power failure, adequate countermeasures, and power recovery
· Backup for problems at related works

## Plant Control System

· Monitoring each equipment
· Automatic start-up and stop of each equipment
· Automatic rate change for each equipment
· Automatic start-up and stop of the calorific value control plant
· Steps in case of power failure

## Disaster Preventing System

· Monitoring gas detectors, low temperature detectors, and other sensors
· Automatic start-up and stop of fire prevention and extinguishing equipment (High Expansion Air Foam Equipment, Dry Chemicals, etc.)
· Control of ITV linking various sensors

## Power Receiving and Distribution System

· Electric power system monitoring and control (including emergency power generation equipment)
· Power-factor and demand monitoring and control
· Operation guidance, and prediction of operation results

## Maintenance System

· LNG pump vibration analyses
· Operation period control of each equipment
· Control of the frequency of each equipment starting and stopping

### CONTROL SYSTEM
### CONFIGURATION AND CONCEPT

In designing the digital instrumentation control system, these points must be taken into consideration.

## Improved Reliability and Efficient Maintenance

· The simplest possible system
· The optimal control station distribution for the equipment and operation systems
· Redundancy in important devices
· Employment of the largest possible amount of packaged software
· Use of backup systems for important loops
· Improved debugging systems

## Promotion of Automation and Labor Saving

Implementation of "MAJOR FUNCTIONS REQUIRED FOR OPERATION CONTROL"

## Improved Operation System

· Centralized monitoring through CRT operation
· Standardized main system and backup system operating procedure
· Employment of an operator supporting system using large scale displays
· Improved training system for operator's console operation

## Provision for Expansion

· Construction of backup systems consisting of distributed computer control systems
· Space to expand in the centralized control room (CCR) and computer room

## Disaster Prevention

· Installation of a backup CCR
· Division of the computer room, the power supply room, and other rooms for fire prevention
· Duplication of optical-fiber cable along separate routes

Figure 2 shows the LNG terminal control system designed based on these considerations. Figure 3 shows the major functions of each sub-system.

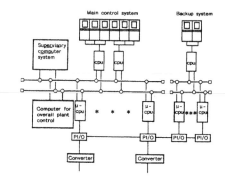

Fig. 2. System configuration.

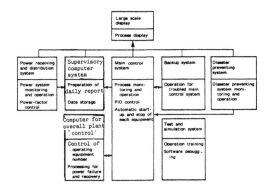

Fig. 3. Functions of each sub-system.

### CHARACTERISTICS OF THE
### CONTROL SYSTEM CONFIGURATION

The LNG terminal control system configuration is characterized below.

## Hierarchization and Function Distribution

Thorough function distribution and hierarchization have been adopted to minimize the influence of expansion, changes, and shutdown of each system on the hardware as shown in Fig. 3 and 4. For the software configuration, hierarchization and a mutual check system on the application level are shown in Fig. 5. To transmit the control signal to other systems, velocity type output is employed.

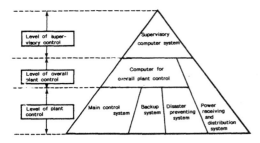

Fig. 4.  System hierarchization.

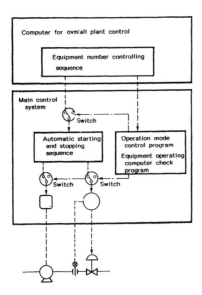

Fig. 5.  Software configuration.

For the efficient automatic operation of each equipment, the operation modes shown in Table 1 are available; a fully automatic mode for full-automatically operating all equipments, a semi-automatic mode for automatically starting and stopping some units, a manual mode for operating all equipments manually, and a maintenance mode for equipment inspection. Each mode can be controlled by checking the normal operation patterns of each equipment.

TABLE 1  Operation Mode Control

| Operation mode | Operation | Computer system |
|---|---|---|
| Full automatic (F-mode) | Equipment number control Processes in case of power failure and recovery | Computer for overall plant control and main control system |
| Semi-automatic (S-mode) | Automatic start and stop for each equipment unit | |
| Manual operation (O-mode) | Manual operation | Main control system |
| Maintenance mode (H-mode) | Equipment maintenance | |

## Digital Backup System

Conventional backup systems for digital instrumentation systems are analog instrumented. Analog backup systems have been expanded with the disadvantages of increased system space and expense, and hardware changes required for the system expansion. For operators familiar with CRT operations, it is likely that they will be more confident with CRT operation than with panel operation. For these reasons, we have employed a digital backup system. As a result, most of the main control system software can be used for the backup system, thus reducing the number of design processes and expense.

Figure 6 shows the digital backup system hardware configuration.

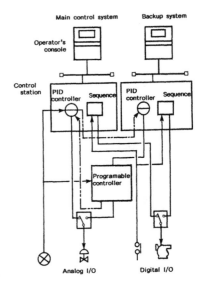

Fig. 6. Digital backup system.

## Test and Simulation System

The complete debugging of the digital instrumentation control system has usually been impossible
because it required so much labor.

A special debugging system was introduced, shown
in Fig. 7. This test and simulation system checks
the practical control program using a prepared
simple plant model. Thanks to this system, nearly
100% of debugging has become possible, and the
application software has become very reliable.
Operators can also be trained using this system as
a plant simulator.

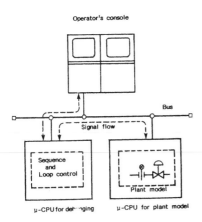

Fig. 7.  Test and simulation system.

## Large Scale Display.

The conventional large graphic panels for overall
monitoring of the entire plant required changes in
the hardware when expanding or altering the plant.
Here, large scale displays have been used instead
of the large graphic panels, and provide operation
guidance for all personnel, ITV displays and TV
conference (Photo 1).

Photo 1  CCR in Senboku Works I

## CONCLUSION

This large scale distributed computer control
system has been installed at Himeji Works, Senboku
Works I of Osaka Gas (Photo 1). Replacement of the
system in the northern area of Senboku Works II now
being planned will be completed in 1993. This will
complete a fully integrated control system for the
Senboku District plants. In addition, an optimal
operation system applicable to all production
plants is under development. The on-line operation
of this optimal operation system and the control
system in each production plant will create the
optimal town gas production.

## REFERENCES

Fujii, T., S. Matsuoka, and K. Murata (1988).
Digital instrumentation system construction promotion under factory operation in LNG terminals.
Instrumentation and Control Engineering, 31-7,
67-71.

Fujii, T., R. Saino, M. Nogami, and others (1988).
A replacement of the computer systems for LNG plant
instrumentation. Yokogawa Technical Report, 32-3,
173-178.

# MULTIVARIABLE NONLINEAR CONTROL
# SYSTEM DESIGN FOR MULTISTAND
# ROLLING MILL

## K. Mizuno,* Y. Morooka** and Y. Katayama**

*Hitachi Ltd., 6 Kanda-Surugadai 4 chome Chiyoda-ku, Tokyo, Japan
**Hitachi Research Laboratory, Hitachi Ltd., 4026 Kuji-cho, Hitachi-shi,
Ibaraki-ken, Japan

Abstract. To satisfy various requirements for a multivariable nonlinear control
system, such as increasing rolling mill gauge accuracy, a new control system using
the blocked decoupling control method has been developed for rolling mill
processes. To design the control system, an interactive simulation system was
developed. This simulation system can interpret nonlinear block diagrams.
Simulation conditions are entered interactively through the graphic terminal. The
newly developed control system has been analyzed using this simulation system, and
has been shown to be useful for multistand rolling mill processes.

Keywords. Decoupling control; Nonlinear system; Rolling mills; Control system
synthesis; Computer control

## I. INTRODUCTION

Rolling mill control systems have traditionally
been discussed as typical multivariable, nonlinear
control processes. An actual complex system
combines one-input one-output control loops that
individually control variables such as strip
thickness, tension, or shape. Rolling of high
value-added semiconductor materials and very thin
materials requires a precise control system which
should readily accept changes of system
configuration and number of rolling stands
corresponding to changes of rolling conditions.
To satisfy these requirements, a multiinput,
multioutput control system is required, and a
blocked decoupling system that can deal with
system change must be developed. This report
describes the blocked decoupling control system
and the state of the art in the design for such a
complex system which increases the practical use
of multivariable control techniques.

## II. STATE EQUATION FOR A MULTISTAND
## ROLLING MILL

The problem in applying multivariable control
theory is developing a mathematical model for
describing the state variables of the system to be
controlled. This research considers the roll gap
and the roll speed, shown in Fig. 1, as the
factors to be controlled.
The state equation is derived by linearizing the
system using the rolling results (i.e., strip
thickness, tension and shape) as the control
inputs based on the command values of roll gap and
roll speed (output of the multivariable
controller).

This section outlines the rolling theory that is
used for finding the state equation. Eq. (1)
shows the relationship between rolling force P and
the strip thickness, tension, material hardness
(deformation resistance), etc.

The basic expression is:

$$P = bK_pQ_p \chi \sqrt{R' (H - h)} \qquad (1)$$

Where b is Width, $K_p$ is Deformation resistance, $Q_p$
is the Coefficient function of rolling force, $\chi$ is
Coefficient based on tension, R' is the Flat roll
radius, H is the Entry thickness, and h is the
Exit thickness

The relationship between roll gap S and thickness
h, after approximating the rolling mill as a
spring system, is shown by the following
expression according to Hooke's law:

$$h = S-So + \frac{P}{K} \qquad (2)$$

Where h is the Exit thickness, S is the Measured
roll gap, So is the Measured roll gap at roll
force zero, S-So is the Roll gap without rolling
force, P is the Rolling force, and K is the Mill
spring constant.
The relationship between roll speed and tension is
shown by the expression (3). When the stand
distance of the tandem rolling mill is 'L' and the
total tension on rolled material is 'T', the strip
speed of the rolled material is 'v' at the exit
point and 'V' at the entry point.

$$\frac{dT}{d\tau} = \frac{E}{L} (V - v) \qquad (3)$$

where $\tau$ is Time and E is Young's coefficient, the
following relationship between exit speed 'v' and
roll speed '$V_R$' is obtained.

$$v = v_R (1 + f) \qquad (4)$$

Parameter 'f' indicates the slip between the roll and the rolled material. The slip 'f' is found by rolling theory using the following expression for a cold rolling mill:

$$f = a_0 + a_1 r + a_2 r^2 + (a_3 + a_4 r + a_5 r^2)\sqrt{\frac{h}{R'}}$$

(5)

Entry speed 'V' is found by the following constant volume theorem which indicates that the volume of the rolling material entering the mill in unit time is equal to the outgoing volume:

$$b\,V\,H = b\,v\,h$$

(6)

The previous section described the mathematical model related to rolling process, but finding the state equation for the control system requires a dynamic model of the roll gap and roll speed controllers. These controllers have higher-order and non-linear transfer functions between the command value and the control output to perform fast response and precise control. Microcomputers have recently been used for digital control. However, the control system is still executed by the one-input one-output feedback control system. The process control system for strip thickness, tension, etc., described in this report allows approximations to be calculated using a low-order model. A first-order lag approximation is used in this report.

If shape is considered, the expression for tension or strip thickness is applied at three to five locations in the direction of the strip width. The results of the expression are related using the roll bending state.

The relationship between parameters depends on variables that influence each other via differential operators (showing the presence of an integrator) and those that influence each other via algebraic operators. If expressions using these variables are adjusted and a deviation equation is produced by Tailor's expansion, a set of linear differential equations can be formulated from Eq. (1) - (6). Sufix $i$ means stand number.

$$Ts\Delta\dot{S} = \Delta Sp - \Delta S$$

(7)

$$Ts(Ph\cdot\Delta\dot{h}_i - P_{tf}\cdot\Delta\dot{t}_{fi} - P_{tb}\cdot\Delta\dot{t}_{bi}) = -P_h\cdot\Delta h_i$$
$$+ P_{tf}\cdot\Delta t_{fi} + P_{tb}\cdot\Delta t_{bi} + \Delta S_{pi} + \frac{1}{K}\Delta Q_i + PH(\Delta H_i$$
$$+ Ts\cdot\Delta\dot{H}_i)$$

(8)

$$\Delta\dot{T}_i = t_{fi}\cdot\Delta\dot{h}_i + h_i\cdot\Delta\dot{t}_{fi} = \frac{E}{L}(\Delta U_{i+1} - \Delta U_i)$$

(9)

$$\text{where} \quad \frac{\Delta U_i}{U_i} = \frac{\Delta VR_i}{VR_i} + \frac{\Delta h_i}{h_i} + \frac{1}{1+f_i}(f_h\cdot\Delta h_i + f_{tf}\cdot\Delta t_{fi}$$

$$= f_{tb}\cdot\Delta t_{bi} + f_H\cdot\Delta H_i)$$

(10)

$$TN\cdot\Delta\dot{VR}_i = Vp_i - VR_i$$

(11)

where coefficient $Px = (\frac{\partial P}{\partial X_i})$, $fx = (\frac{\partial f}{\partial X_i})$

If the output state variable of the controlled system is $X = [\Delta h_i, \Delta t_{fi}, \Delta VR_i]^T$ and the output variable of actuator is $U = [\Delta S_{pi}, \Delta V_{pi}, \Delta Q_i]^T$, then:

$$\dot{X} = AX + BU$$

(12)

### III. DECENTRALIZATION BY BLOCKED DECOUPLING METHOD

From the state equation in the previous section, mutual interaction between rolling stands occurs via tension $T_i$, and, if the tension is fixed at a constant value, the mutual interaction between the stands can be decoupled and a stand control system can be configured. In conventional systems, the gauge control system for each stand is built on the successive speed control of all stands and/or the constant tension control. However, delays in the tension control or errors in the speed control caused tension fluctuations so that precise control was not possible. This report introduces the blocked decoupling control method which separates the system into stand-unit subsystems to reduce the number of dimensions of control operation within the concept of the whole system[1].
When the raw vector $x_1$ is the state variable, plate thickness, tension, etc. of the first rolling stand, the preset values (output of the process controller) of the roll gap and the roll speed are represented by string vector $u_1$, and the entry strip thickness and the observable disturbance are represented by $d_1$, the following state equation can be formulated from the model described in the previous section using the state vector of all stands $X = [x_1, x_2, x_3 ... xn]^T$, operation vector $U = [u_1, u_2, u_3 ... un]^T$, and disturbance vector $D = [d_1, d_2, ... dn]^T$.

$$\dot{X} = \begin{bmatrix} A_{11} & A_{12} & 0 & 0 & \cdots & \cdots & \cdots & 0 \\ A_{21} & A_{22} & A_{23} & 0 & \cdots & \cdots & \cdots & 0 \\ 0 & A_{32} & A_{33} & A_{34} & 0 & \cdots & \cdots & 0 \\ \hline & & & & & & A_{nn-1} & A_{nn} \end{bmatrix} X$$

$$+ \begin{bmatrix} B_{11} & 0 & 0 & | & 0 \\ 0 & B_{22} & 0 & | & 0 \\ 0 & 0 & B_{33} & | & 0 \\ \hline 0 & \cdots & \cdots & \cdots & | & B_{nn} \end{bmatrix} U$$

$$+ \begin{bmatrix} E_{11} & E_{12} & 0 & 0 & | & 0 \\ 0 & E_{22} & E_{23} & 0 & | & 0 \\ 0 & 0 & E_{33} & E_{34} & | & 0 \\ \hline 0 & \cdots & \cdots & \cdots & \cdots & | & E_{nn} \end{bmatrix} D$$

(13)

The method of decoupling the system was suggested by Wonham et al. In this method, the relationship between input and output is mathematically transformed to decoupled form, so that the separated blocks have no meaning with the actual physical process configuration.
The newly developed blocked decoupling method focuses on the physical process structure, and separates it into partial state equations consisting only of a minor coefficient matrix corresponding to state variables in the blocks. When the feedback control function

$$U = FX \qquad (14)$$

is assumed in Eq. (13), Eq. (15) can be formulated for stand '$i$'. $D = 0$ is assumed for simplification.

$$\begin{aligned}
x_i &= A_{i,i-1} x_{i-1} + A_{ii} x_i + A_{i,i+1}\ x_{i+1} + B_{ii} \\
&\quad (f_{i1} x_1 + f_{i2} x_2 + \cdots + f_{i,i-1}\ x_{i-1} + f_{ii}\ x_i \\
&\quad + f_{i,i+1} x_{i+1} \cdots + f_{in} x_n) \\
&= B_{ii} f_{i1} x_1 + B_{ii} f_{i2} x_2 + \cdots + (A_{i,i-1} + B_{ii} \\
&\quad f_{i,i-1}) x_{i-1} + (A_{ii} + B_{ii} f_{ii}) x_i + (A_{i,i+1} \\
&\quad + B_{ii} f_{i,i+1})\ x_{i+1} + \cdots + B_{ii} f_{in} x_n \qquad (15)
\end{aligned}$$

If the coefficient of $X_i$ ($i \neq 1$) is zero in Eq. (14), the state equation considers only x and y. That is,

$$\begin{aligned}
f_{i1} &= f_{i2} = \ldots = f_{i,i-2} = 0 \\
A_{i,i-1} &+ B_{ii} f_{i,i-1} = 0 \\
A_{i,i+1} &+ B_{ii} f_{i,i+1} = 0 \\
f_{i,i+2} &= \ldots \ldots = f_{in} = 0 \qquad (16)
\end{aligned}$$

If $B_{ii}$ is regular, the following equation can be derived:

$$\begin{aligned}
f_{i,i-1} &= - B_{i,i-1} A_{i,i-1} \\
f_{i,i+1} &= - B_{i,i-1} A_{i,i+1} \qquad (17)
\end{aligned}$$

For rolling mill control, Bii can be a regular diagonal matrix.

The feedback matrix '$f_{ii}$' corrects the operation value $u_i$ of stand '$i$' using the '$i$'-stand's state variable $x_1$ to non-interact between '1' stands. At this time, Eq. 15 is simplified to:

$$x_i = (A_{ii} + B_{ii} f_i)\ x_i \qquad (18)$$

'$f_i$' is the feedback coefficient for optimization.

Based on this result, Fig. 2 shows the system configuration when applying the blocked decoupling control to a four-stand tandem rolling mill. With this blocked decoupling control method, each stand will be controlled independently and parallelly. This result is verified using a simulation with a CAD system.

## IV. CONFIGURATION OF THE INTERACTIVE SIMULATION SYSTEM

A schematic diagram of the interactive simulation system is shown in Fig. 3. This system consists of intelligent graphic terminals and a mainframe computer. To reduce the load on the Central Processing Unit (CPU) caused by frequent man-machine communication, the intelligent graphic terminal provides command input functions and pre- and post-command processing functions. The former accept commands selected from menu on display, assemble command and parameter data such as the position of a CRT, and send data to the host computer. The latter can process several commands, such as rotation, extension and refresh,

within the terminal and can interpret and display graphical data sent from the host computer on the CRT. The interactive simulation system consists of basic software, a block diagram input subsystem, a time response analysis subsystem, a frequency response analysis subsystem, and a data base manager. The basic software controls the data flow between the terminal and the host computer, and manages utility programs such as graphic subroutines. The data base manager, especially developed for this simulator, employs a simple data base using a tree structure.

## V. BLOCK DIAGRAM FOR DESCRIBING COMPLEX SYSTEM

The block diagram input subsystem can construct block diagrams on the CRT according to command data containing information about blocks and connections between blocks.

Block diagram elements which can be used in this simulation system are shown in Table 1. To describe a continuous time system corresponding to the controlled process, dynamic elements, linear algebraic elements, and nonlinear algebraic elements can be used. In addition to these elements, logic elements, unit delay elements, and A/D converters are provided to accommodate digital controllers. These elements are referred to as element information. In many cases, a control system is described by more than one block diagram. In order to allow simulator applications to large-scale control systems, the block diagram input subsystem is designed to allow up to 500 elements to be used in the description of one control system.

## VI. TIME RESPONSE ANALYSIS OF DIGITAL CONTROL SYSTEM

Difficulties in analyzing the digital control system result from its being a mixed system (i.e., a discrete-continuous time system). To compute dynamic behavior efficiently, two methods are combined as shown in Fig. 4. Since the block diagram of the continuous time system is equivalent to a set of ordinary differential equations and algebraic equations, it becomes important to conduct a stable numerical computation. In this study, a tableau method, which is a stable computation technique for nonlinear block diagrams, is applied to the continuous time system. For digital controllers, however, selection of the computation method does not affect the simulation results, since behavior of the controller is essentially expressed by difference equations. Therefore, an easily realized sequential computation method is applied to the digital controllers.[2]

An outline of the tableau approach for nonlinear block diagrams of the controlled process is shown in Fig. 5. The state variables are assigned to all the operation blocks of the block diagrams. Input-output equations are derived easily. Dynamic elements are described as

$$x_i(t) + a_{1i} x_i(t) + a_{2i} x_{\alpha_1}(t) + a_{3i} x_{\beta_1}(t) \qquad (19)$$

Where $d\ a_{1i}$, $a_{2i}$ and $a_{3i}$ are constants. The subscripts $\alpha_i$, $\beta_i$ denote input elements.

The following equation is used for linear algebraic elements

$$x_j(t) = a_{2j}x_{\alpha_2}(t) + a_{3j}x_{\beta_2}(t) + a_{4j}x_{\gamma_2}(t) + \delta_j \quad (20)$$

$\delta_j$ represents process input. When the $j$th element is not a process input, $\delta_j$ is set to zero. The nonlinear algebraic elements are expressed as

$$x_\iota(t) = f_\iota(x_{\alpha_3}(t), x_{\beta_3}(t)) \quad (21)$$

where $f_\iota$ represents a nonlinear function corresponding to the type of element. When the one-step linear formula known as the stable computation method for differential equations is applied to Eq. (19), the difference equation below is obtained.

$$(1-h(1-\theta)a_{1\iota})x_\iota(k+1) - h(1-\theta)a_{2\iota}x_{\alpha 1}(k+1)$$
$$-h(1-\theta)a_{3\iota}x_{\beta 1\iota}(k+1)$$
$$= (1+h\theta a_{1\iota})x_\iota(k) + h\theta a_{2\iota}x_{\alpha 1}(k) + h\theta a_{3\iota}x_{\beta 1}(k) \quad (22)$$

where $\theta$ is the weighting factor ($0 \leq \theta \leq 1$). The interval $(0, T)$ is divided by the uniform time step h into discrete time points 0, $t_1$, $\cdots$ $t_n\cdots$ $x(k)$ denotes $x(t_k)$. The Newton-Raphson iteration is applied to Eq. (22), starting from the known values, $x_\iota(k)$, $x_{\alpha_3}(k)$ and $x_{\beta_3}(k)$.

$$x_\iota(k+1) = (\partial f_\iota/\partial x_{\alpha_3})x_{\alpha_3}(k+1) - (\partial f_\iota/\partial x_{\beta_3})x_{\beta_3}(k+1)$$
$$= x_\iota(k) - (\partial f_\iota/\partial x_{\alpha_3})x_{\alpha_3}(k) - (\partial f_\iota/\partial x_{\beta_3})x_{\beta_3}(k) \quad (23)$$

The partial derivatives $\partial f_\iota/\partial x_{\alpha_3}$ and $\partial f_\iota/\partial x_{\beta_3}$ are computed according to the type of nonlinear elements and the values of $x_{\alpha_3}(k)$ and $x_{\beta_3}(k)$. Since the input variables $x_{\alpha_1}$, ....and $x_{\beta_3}$ of a block are the outputs of other elements, the input-output equations can be arranged in a tableau.

$$A(k)X(k+1) = B(k) \quad (24)$$

where X is the N dimensional state vector ($x_1$, $x_2$, ...$x_N$) is the N is the number of operation blocks, $A(k)$ is the $N\times N$ coefficient matrix and $B(k)$ is the N dimensional vector. The generation and solution of Eq. (24) yield $x(k+1)$ for the value of the state variable at the $(k+1)$th point.

## VII. RESULTS OF SIMULATION TESTS

A simulation was performed to evaluate the performance of the newly developed control system on a four-stand tandem rolling mill. Fig. 6 shows a block diagram of a single stand for computer simulation. The roll gap controllers and speed controllers on the stands were approximated using first order lag.
First-order lag time constants were set to 20 ms. and 10 ms. for all stands.

One disturbance cycle of a 60-$\mu$m (p – p value) 2-Hz sine wave was applied to the raw material. This one-cycle wave, which is comparatively larger and of higher frequency than the usual disturbance, was applied to simplify the understanding of the phenomena.

Fig. 7 shows the rolling mill simulation results using a conventional control method, so-called "Gauge meter method" Automatic Gauge Control (AGC), under the following conditions: 2.8 mm raw material is passed through a four-stand mill to produce a 0.8 mm steel strip, and a 60 $\mu$m disturbance is superimposed on the raw material. This figure illustrates the process whereby the disturbance applied to the stands is sequentially transferred from the first stand to the fourth stand. The first deviation at the second stand shows the influence caused by the working tension between the first stand and second stand when the disturbance was applied to the first stand. In this simulation, the strip thickness deviation was found to be 6.1 $\mu$m and 0.8% of the target strip thickness. Fig. 8 shows the simulation results using the blocked decoupling control. Because control was separated for each stand, the influence of the first stand on the second stand, seen in Fig. 7, disappeared. The simulated distributed control of each stand produced a strip, thickness deviation of 1.9 $\mu$m at the last stand. Consequently, the simulation results of the blocked decoupling control system forecast a deviation reduction of one third compared to the conventional mill control.

Performing blocked decoupling control can simplify the complicated operation of stands caused by material tension. It can also improve the control characteristics of a single-stand rolling mill. For example, an eccentricity frequency can easily be detected in roll control to improve control performance.

## VIII. CONCLUSION

We have introduced the outline of the blocked decoupled optimal control system which is adopted in multistand rolling mills. Simulation results revealed that this control system design technique is useful for the nonlinear, multivaliable interacting processes.
Therefore we intend to apply this technique to various processes and to develop a precise control system.

## REFERENCES

[1] Katayama, Y. and Morooka, Y. (1988). Advanced Cooperative Technology for Mill Control. IEE Japan Trans., 108-D, pp. 540-543
[2] Tanuma, M. Morooka, Y. and Takano, T. (1982). Interactive Simulation Method Using Tableau Approach For Nonlinear Control-System. Second IFAC Symposium Computer Aided Design of Multivariable Technological System, 1982-9 (Purdue University)

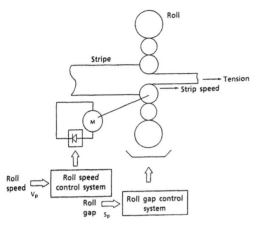

Fig. 1 Rolling mill diagram

## Table 1  Elements of block diagram

| kind | Elements | kind | Elements |
|---|---|---|---|
| Dynamic | · Integral<br>· First time lag<br>· Second time lag<br>· Lead lag<br>· Rate<br>· P1 (proportional plus integral) | Special | · On-off<br>· Sampler<br>· Super time delay |
| Linear algebraic | · Gain<br>· Comparator<br>· Input from another page<br>· Output to another page<br>· Output<br>· Process input<br>  step   ramp<br>  (pulse  sine )<br>  constant<br>  free wave | Digital | · Unit time delay<br>· A/D converter<br>· D/A converter<br>· AND<br>· OR<br>· HAND<br>· EOR<br>· Flip/flop<br>· Level comparater<br>· Switch |
| Nonlinear algebraic | · Product  · Power  · Square root<br>· Division  · Sine  · Absolute<br>· Saturation  · Cosine  · Exponential<br>· Dead band  · Natural logarithm  · Selection of maximum value<br>  · Common logarithm  · Selection of minimum value | | |

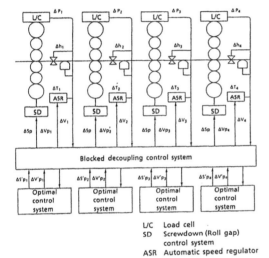

L/C   Load cell
SD   Screwdown (Roll gap) control system
ASR   Automatic speed regulator

Fig. 2 Block diagram of contorl system for multistand rolling mill

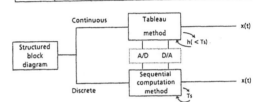

h means that computation is done at interval h

Fig. 4 Outline of time response analysis

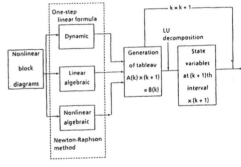

Fig. 5  Basic procedure for the tableau approach

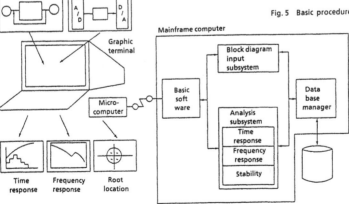

Fig. 3 Configuration of interactive simulation system for digital control system

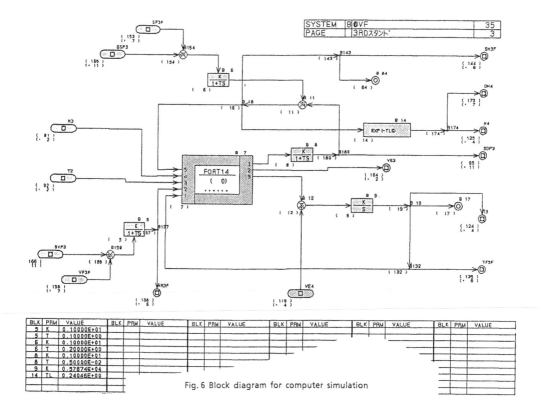

| BLK | PRM | VALUE | BLK | PRM | VALUE | BLK | PRM | VALUE | BLK | PRM | VALUE | BLK | PRM | VALUE | BLK | PRM | VALUE |
|-----|-----|-------|-----|-----|-------|-----|-----|-------|-----|-----|-------|-----|-----|-------|-----|-----|-------|
| 5 | K | 0.10000E+01 | | | | | | | | | | | | | | | |
| 5 | T | 0.10000E+00 | | | | | | | | | | | | | | | |
| 6 | K | 0.10000E+01 | | | | | | | | | | | | | | | |
| 6 | T | 0.20000E+00 | | | | | | | | | | | | | | | |
| 8 | K | 0.10000E+01 | | | | | | | | | | | | | | | |
| 8 | T | 0.50000E-02 | | | | | | | | | | | | | | | |
| 9 | K | 0.57874E+04 | | | | | | | | | | | | | | | |
| 14 | TL | 0.24046E+00 | | | | | | | | | | | | | | | |

Fig. 6 Block diagram for computer simulation

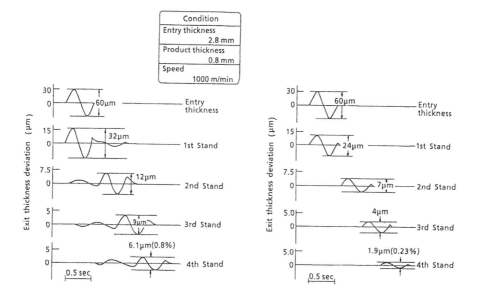

| Condition | |
|-----------|--|
| Entry thickness | 2.8 mm |
| Product thickness | 0.8 mm |
| Speed | 1000 m/min |

Fig. 7    Simulation results by usual control system        Fig. 8    Simulation results by developed control system

Copyright © IFAC Production Control in the
Process Industry, Osaka, Japan 1989

# AN INTEGRATOR DECOUPLING METHOD
# TO THE APPLICATION PROBLEM
# IN MULTIVARIABLE FEEDBACK
# SYSTEM CONTROL

## S. Kawashima

*The Special Engineering College of Kogakuin University, Tokyo, Japan*

Abstract- A development of techniques for the design of multivariable
control systems is of considerable practical importance. Recently the
systems have a great variety of control methods, i.e., the optimization
control, the pole assignment,etc.
    This paper reports some problems, accompanied by the application of
the integrator decoupling method. Furthermore, the decoupling method is
applied to a simple model and a partial system control. Study of the
decoupling controller structure allows one to suggest new feasible
control methods.

Keywords. Decoupling; Eigenvalues; Pole placement; Power control;
Transfer functions.

## INTRODUCTION

The development of techniques for the
design of multivariable control systems is of
considerable practical importance. A particular design approach involves the use of feedback to achieve closed loop control system
stability. In conjunction with this approach
it is often of interest to know whether or
not it is possible to have inputs control
outputs independently, i.e., a single input
influences a single output. The problem of
decoupling a time invariant linear system by
state variable feedback was first considered
by Morgan. The necessary and sufficient condition for decoupling was proposed by Falb
and Wolovich and they have also made definite contributions to the synthesis problem.
Significant contributions in generalising the
decoupling by state feedback were also made by
E.G.Gilbert. The problem of partial decoupling
was considered by W.M.Wonham.
    Fairly large number of example problems
have been worked on. These problems include
flight control by Falb and Wolovich, aircraft
control and the control of binary distillation
column by Gilbert and Pivnichny. The decoupling
technique was applied for power control loops by
K.R.Reddy et al. The decoupling technique was
also applied for Synthesis Problem of Voltage
Transfer Function Matrix by Tajima, Nagase and
Takahashi.
    The decoupling of active power and reactive
power or the voltage and current will have a
considerable beneficial effect on the system
performance and design. K.R.Reddy and M.P. Dave
toke the problem of decoupling the control loops
of active and reactive power as this decoupling
will have better practical application.
    Usually the active power is controlled by
the prime mover input and the reactive power is
controlled through excitation. This can be regarded as the multi-input-multi-output control
system, with the control inputs as the changes
in the excitation and changes in the prime mover
input, and the two outputs as the changes in
reactive power and active power. Any change in
one of the control inputs will in general
affect both the outputs, i.e., there is coupling between the two control loops of P and Q.
    In the paper K.R.Reddy and M.P.Dave achieve
the decoupling between the two control loops of
P and Q by statevariable feedback. The state
feedback used for decoupling is also utilized
for locating the poles at desired locations,
thus achieving the decoupling along with the
pole placement. It is found that a system without exciter needs a precompensator at the input
terminals for decoupling. If the system includes the exciter, the decoupling is possible
without the precompensator.
    The first order of business is development
of a power system model. The following analysis
is based on the assumption that the electrical
interconnections within each individual control
area are so strong, at least in relation to the
ties with the neighboring areas, that the whole
area can be characterized by a single frequency
only.
    A noninteracting (decoupling) control of
interconnected electric power system applied by
the matrix methods of multipole system feedback
control theory had a considerable beneficial
effect on transient analysis of system and design of frequency and tie-line power regulators.
It was not a general approach with interconnection conditions that was treated the case of the
n-area with stiff interconnections.
    The decoupling of a control system with a

general interconnection condition will have a considerable beneficial effect on system performance and design. In the paper I consider the problem of decoupling of control loops of power system control as this decoupling will have better practical applications.

In this paper I achieve the decoupling between the two-area control systems by state variable feedback. The state feedback used for decoupling is also utilized for locating the poles at desired locations, thus achieving the decoupling along with the pole placement. Study of the decoupling controller structure allows one to suggest new feasible control methods.

Usually the power systems with general interconnection conditions have woven a great variety of control systems into their own systems. The partial area power systems control method of state feedback decoupling involves the use of state feedback decoupling method to apply a subsystem and a cross coupling dynamic response subsystem. The state feedback used for the decoupling subsystem is also utilized for locating the poles at desired locations, thus achieving the decoupling along with the pole placement. These methods could give improved dynamic response and wider stability margins to the control systems.

## DECOUPLING OF MULTIVARIABLE CONTROL SYSTEMS

The statement of the problem in decoupling control theory form is as follows: Given the linear time invariant system represented by the state variable differential equation

$$\dot{x}=Ax+Bu$$

$$y=Cx \qquad (1)$$

where x type indicates the time derivative d/dt, x is an $n\times1$ state vector, y is an $m\times1$ output vector, u is an $m\times1$ control vector, A is an $n\times n$ state distribution matrix, B is an $n\times m$ control distribution matrix, C is an $m\times n$ output matrix.

The decoupling of system Eq.(1) is achieved using a control law.

$$u=Fx+Gv. \qquad (2)$$

If the closed loop transfer function

$$H(s,F,G)=C(sI-A-BF)^{-1}BG \qquad (3)$$

is diagonal and nonsingular.
Let us define

$$B^{*}=[C_iA^{d_i}B],\ i=1,2,\cdots,m. \qquad (4)$$

The necessary and sufficient condition for decoupling is that $B^{*}$ should be nonsingular. Falb and Wolovich have given a synthesis procedure for directly obtaining feedback matrix F, whose elements are so determined as to yield a desired closed loop pole structure. The feedback matrix F which yeilds the desired closed loop poles is given by

$$F=B^{*-1}[\sum_{k=0}^{\delta} M_kCA^{k}-A^{*}]$$

$$G=B^{*-1} \qquad (5)$$

where $\delta$=max.$d_i$ and $M_k$, k=0,1,$\cdots,\delta$ are $m\times m$

diagonal matrices. Thus the decoupled system is given by

$$\dot{x}=(A+BF)x+BGv$$

$$y=Cx. \qquad (6)$$

The feedback decoupled system diagram is given by

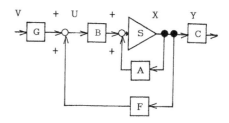

Fig.1. A general decoupling by state feedback control system.

## PARTIAL AREA SYSTEMS CONTROL METHOD OF STATE FEEDBACK DECOUPLING

Consider a subsystem S' of linear dynamical system S defined by

$$\dot{x}=Ax+B'u'$$

$$y'=C'x \qquad (7)$$

where x is a real n-state vector, u' is an m'-vector $m'\leqq m$, y' is an m'-vector $m'\leqq m$ and the submatrices A,B' and C' are real constant matrices. The decoupling of S is achieved by using a control law.

$$u'=F'x+G'v'. \qquad (8)$$

If the closed loop transfer function

$$H'(s,F',G')=C'(sI-A-B'F')^{-1}B'G'$$

$$=Y'(s)/V'(s) \qquad (9)$$

is diagonal and nonsingular, here, v' is the m'-new-input vector, and F' and G' are constant matrices of order $m'\times n$ and $m'\times m'$.
Let us define

$$B'^{*}=[c'_iA^{d_i}B'],\ i=1,2,\cdots,m' \qquad (10)$$

where $C'_i$ is the ith row of C',

$d_i$=min j:$C'_iA^jB'\neq0$, j=0,1,$\cdots,$n-1

=n-1 if $C'_iA^jB'$=0 for all values of j

$$A'^{*}=[C'_iA^{d_{i+1}}],i=1,2,\cdots,m'. \qquad (11)$$

The necessary and sufficient condition for decoupling is that $B'^{*}$ should be nonsingular. If det.$B'^{*}\neq0$, (the system is said to have no inherent coupling) then the decoupling of S' is achieved by the following F' and G' matrices.

the used power system is expressed in the state space form Eq.(17) and Eq.(18).

The new partial area systems control method is applied to the system 1 of the interconnected two-area system's one side. Then the output matrix is given by

$$[y1]=[1\ 0\ 0\ 0\ 0]\begin{bmatrix}\triangle f1\\\triangle G1\\\triangle f2\\\triangle G2\\\triangle T\end{bmatrix}.\qquad(19)$$

The $B'^*$ matrix Eq.(10) is calculated as

$$B'^*=[C'iA^{di}\ B']=[C'1AB']=[-0.05].\qquad(20)$$

The $B'^*$ matrix is nonsingular and hence the system can be decoupled with the control law Eq.(8). Thus the system can be controlled with the decoupling control law.

Since m'+d1=2<n, all the 2 poles can be placed at desired locations. The natural system Eq.(7) has eigenvalues at $-0.475,-0.063\pm j3.12$ and $-0.300\pm j0.673$. The partial area decoupled system Eq.(12) has eigenvalues at $0.0,0.0,$ $-0.453,-0.072\pm j2.26$. The partial area decoupled system will have the following transfer function

$$H'(s,F'^*,G')=[1/s^2].\qquad(21)$$

The M'0 and M'1 matrices (this exemplar's M'0 and M'1 are real constants) of Eq.(14) are given by

$$M'0=[-5.0],M'1=[-10.0].\qquad(22)$$

The F' matrix using Eq.(14) for the above Eq.(22) is calculated as

$$F'=[-22.3\ 18.8\ 92.5\ 0.0\ -19.8].\qquad(23)$$

The partial area decoupled system for the F' matrix Eq.(23) has eigenvalues at $-0.528,$ $-9.472$ (which approximately has the original system response with increased damping) and three other eigenvalues at $-0.453,-0.074\pm j2.26$ (which approximately has the original system response). Thus the partial area decoupled system makes the system stable.

### COMPUTER RESULTS

The decoupled and partial area decoupled systems were simulated on the digital computer. 6 cases were tried, for the two-area problem I considered, the A,B,M and M' matrices did not change. The B' and C' matrices changed for the two-area system input-output sets. The C and C' matrices were varied to show output matrices effect on the system response. The following output matrices were used:
case 1

$$C=\begin{bmatrix}1\ 0\ 0\ 0\ 0\\0\ 0\ 1\ 0\ 0\end{bmatrix}\qquad(24)$$

case 2

$$C=\begin{bmatrix}1\ 0\ 0\ 0\ 0.2\\0\ 0\ 1\ 0-0.2\end{bmatrix}\qquad(25)$$

case 3

$$C=\begin{bmatrix}0\ 0\ 0\ 0\ 1\\0\ 0\ 1\ 0\ 0\end{bmatrix}\qquad(26)$$

case 4

y1=$\triangle$f1

$$C'=[1\ 0\ 0\ 0\ 0]\qquad(27)$$

case 5
y1=$\triangle$f1+0.02$\triangle$T

$$C'=[1\ 0\ 0\ 0\ 0.02]\qquad(28)$$

case 6
y1=$\triangle$T

$$C'=[0\ 0\ 0\ 0\ 1].\qquad(29)$$

The M0,M1 and M2 matrices of Eq.(5) are given by

$$M0=\begin{bmatrix}-5.0&0\\0&-5.0\end{bmatrix},M1=\begin{bmatrix}-10.0&0\\0&-10.0\end{bmatrix},$$

$$M2=\begin{bmatrix}-10.0&0\\0&0.0\end{bmatrix}.\qquad(30)$$

The M'0,M'1 and M'2 matrices (this exemplar's M'0 and M'1 and M'2 are real constants) of Eq.(14) are given by

$$M'0=[-5.0],M'1=[-10.0],M'2=[-10.0].\qquad(31)$$

The decoupled system for case 1 has eigenvalues at $-0.528,-0.528,-9.472,-9.472$( which approximately has the original system response with increased damping) and one other eigenvalue at 0.0 (this eigenvalue is placed at the original point). All the 4 poles can be placed at desired locations (since m+d1+d2=4<n). Thus the decoupled system makes the system unstable. The decoupled system for case 2 has eigenvalues at $-18.50,-9.472,-0.528,-9.472,-0.528$ (which approximately has the original system response with increased damping). Thus the decoupled system makes the system stable. The decoupled system for case 3 has the 5 poles at desired locations (since m+d1+d2=5=n). Thus the decoupled system makes the system stable. The partial area decoupled system for case 5 has eigenvalues at $-9.472,-0.528,-0.543,-0.491\pm j2.226$. Thus the decoupled system makes the system stable. The partial area decoupled system for case 6 has eigenvalues at $-8.975,-0.307\pm j2.365,-0.506\pm j0.158$ (which approximately has the orignal system response with increased damping). Thus the decoupled system makes the system stable.

A five percent per unit increase step input during five seconds is applied to area 1 and the subsequent variations of $\triangle$f1,$\triangle$f2 and $\triangle$T are studied.

Figs.3-10 show the results as recorded on an digital computer. I note the following important features;
perdecoupled case
1) the two frequency errors will be equal after steady state is reached;
2) the system is oscillatory but stable.

$$G'=B'^{*-1}$$

$$F'^{*}=-B'^{*-1}A'^{*}. \tag{12}$$

The closed loop transfer function after decoupling will be of the form

$$H'(s,F'^{*},G')=diag(1/s^{di+1}),$$

$$i=1,2,\cdots,m'. \tag{13}$$

Thus the decoupled system has multiple poles at the origin which makes the system unstable. Falb and Wolovich have given a synthesis procedure for directly obtaining feedback matrix F' whose elements are so determined as to yield a desired closed loop pole structure. The feedback matrix F' which yields the desired closed loop poles is given by

$$F'=B'^{*-1}[\sum_{k=0}^{\delta}M'kC'A^{k}-A'^{*}] \tag{14}$$

where $\delta$=max.di and M'k, k=0,1,$\cdots$,$\delta$ are m'$\times$m' diagonal matrices. The decoupled system's transfer function will have the denominator of form

$$s^{di+1}+m'^{i}_{di}s^{di}+m'^{i}_{di-1}s^{di-1}+\cdots+m'^{i}_{0}. \tag{15}$$

The required closed loop poles are obtained by a suitable choice of M'k. But only m'+d1+d2+$\cdots$+dm' of closed loop poles can be located arbitrarily. The pole-zero cancellations's closed loop transfer function is given in references.

### APPLICATION OF A DECOUPLING METHOD TO PARTIAL AREA SYSTEMS

Let us consider a simple power control of two-area system. The performance equations of the system together with the expressions for a block diagram of two-area perturbation model are as given in references. This paper used a block diagram of a two-area perturbation model given below-

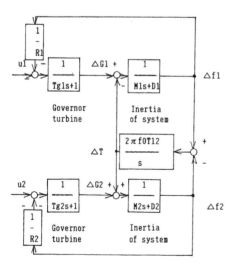

Fig.2. Block diagram of two-area perturbation model.

The result of the two-area linear perturbation relationship may be expressed in the state space form Eq.(1). The states, the inputs, the constant matrices A and B of the state space form Eq.(1) are given by

$$\frac{d}{dt}\begin{bmatrix}\triangle f1\\\triangle G1\\\triangle f2\\\triangle G2\\\triangle T\end{bmatrix}=\begin{bmatrix}-D1/M1 & 1/M1 & 0 & 0 & -1/M1\\-1/Tg1R1 & -1/Tg1 & 0 & 0 & 0\\0 & 0 & -D2/M2 & 1/M2 & 1/M2\\0 & 0 & -1/Tg2R2 & -1/Tg2 & 0\\2\pi f0T12 & 0 & -2\pi f0T12 & 0 & 0\end{bmatrix}\begin{bmatrix}\triangle f1\\\triangle G1\\\triangle f2\\\triangle G2\\\triangle T\end{bmatrix}$$

$$+\begin{bmatrix}0 & 0\\-1/Tg1 & 0\\0 & 0\\0 & -1/Tg2\\0 & 0\end{bmatrix}\begin{bmatrix}u1\\u2\end{bmatrix}. \tag{16}$$

The procedure and the results are explained with the help of the following example. The equation (16) is calculated at the nominal operating point with the following system parameters. Tg1=Tg2=2(seconds), R1=R2=0.1((rad/sec)/MW), M1=M2=10(seconds), D1=D2=1 (MW/(rad/sec)), T12=0.14722(MWsec/(rad/sec)), f0=50[Hz],Base MVA=1000MVA, All the values are given in per uint.

The equation (16) is given by

$$\frac{d}{dt}\begin{bmatrix}\triangle f1\\\triangle G1\\\triangle f2\\\triangle G2\\\triangle T\end{bmatrix}=\begin{bmatrix}-0.1 & 0.1 & 0 & 0 & -0.1\\-5.0 & -0.5 & 0 & 0 & 0\\0 & 0 & -0.1 & 0.1 & 0.1\\0 & 0 & -5.0 & -0.5 & 0\\46.25 & 0 & -46.25 & 0 & 0\end{bmatrix}\begin{bmatrix}\triangle f1\\\triangle G1\\\triangle f2\\\triangle G2\\\triangle T\end{bmatrix}$$

$$+\begin{bmatrix}0 & 0\\-0.5 & 0\\0 & 0\\0 & -0.5\\0 & 0\end{bmatrix}\begin{bmatrix}u1\\u2\end{bmatrix}. \tag{17}$$

The following usual output matrix of the outputs of the frequency deviation $\triangle$f1 and $\triangle$f2 is used

$$\begin{bmatrix}y1\\y2\end{bmatrix}=\begin{bmatrix}1 & 0 & 0 & 0 & 0\\0 & 0 & 1 & 0 & 0\end{bmatrix}\begin{bmatrix}\triangle f1\\\triangle G1\\\triangle f2\\\triangle G2\\\triangle T\end{bmatrix}. \tag{18}$$

I developed a decoupling method by applying state feedback to the power system control. Then

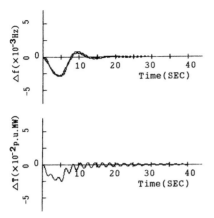

Fig.3. System dynamic response of original
system to step-input increase in area 1.

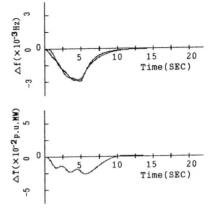

Fig.6. System dynamic response, Case 3;
m+d1+d2=5=n.

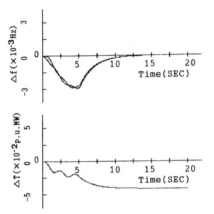

Fig.4. Uncontrolled response of tie-line
deviation to step-input increase in
area 1, Case 1.

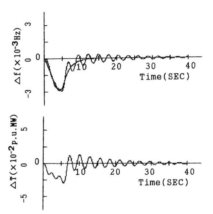

Fig.7. Swing response of decoupled system tie--
line and frequency deviation to step--
input increase in area 1, Case 4.

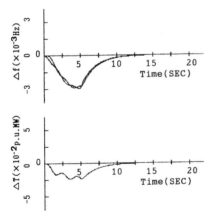

Fig.5. System dynamic response, Case 2.

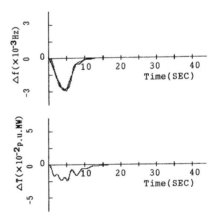

Fig.8. Good damping swing response of decoupled
system tie-line and frequency deviation
to step-input increase in area 1, Case 5.

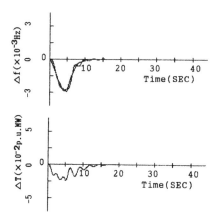

Fig.9. Good damping swing response of desired
       locations the three poles decoupled
       system tie-line and frequency deviation
       to step-input increase in area 1, Case 6.

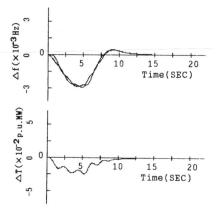

Fig.10. Coupling response by restricted control
        input, Case 2; $|u| \leqq 0.01$, sampling period
        T=0.05sec.

decoupled case 1
1) the two frequency errors will be equal after
steady state is reached;
2) $\Delta T$ variable has nonzero errors;
3) the system is nonswing.
decoupled case 2 and case 3
1) the two frequency errors will be equal after
steady state is reached;
2) the system is nonswing, stable and good
damping.
partial area decoupled case 4
1) the two frequency errors will be equal after
steady state is reached;
2) the system is swing, stable and about the
original system's damping.
partial area decoupled case 5 and case 6
1) the two frequency errors will be equal after
steady state is reached;
2) the system is swing, stable and good damping.
    A restricted input decoupling control
system has a cross coupling dynamic response
when over the input level. But the restricted
system has the decoupled dynamic response when
under the input level. The cross coupling dynamic response is the original system dynamic
response. The original system dynamic response
is the response with governor action (R1=R2=0.1

((rad/sec)/MW)).
    Figs.5-6,8-10 illustrate the improvement
in damping given by the decoupled and the partial area decoupled controller to the system.

CONCLUSION

    Decoupling control techniques have been ap-
plied to develop decoupling controllers for the
simple system that significantly improve transient response to perturbations in step input. The
form of the decoupling controller has shown that
1) the state feedback used for decoupling and
partial area decoupling also helps to place the
closed loop poles at desired locations;
2) the only desired system output is affected by
the state feedback control;
3) by using all information about the system in
the controller, significant improvements in
system response can be achieved.

REFERENCES

B.S.Morgan. (1964). The synthesis of multivaria-
    ble systems by state feedback. IEEE Trans.
    Vol.AC-9,pp405-411.
P.L.Falb and W.A.Wolovich. (1967). Decoupling in
    the design and synthesis of multivariable
    control systems. IEEE Trans. Vol.AC-12,
    pp.651-659.
E.G.Gilbert. (1969). Decoupling of multivariable
    systems by state feedback. J.SIAM Control,
    Vol.7,pp.50-63.
L.M.Silverman. (1970). Decoupling with state
    feedback and precompensation. IEEE Trans.
    Vol.AC-15,pp.487-489.
E.G.Gilbert and J.R.Pivnichny. (1969). A Com-
    puter program for the synthesis of decou-
    pled multivariable feedback systems. IEEE
    Trans. Vol.AC-14,pp.652-659.
G.Quazza. (1966). Noninteracting controls of
    interconnected electric power systems. IEEE
    Trans. Vol.PAS-85,pp.727-741.
K.Raja,M.P.Dave. (1975) Decoupling of active
    and reactive power control loops by state
    feedback. IEEE PES Summer Meeting. San
    Francisco,Calif.July 20-25. A 75 4416.
O.I.Elgerd,C.E.Fosha. (1970). Optimum megawatt--
    frequency control of multiarea electric
    energy systems. IEEE Trans. Vol.PAS-89,
    pp.556-563.
C.E.Fosha,O.I.Elgerd. (1970). The megawatt-fre-
    quency control problem: A new approach via
    optimal control theory. IEEE Trans. Vol.PAS
    -89,pp.563-577.
A.J.Wood,B.F.Wollenberg. (1986). Power genera-
    tion, operation, and control. John wiley &
    Sons,New York.
P.C.Chandrasekharan. (1971). Observability and
    decoupling. IEEE Trans. Vol.AC-16,pp.
    482-484.
J.Tajima,H.Nagase and S.Takahashi. (1978).
    Synthesis problem of voltage transfer
    function matrix. Trans. IEcE Japan.
    Vol.J61-A NO.5,pp.433-440.
W.M.Wonham. (1985). Linear multivariable
    control. 3rd ed. Springer-verlag. New York.
    pp.221-275.

# APPLICATION OF THE FUZZY EXPERT
# SYSTEM TO FERMENATION PROCESSES

## M. Kishimoto,* T. Yoshida** and M. Moo-Young***

*Department of Biological Science and Technology, Science University of Tokyo,
Noda, Chiba, Japan
**International Centre of Cooperative Research in Biotechnology,
Japan Faculty of Engineering, Osaka University, Suita, Osaka, Japan
***Industrial Biotechnology Centre, University of Waterloo, Waterloo, Ontario,
Canada N2L 3G1

Abstract. The optimal operation of fermentation processes was inferred by using
the fuzzy expert system. The system contains two parts: a fuzzy inference
engine and a knowledge data base. The knowledge data base contains fuzzy
relations and arithmatic ones. The fuzzy relation shows a relationship between
variables, the same as a relationship on a normal graph. The fuzzy inference
engine consisted of searching procedure and several kinds of fuzzy arithmatic
methods. The expert system was applied to maridomycin production, and it can be
easily applied to other fermentation processes, because we need not alter the
inference engine but only the input of the kowledge data base for any variation
in the process considered. The construction of the data base from the
experimental data or available semi-quantitative information can be carried out
semi-automatically by using a computer graphic tool.

Keywords. Optimization; fermentation process; expert system; fuzzy
inference; maridomycin production

## INTRODUCTION

Recently, computers have made a significant
input into the field of bioprocess control and
several sensors such as those of pH, DO,
temperature, turbidity and ORP are available
for monitoring bioreaction processes. However,
it is difficult to construct kinetic models
which can describe the behaviour of organisms
in a bioreactor because the bioreactions,
including many kinds of control mechanisms at
genetic and enzymatic levels, are much more
complicated than ordinary chemical reactions.
In addition microbial characteristics can be
easily changed and the improvements of them
are often attempted in industrial production.
So a fermentation model can only work for a
limited period.

Models of fermentation processes, therefore,
should be constructed quickly. From this point
of view, we considered that the use of a
perfectly identified model was not practical
and first proposed a statistical procedure
which is based on the actual data of the
present fermentation process. The procedure
need not construct precise models and not
identify the process perfectly. However, it
needs some amount of data which increased a
lot if the number of state variable increased
or the dimension of system equations became
high. The procedure can select the independent
variables, but we need the previous candidates
and much more data for the selection.

In order to overcome the defects of the sta-

tistical procedure, we need to investigate the
fermentation process with the aid of various
information in literature or discussion with
experienced operators. The conclusion usually
are not precise one, and some one might be
unclear. But this inference result (even an
unclear one) can play an important role for
the statistical procedure or the construction
of the mathematical model.

Human inference consumes a lot of time and
varies according to each person. Then, as a
substitute of some part of the human
inference, a fuzzy expert system was developed
and applied to the fermentation process.

### FUZZY EXPERT SYSTEM

Almost all fuzzy inferences are classified
into two groups as follows.

1. Inference based on the fuzzy production
   rules.
2. Inference based on the handling of fuzzy
   relations.

In the present research, the inference method
based on the handling fuzzy relations was
applied to the construction of the expert
system of the fermentation process. This
approach is favoured because operator or human
imagination for the control or prediction of
progress of the fermentation process is
similar to the inference using fuzzy relations,
which was used to describe the relationship
between the independent variables and depend-

ent variables which might be operative variables, state variables, or objective function of the process.

The inference engine component of the fuzzy expert system was constructed in the previous work(Kishimoto and Moo-Young) and included the fuzzy arithmatic techniques shown as follows.

If we have two fuzzy relations, $R_1 \subset X \times Y$ and $R_2 \subset Y \times z$, we define the min-max composition of $R_1$ and $R_2$ such that

$$\mu_{R_2 R_1}(x,z) = \bigvee_y \left[ \mu_{R_1}(x,y) \wedge \mu_{R_2}(y,z) \right]$$
$$= \underset{y}{MAX} \left[ MIN \left( \mu_{R_1}(x,y), \mu_{R_2}(y,z) \right) \right]$$
$$\forall \ x, \ y, \ z \in R^+ \tag{1}$$

We also used other fuzzy arithmatic techniques for operation based on two fuzzy relations. The following operation in $R^+$ are called the min-max convolution.
$\forall \ x, \ y, \ z \in R^+$

$$\mu_{A(+)B}(z) = \underset{z=x+y}{\bigvee}(\mu_A(x) \wedge \mu_B(y)) \tag{2}$$

$$\mu_{A(-)B}(z) = \underset{z=x-y}{\bigvee}(\mu_A(x) \wedge \mu_B(y)) \tag{3}$$

$$\mu_{A(\times)B}(z) = \underset{z=x \times y}{\bigvee}(\mu_A(x) \wedge \mu_B(y)) \tag{4}$$

$$\mu_{A(\div)B}(z) = \underset{z=x \div y}{\bigvee}(\mu_A(x) \wedge \mu_B(y)) \tag{5}$$

The system can handle the algebraic relations for the rearrangement of the fuzzy relations in the database as follows.

Consider the two equations which represent algebraic relations in the database.

$$Z_1 = Y_1 + Y_2 \tag{6}$$

$$Y_1 = Y_3 + Y_4 \tag{7}$$

Eq.(7) is substituted into (6) and the following equation is constructed automatically.

$$Z_1 = (Y_3 + Y_4) + Y_2 \tag{8}$$

By combining these techniques, we can use more complicated relationships, which might include material balances, system equations and so on, for the fuzzy inference of the expert system. Therefore, if the database included the fuzzy relations, $Y_3$ vs. X, $Y_4$ vs. X, $Y_2$ vs. X, the inference engine would infer the fuzzy relation $Z_1$ vs. X by combination of the functions and the fuzzy arithmatic techniques mentioned before.

A scheme of the procedure for the fuzzy inference is shown in Fig.1 and the main structure of the fuzzy inference engine is as follows.
1. rearrangement of the database for application to the inference procedure (cf. rearrangement of eq.(6) and eq.(7) into eq.(8)).
2. Set I = 1, I : The depth of Searching

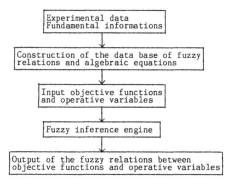

Fig. 1. Schematic diagram of the procedure for the fuzzy inference

3. Search for the same dependent variable of the fuzzy relation as the objective function (I=1) or the independent variable of the pre-stage inference(I>1).
4. If the same variable was found, calculate the fuzzy relation between the objective function and the new independent variable by the fuzzy composition procedure described in Eq.(1). If I = 1, the fuzzy composition procedure is not needed; i.e., the new fuzzy relation equals the searched one.
5. Search the same dependent variable of the algebraic relation as the objective variable (I = 1) or the independent variable of the pre-stage inference (I > 1).
A simple example of the algebraic relation in the database is developed to give:
  Production rate = Specific production rate
                × cell mass
6. If the same dependent variable of algebraic relation exist in the data base, calculate a new fuzzy relation between the objective function and the new independent variable by the fuzzy arithmatic techniques (fuzzy convolution described in Eq.(2-5) and fuzzy composition procedure).
7. If the independent variables of the new fuzzy relations derived from step 3 to step 6 are the same as the operative condition, the relationships would be memorized in the computer memory.
8. I = I + 1, and if I < IM, return to the step 3, or else, go to the next step.
9. Summarize the fuzzy relations between the operative condition and the objective function by using a maximizing method with the memorized data of the fuzzy relations at the step 7.
10. Output the summarized fuzzy relation.

## Example of the inference by using the fuzzy expert system

Let us consider a very simple example of the application of the expert system to glutamic acid production by fermentation process. It is well known that one of the most important fractors for production is the control of the permiability of glutamic acid through the cell membrane. Penicillin addition drastically affects the cell wall and the cell membrane. Several hours after penicillin addition, the membrane cannot keep high concentrations of glutamic acid inside the cell, and a large amount of glutamic acid leaks into the medium and "overproduction" of glutamic acid begins at this time.

A detailed mechanism of the penicillin effect on cell wall and membrane is not available, and it is difficult to construct a deterministic model for the process. However, we can readily develop a fuzzy relationship :

> If penicillin addition was too early, the cell growth would be severely inhibited, the cell membrane would become weak, and glutamic acid would be produced slowly.
> If penicillin addition was too late, the growth would be high, but cell membrane would not be effected so much and glutamic acid would not be produced.

From this imagination, we can easily propose a fuzzy relation as shown in Fig. 2a and 2b.

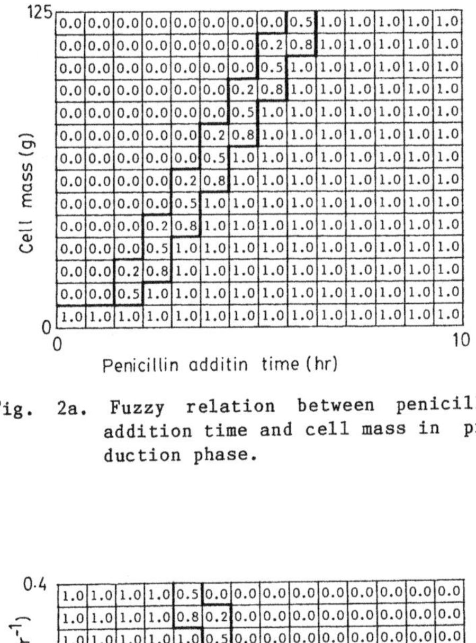

The values of membership functions, which are the components of the matrices (fuzzy relations), represent the possibility of the culture state or culture condition which is partially determined by the coordinates in the table. Each fuzzy relation represents the relationship between two variables the same as a relationship on a normal graph.

For simplicity, we assume a simple shape of the fuzzy relation. In an actual inference, we would modify them with information obtained from experimental data of the fermentation process.
The following algebraic relation was also used for the inference.

$$\text{production rate} = \text{cell mass} \times \text{specific production rate} \qquad (9)$$

The fuzzy relation in Fig. 2a and 2b and the algebraic relation (9) was input into the knowledge database of the expert system, and the expert system would infer the effect of the operative condition (penicillin addition time) on the objective function (production rate), and give us the result shown in Fig. 3.

### Fig. 2a

Cell mass (g), max 125; Penicillin addition time (hr) 0 to 10.

| | | | | | | | | | | | | | | | |
|---|---|---|---|---|---|---|---|---|---|---|---|---|---|---|---|
|0.0|0.0|0.0|0.0|0.0|0.0|0.0|0.0|0.0|0.0|0.5|1.0|1.0|1.0|1.0|1.0|
|0.0|0.0|0.0|0.0|0.0|0.0|0.0|0.0|0.0|0.2|0.8|1.0|1.0|1.0|1.0|1.0|
|0.0|0.0|0.0|0.0|0.0|0.0|0.0|0.0|0.0|0.5|1.0|1.0|1.0|1.0|1.0|1.0|
|0.0|0.0|0.0|0.0|0.0|0.0|0.0|0.0|0.2|0.8|1.0|1.0|1.0|1.0|1.0|1.0|
|0.0|0.0|0.0|0.0|0.0|0.0|0.0|0.0|0.5|1.0|1.0|1.0|1.0|1.0|1.0|1.0|
|0.0|0.0|0.0|0.0|0.0|0.0|0.0|0.2|0.8|1.0|1.0|1.0|1.0|1.0|1.0|1.0|
|0.0|0.0|0.0|0.0|0.0|0.0|0.5|1.0|1.0|1.0|1.0|1.0|1.0|1.0|1.0|1.0|
|0.0|0.0|0.0|0.0|0.0|0.2|0.8|1.0|1.0|1.0|1.0|1.0|1.0|1.0|1.0|1.0|
|0.0|0.0|0.0|0.0|0.0|0.5|1.0|1.0|1.0|1.0|1.0|1.0|1.0|1.0|1.0|1.0|
|0.0|0.0|0.0|0.0|0.2|0.8|1.0|1.0|1.0|1.0|1.0|1.0|1.0|1.0|1.0|1.0|
|0.0|0.0|0.0|0.5|1.0|1.0|1.0|1.0|1.0|1.0|1.0|1.0|1.0|1.0|1.0|1.0|
|0.0|0.0|0.2|0.8|1.0|1.0|1.0|1.0|1.0|1.0|1.0|1.0|1.0|1.0|1.0|1.0|
|0.0|0.0|0.5|1.0|1.0|1.0|1.0|1.0|1.0|1.0|1.0|1.0|1.0|1.0|1.0|1.0|
|1.0|1.0|1.0|1.0|1.0|1.0|1.0|1.0|1.0|1.0|1.0|1.0|1.0|1.0|1.0|1.0|

Fig. 2a. Fuzzy relation between penicillin addition time and cell mass in production phase.

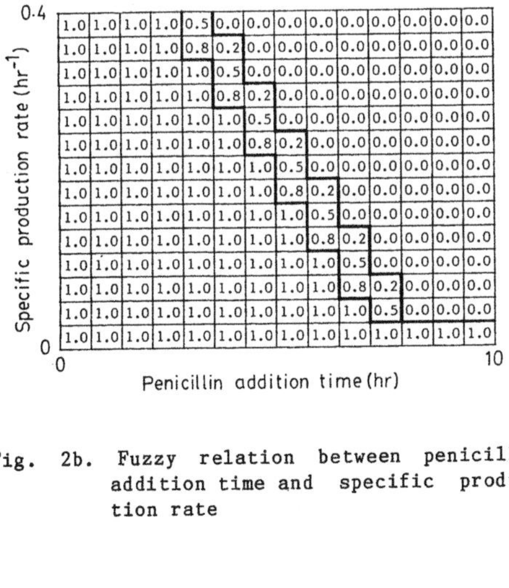

### Fig. 2b

Specific production rate (hr$^{-1}$), max 0.4; Penicillin addition time (hr) 0 to 10.

| | | | | | | | | | | | | | | | |
|---|---|---|---|---|---|---|---|---|---|---|---|---|---|---|---|
|1.0|1.0|1.0|1.0|1.0|0.5|0.0|0.0|0.0|0.0|0.0|0.0|0.0|0.0|0.0|0.0|
|1.0|1.0|1.0|1.0|1.0|0.8|0.2|0.0|0.0|0.0|0.0|0.0|0.0|0.0|0.0|0.0|
|1.0|1.0|1.0|1.0|1.0|1.0|0.5|0.0|0.0|0.0|0.0|0.0|0.0|0.0|0.0|0.0|
|1.0|1.0|1.0|1.0|1.0|1.0|0.8|0.2|0.0|0.0|0.0|0.0|0.0|0.0|0.0|0.0|
|1.0|1.0|1.0|1.0|1.0|1.0|1.0|0.5|0.0|0.0|0.0|0.0|0.0|0.0|0.0|0.0|
|1.0|1.0|1.0|1.0|1.0|1.0|1.0|0.8|0.2|0.0|0.0|0.0|0.0|0.0|0.0|0.0|
|1.0|1.0|1.0|1.0|1.0|1.0|1.0|1.0|0.5|0.0|0.0|0.0|0.0|0.0|0.0|0.0|
|1.0|1.0|1.0|1.0|1.0|1.0|1.0|1.0|0.8|0.2|0.0|0.0|0.0|0.0|0.0|0.0|
|1.0|1.0|1.0|1.0|1.0|1.0|1.0|1.0|1.0|0.5|0.0|0.0|0.0|0.0|0.0|0.0|
|1.0|1.0|1.0|1.0|1.0|1.0|1.0|1.0|1.0|0.8|0.2|0.0|0.0|0.0|0.0|0.0|
|1.0|1.0|1.0|1.0|1.0|1.0|1.0|1.0|1.0|1.0|0.5|0.0|0.0|0.0|0.0|0.0|
|1.0|1.0|1.0|1.0|1.0|1.0|1.0|1.0|1.0|1.0|0.8|0.2|0.0|0.0|0.0|0.0|
|1.0|1.0|1.0|1.0|1.0|1.0|1.0|1.0|1.0|1.0|1.0|0.5|0.0|0.0|0.0|0.0|
|1.0|1.0|1.0|1.0|1.0|1.0|1.0|1.0|1.0|1.0|1.0|1.0|1.0|1.0|1.0|1.0|

Fig. 2b. Fuzzy relation between penicillin addition time and specific production rate

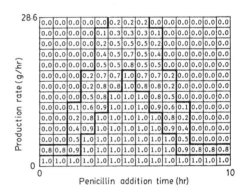

Fig. 3. Inference result which show the fuzzy relations of glutamic acid production vs. penicillin addition time.

This inference result means that the middle time addition is the best for the glutamic acid production, and the result of the primitive inference is almost the same as the trend of the experimental result in the previous work by Kishimoto and Moo-Young(submitted) and is also coincident with the common view for glutamic acid production. Therefore, the fuzzy expert system could suggest a suitable policy for penicillin addition roughly but easily from the semi-quantitative information available.

## APPLICATION OF THE FUZZY EXPERT SYSTEM TO THE MARIDOMYCIN PRODUCTION

Maridomycin is one of the macroride antibiotics, which is active against Gram positive bacteria and some Gram negative bacteria such as _Neisseria gonorroheae_ and _Vibrio cholera_. Asai et al.(1978) showed that how to secure

the critical value of dissolved oxygen level
and to avoid mechanical damage to microbial
cells were important subject for the scale up
of maridomycin fermentation. At present, it
is difficult to construct system equations for
complete description of the fermentation
because of the lack of the information related
with the mechanism of maridomycin production.
As the expert system can handle the fuzzy
relation which represents the trend of experimental data of the process and semi-
quatitative information found in the
literature, we tried to apply this method to
find suitable operative conditions (DO level
and agitation speed)

## Experimental condition

Streptomyces hygroscopicus B5050AV, which was
kindly supplied by Takeda Chemical Industry
Ltd., Osaka, was used in this study.

The composition of the fermentation medium is
as follows: 3% glycerol, 3.5% corn steep
liquor, 1% Proflo (protein hydrolysate, Takede
Pharm. Ind. Co. Ltd.), 0.1% $K_2HPO_4$, 0.02%
$MgSO_4.7H_2O$, 0.05%$MnSO_4.7H_2O$, 0.05% $FeSO_4.7H_2O$,
0.05% $ZnSO_4.7H_2O$, 0.12% $K_2HPO_4$, 0.1% DL-
methionine, 0.05% DL-alanine, and 1.2% $CaCO_3$.
$CaCO_3$ was separately sterirized.

Fermentation was carried out at 28 °C for 8
days in 500-ml Erlenmeyer flasks and in a 5-ℓ
jar fermentor where dissolved oxygen level was
automatically controlled at the set level and
the Aeration rate was 1 vvm.

Maridomycin concentration was estimated by a
colorimetric method using picric acid (Uchida,
et al. 1979). Glucose concentration was analyzed by a glucostat method, and the microbial
cell concentration was determined by Lowry's
Method after sonication(Sawada, et al., 1978).

## Experimental results

In order to investigate the effect of aeration
on the production of maridomycin, the flask
cultures were carried out under the various
working volume of 40 ml, 50 ml, 100 ml, and
150 ml. It was observed that the maximum
maridomycin 0.26 g/l was produced under a 40
ml working volume(Fig. 4) and the increase of
volume led to a decrease of maridomycin
production. The supply of oxygen might be an
important factor for the production. Several
batch cultures were carried out using the jar
fermentor in order to investigate the optimal
operative conditions and the results of some
of them are shown as follows.

A batch culture of S. hygroscopicus B-5050AV
in the fermentor was carried out with the
control of the dissolved oxygen level(DO) at
10% saturated with air throughout the
cultivation (Fig. 5). The maridomycin production was low (0.605g at 90hr), and a low DO
level was not favorable for the production of
maridomycin.

A similar batch culture was conducted with 40%
DO level(Fig.6). The maximum concentration of

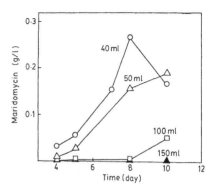

Fig. 4. Effect of culture volume in 500 ml
flask culture on the production of
maridomycin.

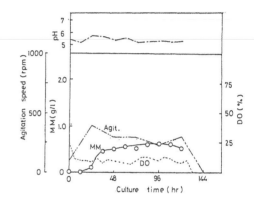

Fig. 5. Time course of maridomycin prodiction
with 10% DO level.

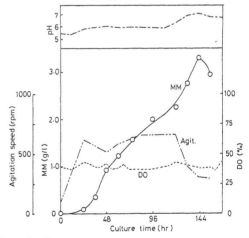

Fig. 6. Time course of maridomycin production
with 40% DO level

the product in this cultivation was much
higher than that obtained with 10% DO control
(Fig. 5) and in the flask cultures.

Fig. 7 shows the time courses of maridomycin production with the control of DO at 70% by using the aeration of air only, or of oxygen enriched gas. It was found that the production with the aeration using the mixed gas was less than a half of that using only the air. A high partial pressure of oxygen might have a negative effect on the production, and reduced to the production with 10% DO level. In this experiment, 70% DO level was maintained automatically varying agitation speed. The maximum agitation speed during the cultivation using air was 840 rpm while it was 500 rpm in the case that the enriched gas was used. Even if agitation speed was kept low to avoid a possible mechanical damage on cells, the production was low, when an oxygen enriched gas was used.

From the above results, it can be conclude that a high level of DO beyond 40% could not increase the production of maridomycin.

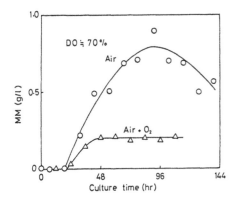

Fig.7.  Maridomycin production with 70% DO level by the aeration of air only, or of mixed gas of $O_2$ and air.

The construction of the database of fuzzy expert system.

A fuzzy expert system was used to optimize the operational condition of maridomycin production. This fuzzy expert system consisted of two parts : a fuzzy inference engine and a knowledge database. The inference engine was the the same as that applied to the glutamic acid production in the previous section "Example of the application". The different knowledge database, however, should be constructed for the application to the different process. Almost all part of the database was consisted of fuzzy relations, which are the tables of membership functions. In the case of a normal graph, the relationship is presented as a line or a curve, whereas it is presented as bands or zones, for example, a possible zone, an unclear zone and impossible zone which are defined in a fuzzy relation. The meaning of a possible zone is that the relationships between two variables are clearly possible in this zone and all the

values of the membership functions are 1.0 in this zone. The unclear zone implies that the possibility of the ralationship is unclear in this zone and the values of the membership function fall between 0.1 and 0.9. The impossible zone reveals that the relationship defined by the coodinates cannot be existed in this zone and values of the membership functions are always zero.

If all of the fuzzy relations were input by using the keyboard, we would waste time and input mistake can lead to a serious error of inference. Therefore the graphic input system was made in order to eliminate the error from the complexity of the input procedure of the membership functions.

1. Operator input the experimental data by using a key board

2. The frame of the graph and the symbols for the experimental data were drawn automatically.

3. Operator draws the two borderlines on the graph which was already drawn by the computer at step 2, one borderline between the possible zone and unclear zone, and the other between the unclear zone and the impossible zone, by a graphic tool, e.g. , a mouse.

4  Computer calculate the value of the membership function for every component of the fuzzy relation and construct the database for the fuzzy relations.

In the fuzzy expert system coupled with such a graphic input system, the operator need only to input experimental data, and draw the two curves using a mouse for the construction of the database.

Fig. 8 show the graphic example of the construction of fuzzy relations which are the main part of the knowledge data base. The experimental data is shown as circles. Asai et al.(1978) reported that too high agitation speed might decrease the production because of mechanical damage to the cells. But a low agitation may not cause any damage to the cells. Therefore it is not clear whether a low agitation decreases the production or not.

We drew the curves by the computer graphic tool "mouse" which represented the border line. The arrow represents the position of the mouse and the line was drawn by moving the mouse position. After the curves were drawn, the possible, unclear and impossible zones, are automatically defined by the computer as shown in Fig.8(B).

The matrix of the membership functions are shown in Fig. 9. The assignments of the values were also carried out automatically by the computer.

The fuzzy relation between production and agitation (A), another one between production and dissolved oxygen (B), and the other one between dissolved oxygen and agitation (C) were input to the knowledge database. A new

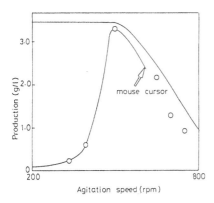

Fig. 8a. First step of the construction of the fuzzy relation

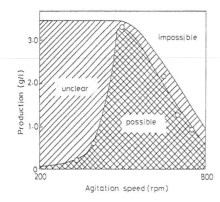

Fig. 8b. Second step of the construction of the fuzzy relation

3.5

| 0.0 | 0.0 | 0.0 | 0.0 | 0.0 | 0.0 | 0.0 | 0.0 | 0.0 | 0.0 | 0.1 | 0.3 | 0.3 | 0.0 | 0.0 | 0.0 | 0.0 | 0.0 |
|---|---|---|---|---|---|---|---|---|---|---|---|---|---|---|---|---|---|
| 0.1 | 0.1 | 0.1 | 0.1 | 0.1 | 0.1 | 0.2 | 0.6 | 0.9 | 0.9 | 0.1 | 0.0 | 0.0 | 0.0 | 0.0 | 0.0 | | |
| 0.2 | 0.2 | 0.2 | 0.2 | 0.2 | 0.2 | 0.4 | 0.9 | 1.0 | 1.0 | 0.4 | 0.0 | 0.0 | 0.0 | 0.0 | 0.0 | | |
| 0.2 | 0.2 | 0.2 | 0.2 | 0.3 | 0.3 | 0.6 | 1.0 | 1.0 | 1.0 | 0.7 | 0.0 | 0.0 | 0.0 | 0.0 | 0.0 | | |
| 0.3 | 0.3 | 0.3 | 0.3 | 0.3 | 0.4 | 0.8 | 1.0 | 1.0 | 1.0 | 0.9 | 0.0 | 0.0 | 0.0 | 0.0 | | | |
| 0.4 | 0.4 | 0.4 | 0.4 | 0.4 | 0.5 | 0.9 | 1.0 | 1.0 | 1.0 | 1.0 | 0.2 | 0.0 | 0.0 | 0.0 | | | |
| 0.5 | 0.5 | 0.5 | 0.5 | 0.5 | 0.6 | 1.0 | 1.0 | 1.0 | 1.0 | 1.0 | 0.5 | 0.0 | 0.0 | 0.0 | | | |
| 0.5 | 0.5 | 0.5 | 0.5 | 0.6 | 0.7 | 1.0 | 1.0 | 1.0 | 1.0 | 1.0 | 0.8 | 0.2 | 0.0 | | | | |
| 0.6 | 0.6 | 0.6 | 0.6 | 0.7 | 0.8 | 1.0 | 1.0 | 1.0 | 1.0 | 1.0 | 1.0 | 0.5 | 0.1 | | | | |
| 0.7 | 0.7 | 0.7 | 0.7 | 0.7 | 0.9 | 1.0 | 1.0 | 1.0 | 1.0 | 1.0 | 1.0 | 0.9 | 0.4 | | | | |
| 0.8 | 0.8 | 0.8 | 0.8 | 0.8 | 0.9 | 1.0 | 1.0 | 1.0 | 1.0 | 1.0 | 1.0 | 1.0 | 0.8 | | | | |
| 0.8 | 0.8 | 0.8 | 0.8 | 0.9 | 1.0 | 1.0 | 1.0 | 1.0 | 1.0 | 1.0 | 1.0 | 1.0 | 1.0 | | | | |
| 0.9 | 0.9 | 0.9 | 0.9 | 1.0 | 1.0 | 1.0 | 1.0 | 1.0 | 1.0 | 1.0 | 1.0 | 1.0 | 1.0 | | | | |
| 1.0 | 1.0 | 1.0 | 1.0 | 1.0 | 1.0 | 1.0 | 1.0 | 1.0 | 1.0 | 1.0 | 1.0 | 1.0 | 1.0 | | | | |

0
200    Agitation speed (rpm)    800

Maridomycin (g/l)

Fig. 9. Data of fuzzy relation

fuzzy relation (D) between production and agitation was calculated by the inference engine from the fuzzy relations (B) and (C) using the combination procedure (Eq.(1)). The fuzzy relations (A) and (D) were summarized by the comparizon of the membership function of the fuzzy relations. More unclear informations were rejected and clear data remained in the inference result.

3.5

| 0.0 | 0.0 | 0.0 | 0.0 | 0.0 | 0.0 | 0.0 | 0.0 | 0.0 | 0.0 | 0.1 | 0.3 | 0.3 | 0.0 | 0.0 | 0.0 | 0.0 | 0.0 |
|---|---|---|---|---|---|---|---|---|---|---|---|---|---|---|---|---|---|
| 0.0 | 0.0 | 0.0 | 0.0 | 0.1 | 0.1 | 0.1 | 0.2 | 0.6 | 0.9 | 0.9 | 0.9 | 0.1 | 0.0 | 0.0 | 0.0 | 0.0 | 0.0 |
| 0.0 | 0.0 | 0.0 | 0.0 | 0.1 | 0.2 | 0.2 | 0.4 | 0.9 | 1.0 | 1.0 | 1.0 | 0.4 | 0.0 | 0.0 | 0.0 | 0.0 | 0.0 |
| 0.0 | 0.1 | 0.2 | 0.2 | 0.3 | 0.3 | 0.6 | 1.0 | 1.0 | 1.0 | 1.0 | 0.7 | 0.0 | 0.0 | 0.0 | 0.0 | 0.0 | |
| 0.0 | 0.1 | 0.3 | 0.3 | 0.3 | 0.4 | 0.8 | 1.0 | 1.0 | 1.0 | 1.0 | 0.9 | 0.0 | 0.0 | 0.0 | 0.0 | 0.0 | |
| 0.4 | 0.4 | 0.4 | 0.4 | 0.4 | 0.5 | 0.9 | 1.0 | 1.0 | 1.0 | 1.0 | 1.0 | 0.2 | 0.0 | 0.0 | 0.0 | 0.0 | |
| 0.5 | 0.5 | 0.5 | 0.5 | 0.5 | 0.6 | 1.0 | 1.0 | 1.0 | 1.0 | 1.0 | 1.0 | 0.5 | 0.0 | 0.0 | 0.0 | | |
| 0.5 | 0.5 | 0.5 | 0.5 | 0.5 | 0.6 | 0.7 | 1.0 | 1.0 | 1.0 | 1.0 | 1.0 | 1.0 | 0.8 | 0.2 | 0.0 | | |
| 0.6 | 0.6 | 0.6 | 0.6 | 0.6 | 0.7 | 0.8 | 1.0 | 1.0 | 1.0 | 1.0 | 1.0 | 1.0 | 1.0 | 0.5 | 0.1 | | |
| 0.7 | 0.7 | 0.7 | 0.7 | 0.7 | 0.7 | 0.9 | 1.0 | 1.0 | 1.0 | 1.0 | 1.0 | 1.0 | 1.0 | 0.9 | 0.4 | | |
| 0.8 | 0.8 | 0.8 | 0.8 | 0.8 | 0.8 | 0.9 | 1.0 | 1.0 | 1.0 | 1.0 | 1.0 | 1.0 | 1.0 | 1.0 | 0.8 | | |
| 0.8 | 0.8 | 0.8 | 0.8 | 0.9 | 1.0 | 1.0 | 1.0 | 1.0 | 1.0 | 1.0 | 1.0 | 1.0 | 1.0 | 1.0 | 1.0 | | |
| 0.9 | 0.9 | 0.9 | 0.9 | 1.0 | 1.0 | 1.0 | 1.0 | 1.0 | 1.0 | 1.0 | 1.0 | 1.0 | 1.0 | 1.0 | 1.0 | | |
| 1.0 | 1.0 | 1.0 | 1.0 | 1.0 | 1.0 | 1.0 | 1.0 | 1.0 | 1.0 | 1.0 | 1.0 | 1.0 | 1.0 | 1.0 | | | |

0
200    Agitation speed (rpm)    800

Maridomycin (g/l)

Fig. 10. Result of the fuzzy inference.

The summarized result is presented as the fuzzy relations between agitation speed and production shown in Fig. 9, and shows the suitable agitation speed was about 600 rpm. The result agreed well with our human imagination because high agitation cannot increase the production due to the mechanical damage, and low agitation cannot favour the production because of the low DO, we should keep the moderate agitation.

## CONCLUSION

The fuzzy expert system was applied to the fermentation processes and could infer the suitable operative conditions. The expert system can be readily applied to many kinds of fermentation processes because only change of the knowledge database is needed for the variation of applied processes, and it is easily carried out by using the graphic input system. Another advantage of the fuzzy expert system is that it can effectively use various kinds of biological information, which is often qualitative and imperfect , and cannot be used directly for the simulation or optimization of fermentation processes.

## REFERENCES

Asai, T., H. Sawada, T. Yamaguchi, M. Suzuki, E. Higashide, and M. Uchida (1978). J. Ferment. Technol., 56, 374

Kaufmann, A. (1975). Introduction to the Theory of Fuzzy Subsets, Vol. 1, Academic press, New York. pp46-60

Kaufmann, A. and M. M. Gupta (1985). Introduction to Fuzzy Arithmetic Theory and Application, Van Nostrand Reinhold New York pp14-pp46

Kishimoto, M. and M. Moo-Young (submitted). Bioprocess Engineering

Sawada, H., M. Suzuki, and M. Uchida (1978). J. Ferment. Technol., 57, 248

Uchida, M., M. Suzuki, T. Takayama and N. Suzuki (1979). Agri. Biol. Chem., 43, 847

# CONTROL OF FERMENTATION PROCESSES
# AS VARIABLE STRUCTURE SYSTEMS

## K. B. Konstantinov, M. Kishimoto and T. Yoshida

*International Center of Cooperative Research in Biotechnology, Japan,*
*Faculty of Engineering, Osaka University, Yamada-oka, Suita-shi, Osaka 565, Japan*

Abstract. An essential pending problem in the modeling and control of fermentation
processes is the variability of biological systems, expressed as predictable or unpredictable structural alterations in the control plant. Such phenomena, naturally or
artificially induced, require flexible alteration of control system structure by proper
switching to one or another control strategy. In order to overcome the rigid restriction imposed upon the plant structure by the conventional control approach, we
propose here a two-level hierarchical scheme, which provides the control system with
a kind of structural adaptability. In the second level of the system, which operates
in the structural space of the plant and performs intelligent functions, the current
plant structure is recognized as an element of a finite set of structures defined on
the base of expert knowledge. In the first level, which works in the state space of
the plant, the control strategy relevant to the current plant structure is picked out
from a defined pool and the corresponding control action is calculated. This approach
has been realized as a real-time software system intended for control of various
fermentation processes.

Keywords. Process control; varying structures; large-scale systems; self-adjusting
systems; pattern recognition; artificial intelligence; control engineering computer
applications; fermentation processes; fuzzy sets.

## INTRODUCTION

The plants under the scope of control theory can
be classified into two general groups: those with
constant structure, and those with variable
structure. There are a large number of theoretical achievements and practical applications
dealing with single-structure control plants.
However, probably because of their great complexity, contributions to the analysis and control
of multi-structure plants have been few and the
corresponding theoretical basis is not yet well-
founded. Some results fundamental to analysis
and control of variable structure plants have been
presented by Varshavskii and Vorontsova (1963),
McLaren (1967), Tsetlin (1973), Saridis and
Hofstadter (1974), Young and Kwatny (1981) and
Sworder and Rogers (1983). In this report we
propose a nonstandard approach for control of
such systems, using an heuristic rather than
analytical approach. It is applicable particularly to fermentation processes, which are among the
most difficult plants to control. To the
authors' knowledge, to date there have been no
reports analyzing fermentation processes as
variable structure plants.

An essential pending problem in control of
fermentation processes is the variability of
animate systems, expressed as predictable or unpredictable structural alterations in the control
plant. This phenomenon, naturally or artificially
induced, means changes in the behavioral properties and physiological necessities of cells, and
require proper replacement of the control
strategies, even in the form of physical alteration of state and control vectors. In this case
the control system must track the plant structure
on-line and should be able to respond flexibly to
its variation by proper reconfiguration of its
own structure. However, a conventional approach
to fermentation process control imposes rigid

restrictions upon plant structure by enclosing it
in a single-structure frame unable to handle such
perturbations. In this case a single-structure
controller, which has poor information competence
(Kickert, Bertrand and Praagman, 1978), will be
ineffective in accord with the cybernetic law for
requisite variety (Ashbi, 1956): "only variety in
the regulator can force down the variety in the
plant; only variety can destroy variety". A more
sophisticated multi-structure controller is required, one able to alter its own structure in
response to plant variations. In order to overcome this limitation, we have considered fermentation processes as variable structure systems and,
on this basis, have proposed a novel hierarchical
method for their control. It is composed of the
following tasks, working in an on-line cyclic
sequence: detection of structural variations by
an expert pattern recognition procedure (second
level), switching of the control strategy corresponding to the current plant structure and
calculation of a new control action (first level).
We believe that this approach is more practically
applicable compared with conventional methods to
global modeling of the plant and formal synthesis
of control algorithms which often result in over-
mathematization and overdesign of control systems.
The proposed approach may be useful at more advanced stages of fermentation process investigation, when consistent base of expert knowledge has
been accumulated.

## BASIC DEFINITIONS

We define here several basic terms used in this
paper.

*Structure-altering variables* are those input
variables which, when exceeding certain values,
induce structural alterations in a plant.

*Structure-altering phenomena* are the physiological phenomena responsible for the structural alterations in a plant. Only significant alterations, emerging as a strong behavioral perturbations or as new physiological requirements, are considered.

*Structure-reflecting variables* are those output variables which respond to the alteration of the plant structure in some manner. While the structure-altering variables are the *cause* of the structure-altering phenomena, the structure-reflecting variables are related with its *effect*. In some cases these two types will be jointly referred to as *structural variables*.

*Structural space* is the space defined by the structure-altering variables and/or structure-reflecting variables. The structural space does not coincide with the state space. It includes only coordinates related to the structure of the plant and not variables describing its general state.

A *subspace of constant structure* is any region of the structural space where plant structure remains unaltered.

A *control strategy* is a set of control loops used simultaneously to control the plant in a particular subspace of constant structure.

usually with low time-constants (Sworder and Rogers, 1983; Young and Kwatny, 1981).
fp: Structural transfers are smooth, with broad dynamical range; they may be fast (e.g., temperature gene expression (Sugimoto and colleagues, 1986), moderate (metabolic product inhibition (Shimizu and colleagues, 1988)) or slow (plasmid loss (Siegel and Ryu, 1985)).

Number and characteristics of structures.
ip: Reduced number of structures; structures are known *a priori* or are easily separable.
fp: The number of structures is not well-defined; many structures are not known *a priori*, but with increasing knowledge may be discovered.

Dimension and characteristics of the structural vector.
is: Low, often one-dimensional; structure-altering variables are usually directly measurable.
fp: May be multi-dimensional; structure-altering variables are often unmeasurable and sometimes unknown.

Characteristics of the structure-altering phenomena.
is: They have clear physical, chemical or mechanicl meaning.
fp: Phenomena are badly understood, or of vague physiological origin; often they are entirely unknown.

## FERMENTATION PROCESSES AS VARIABLE STRUCTURE PLANTS

The consideration of the fermentation processes as variable structure plants can be best motivated by some examples. There is a large range of structure-altering phenomena in the living plant, e.g., changes of metabolite pathways (Furukawa, Heinzle and Dunn, 1983; San and Stephanopoulos, 1989), expression or repression of genes in recombinant cells caused by chemical factors (Kawai and colleagues, 1986) or by temperature (Sugimoto and colleagues, 1986), inhibition induced by accumulated medium substances (Shimizu and colleagues, 1988), diauxic growth (Barford, 1981), changes in membrane transport mechanisms, morphological variations (Barford, 1981), plasmid instability of recombinant cultures (Siegel and Ryu, 1985), etc. These phenomena result in alterations of the internal biochemical structure of the plant and of its behavioral characteristics, which in some cases are significant enough to require radically different strategies to control the process.

Physiological transfers may be induced artificially, as well, by proper variation of selected structure-altering variable. Recently, it has become a common practice, in laboratory and industrially, to provoke such phenomena purposefully at certain stages of cultivation, because the induced new structures leads to great improvement in process efficiency. Examples of this include glutamic acid fermentation (in which the structure-altering variable is penicillin concentration), phenylalanine fermentation (structure-alterning variable is tyrosine concentration) and penicillin fermentation (structure-altering variable is glucose concentration).

Structural variability in the control plant has already been observed and commented upon in the case of inanimate systems (Sworder and Rogers, 1983; Young and Kwatny, 1981). However, there are several important differences between the variablity of fermentation plants (fp) and inanimate plants (ip), which can be summarized as follows:

Dynamical characteristics of the structural transfers.
ip: Structural transfers are smooth or abrupt,

## CONSIDERATIONS OF THE CONTROL OF FERMENTATION PLANTS EXPRESSING THE VARIABLE STRUCTURE PHENOMENON

In their paper, "Every good regulator of a system must be a model of that system", Conant and Ashby (1970) have proven analytically, under very broad conditions, that there is a strict isomorphic mapping between plant and controller structures. There is one, and only one controller which can fit best a plant of interest. In the case of variable structure plants, each plant structure therefore requires a unique control structure, or, in the words of Conant and Ashby: "if the plant changes slowly in time, the theorem holds over any period throughout which the plant is essentially constant. As it changes, the isomorphic mapping will change appropriately, so that the best regulator in such a situation will still be a model of the plant,

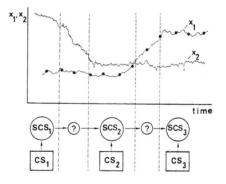

Fig. 1.   Representation of the continuous time-course of the fermentation plant structural variables $x_1$ and $x_2$ by a discrete sequence of subspaces of constant structure (SCS) with the corresponding control strategies (CS). "?" denotes periods of unknown structure.

but a time-varying model will be needed to regulate the time-varying plant". Division of time into "periods throughout which the plant is essentially constant" suggests the possibility of representing the continues course of the plant in the time domain by a discrete sequence of structures, which alternatively replace each other (Fig. 1). Since physiological changes are not induced instantly, there will be certain periods during which the structure of the plant is unclear. Such periods must occur also because of lack of complete knowledge of the plant structural space.

The general scheme of a switching multi-structure controller, able to respond to the internal altera-tions of the plant, is shown in Fig. 2. It is a two-level hierarchical system provided with a kind of structural adaptability which has strictly different functions at each of its levels. Detec-tion of the structural alterations is a task assigned to the second level and operates in the structural space of the plant. At the first level, control strategy relevant to the current plant structure is picked out from a defined pool, and the corresponding control action is calculated within the state space of the plant.

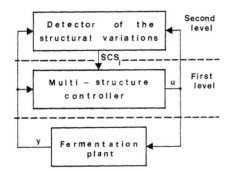

Fig. 2.   Scheme of the proposed multi-structure control system.

Application of the theorem of Conant and Ashby supplies one other reason whether or not to con-sider given fermentation process as a variable structure plant in the case when the existence of structural transformations cannot be demonstrated physiologically. Because of the symmetry of the isomorphic mapping, the theorem may be reversed. The resultant theorem, in a more relaxed form, states that if different control strategies have been shown to work successfully during different periods of time, then the plant must have differ-ing structures during these periods.

A fermentation process should, therefore, be considered and controlled as a variable structure plant in two cases. First, if there are sound physiological justifications for it, or, second, if practical experience in the control has shown that the process cannot be controlled well by a single strategy.

THE SECOND HIERARCHICAL LEVEL:
DETECTION OF STRUCTURE-ALTERING
PHENOMENA

The "periods throughout which the plant is essentially constant" in time correspond to sub-spaces of constant structure in the structural space of the plant (Fig. 3). Obviously, it is a typical recognition problem to add the current position of the plant in its structural space to one or another subspace of constant structure. Therefore, to solve this detection task, a pattern

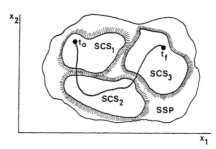

Fig. 3.   Representation of fermentation process in its structural space (SSP), including three subspaces of constant structure (SCS) with uncertain boundaries.

recognition technique was employed.

Typically, the synthesis of control system for fermentation processes is the final part of con-secutive research phases including microbiological, physiological, genetic, biochemical and behavioral studies of the microorganisms of interest. As a result, design of a control synthesis begins after much knowledge has accumulated about the plant, which may have high value in the informal creating of adequate control system. Therefore, we con-sider use of a priori accumulated expert knowledge indispensable and assume it in the synthesis of the pattern recognition procedure.

The basic knowledge required for the synthesis of the recognition task includes the set of struc-tural variables, the set of known structures, and the relations between these variables and structures.

Segmentation of Structural Variables

The recognition procedure requires appropriate segmentation of the scales of properly selected structural variables, translating them into useful qualitative terms which are easy to interpret. Instead of segmentation across time, proposed by Forbus (1987), we have applied fuzzy segmentation in the structural space of the plant. A fuzzy approach was selected to better suit the uncertain character of expert knowledge. It results in generation of fuzzy values $l_{kj}$, $kj = 1, ..., r_j$, across the scale of every structural variable $x_j$, $j = 1, ..., m$, defined quantitatively by their membership functions $\mu_{kj}(x_j)$

$$l_{kj} \equiv \{ \mu_{kj}(x_j) / x_j \} \tag{1}$$

Definition of the Set of Expert Production Rules

The expert production rules can be written in the following linguistic form:

If $x_1$ is $l_{11}$ and $x_2$ is $l_{12}$ and ... $x_m$ is $l_{1m}$, then the structural state belongs to SCS1 with the possibility $M_1 = M_1^\circ$.

If $x_1$ is $l_{21}$ and $x_2$ is $l_{22}$ and ... $x_m$ is $l_{2m}$, then the structural state belongs to SCS2 with the possibility $M_2 = M_2^\circ$.
$$\tag{2}$$

If $x_1$ is $l_{ni}$ and $x_2$ is $l_{n2}$ and ... $x_m$ is $l_{nm}$, then the structural state belongs to SCSn with the possibility $M_n = M_n^\circ$.

where SCS$_i$ is the i-th subspace of constant structure, $i = 1, ..., n$, $l_{ij}$ is a fuzzy value of the j-th structural variable used for the determi-

nation of $SCS_i$, $M_i$ is a real number in the range
$[0, 1]$, which represents the possibility for the
recognition of the current structural state as an
element of $SCS_i$, and the $M_i^o$ is the maximum value
of this possibility. In the general case $M_i^o$ could
be less than one.

The collected expert information should be formal-
ized in a suitable for computer implementation
form. To do this we apply the following norm,
modelling the fuzzy expert rules (2)

$$\sum_{j=1}^{m} w_j \cdot \mu_{ij}(x_j) = M_i, \quad \sum_{j=1}^{m} w_j = M_i^o, \qquad (3)$$

where $w_j$ are weight coefficients expressing the
different importances of the structural variables
$x_j$ in a particular rule.

Application of (3) to the system of fuzzy expert
rules (2) results in a set of equations

$$w_{11} \cdot \mu_{11}(x_1) + w_{12} \cdot \mu_{12}(x_2) + \ldots$$
$$+ w_{1m} \cdot \mu_{1m}(x_m) = M_1$$
$$w_{21} \cdot \mu_{21}(x_1) + w_{22} \cdot \mu_{22}(x_2) + \ldots$$
$$+ w_{2m} \cdot \mu_{2m}(x_m) = M_2 \qquad (4)$$
$$\ldots \ldots \ldots \ldots \ldots \ldots \ldots$$
$$w_{n1} \cdot \mu_{n1}(x_1) + w_{n2} \cdot \mu_{n2}(x_2) + \ldots$$
$$+ w_{nm} \cdot \mu_{nm}(x_m) = M_n,$$

known in pattern recognition theory as a nonlinear
system of decision functions, in which nonlinear
analytical functions are replaced by nonlinear
membership functions $\mu_{ij}(x_j)$. Obviously, (4) can
be easily expanded as new knowledge accumulates.
A specification of a new structural variable will
result in addition of one more column to the
system; definition of a new subspace of a constant
structure will result in addition of one more row.

### Supervised Learning

Learning is the final step of the synthesis of the
recognition procedure. It consists in determination of the unknown weight coefficients $w_{ij}$. Since
equations (4) are linear in respect to $w_{ij}$, they
can be estimated by linear regression, using an
additional data base containing realizations of
the structural vector and the corresponding expert
estimations of $M_i$. For application for on-line
use, the regression procedure was constructed in
proper recursive form:

$$P_i(q) = P_i(q-1) - P_i(q-1)\mu_i(q)[FF+$$
$$\mu_i^T(q)P_i(q-1)\mu_i(q)]^{-1}\mu_i^T(q)P_i(q-1)$$
$$w_i(q) = w_i(q-1) + P_i(q)\mu_i(q)[M_i(q) - \mu_i^T(q)w_i(q-1)]$$
$$w_i(q) = M_i^o \frac{w_i(q)}{\sum_{j=1}^{m} w_{ij}(q)} \qquad (5)$$

with initial conditions $w_i(0) = 0$ and $P_i(0) = C \circ I$,
where $\mu_i(q)^T = [\mu_{i1}(q), \mu_{i2}(q), \ldots, \mu_{im}(q)]$,
$w_i(q)^T = [w_{i1}(q), w_{i2}(q), \ldots, w_{im}(q)]$, $P(q)$ is the
renovation matrix, FF the forgetting factor, C a
large constant and q the iteration number. The
regression (5) permits accounting for the old
values of $w_{ij}$ and normalization of the result
towards $M_i^o$.

### Final Determination of the Current Subspace of Constant Structure

From the set of nonlinear decision functions (4)
with already determined weight coefficients, the
possibilities for existence of every structural
subspace are calculated, using the current value
of the structural vector. The hypothesis with
highest confidence (i.e., highest $M_i$) is extracted,
which is the final result of the recognition
procedure.

### THE FIRST HIERARCHICAL LEVEL: STRATEGY SWITCHING AND CONTROL IN SUBSPACES OF CONSTANT STRUCTURE

According to the theorem of Conant and Ashby
(1970), the subspaces of constant structure are
isomorphically mapped into a set of control
strategies. Thus, the result of the recognition
procedure can be used directly for activation of
a control strategy, corresponding to the current
plant structure.

The *global goal* of the control system is to keep
the plant in this subspace of constant structure,
which corresponds to the currently desired physiological properties of the plant. If structural
transfers to other subspaces are encountered, the
system should force the plant back to the desired
structure by appropriate switching of control
strategies. The *local goal* in every subspace of
constant structure is to realize qualitative
control within this subspace, i.e. to satisfy
given performance index by applying proper algorithms in the loops, implementing the current
control strategy. In this sense, the synthesis of
the first level of the control system is a two-
stage process (Table 1). First, physiologically

TABLE 1    Stages of Synthesis of the First
System Level

| Stage | | Goal | Basis for synthesis |
|---|---|---|---|
| First | Specification of physiologically motivated control strategies: input, outputs and control purposes. | General: control of the plant structure. | Expert knowledge |
| Second | Synthesis of control algorithms, imple- menting these strategies. | Local: control within SCS. | Formal approach |

motivated strategies, meant to realize the global
goal of the system, are to be constructed. This
task is beyond the competence of a formal control
approach and should be solved using expert
knowledge about the plant. At this stage, the
strategies are conceptually specified by their
inputs, outputs and control loops. Second, the
particular algorithms, implementing these strategies, should be synthesized. In every subspace
of constant structure the plant variations are
reduced to simple parametrical perturbations.
Therefore, conventional algorithms can be used,
based on local mathematical models, providing the
required structural morphism with the current
plant structure (Ziegler, 1970).

There is an important peculiarity of this type of
control, ensuing from the diverse characteristics
and physiological requirements of the plant in its

subspaces of constant structure. State variables essential for the plant in one structure may lose their importance in others, in which new variables may become significant. State variables may also physically appear and disappear associated with structural transfers. The same holds true for the control variables, which are subject to physical replacement according to the physiological necessities of the cell population. Therefore, the structural perturbations of the fermentation plant should be understood in a broad sense. They result not only in alteration of the loop structure of the control system, but also in physical re-configuration of state and control vectors, which are not fixed, but flexible and differing in their dimension and composition. In other words, each control strategy can work in its own environment composed by local state space and local control vector.

### SOFTWARE IMPLEMENTATION OF THE CONTROL CONCEPT

Based on the concept proposed above, a real-time software system has been constructed (Konstantinov, Kishimoto and Yoshida, 1988), intended to serve as a modern, intelligent and powerful scientific tool for development and investigation of sophisticated control strategies of fermentation processes. The system is independent of the fermentation equipment used as well as of the fermentation process. It is designed for use with the IBM personal computer series and works under the QNX (Quantum, Otawa) multitasking real-time operating system.

The flow-chart of the system is shown in Fig. 4, where both off-line and on-line modules are presented. Up to 16 subspaces of constant structure can be specified. The corresponding 16 user-defined control strategies may be switched automatically, using the described pattern recog-nition procedure, or manually, on user request. Every strategy represents arbitrary combinations of up to 16 user-defined algorithms. The struc-tural variables can be selected from among 16 directly measured variables and 16 indirectly calculated variables, all user-defined. The system facilitates not only on-line operations, but also procedures for off-line control synthesis. It can accept and handle the required expert

information in its natural fuzzy form and formalize it, using the method described above. The supervised learning procedure can work on-line, making possible improvement of system performance during cultivation. The system also supports a self-reasoning capability.

### CONCLUSION

With the proposed control concent we have tried to approach more closely the complex and variable nature of living plants, creating a realistic method for their control synthesis. This approach is intended to meet unsolved problems of control of fermentation processes subjected to significant structural perturbation. Compared with the conventional 'single-structure' approach, it offers important improvement: instead of globally modeling the fermentation plant, the most vulnerable point in the conventional control syntheiss, it decomposes the structural space of the the plant into several subspaces, in each of which plant characteristics are constant and control, either heuristic or model based,is simplified.

The discussed approach focuses our belief that a control system should be built using knowledge of the physiology and behavioral characteristics of the fermentation plant, rather than with formal mathematical methods alone. The inclusion of expert knowledge is essential for adequate opera-tion of control systems, providing them with a kind of intelligence. In this sense, with this work we have neared our goal - creation of intelligent system for control of fermentation processes, which at present, appears to be the most reasonable and perspicacious way to deal with these complex, variable and always puzzling plants.

Besides fermentation processes, the described control concept may be applied to other techno-logical plants expressing the variable structure phenomenon.

### REFERENCES

Ashby, W.R. (1956). *An Introduction to Cybernetics*, Chapman & Hall, London.
Barford, J.P. (1981). A Mathematical Model for the Aerobic Growth of *Saccharomyces cerevisiae* with a Saturated Respiratory Capacity. *Biotechnol. Bioeng.*, 23, 1735-1762.
Conant, R.C., and W.R. Ashby. (1970). Every good regulator of system must be a model of that system. *Int. J. Systems Sci.*, I, 89-97.
Forbus, K.D. (1987). Interpretting Observations of Physical Systems. *IEEE Trans. Syst., Man., Cybern.*, SMC-17, 349-359.
Furukawa, K., E. Heinzle, and I.J. Dunn. (1983). Influence of Oxygen on the Growth of *Saccharomyces cerevisiae* in Continuous Culture. *Biotechnol. Bioeng.*, 25, 2293-2317.
Harrison, D.E.F. (1976). Making protein from methane. *Chemtech.*, 6, 570-574.
Kawai, S., S. Mizutani, S. Iijima, and T. Kobayashi. (1986). On-off Regulation of the Triptophan Promoter in Fed-Batch Culture. *J. Ferment. Technol.*, 64, 503-510.
Kickert, W.J.M., J.W.M. Bertrand, and J. Praagman. (1978). Some Comments on Cybernetics and Control. *IEEE Trans. Syst., Man, Cybern.*, SMC-8, 805-809.
Konstantinov, K.B., M. Kishimoto, and T. Yoshida. (1988). Construction of a Flexible Software System for Development of Control Strategies of Fermentation Processes. In H. Okada (Ed.), *Annual Reports of ICBiotech*, ICBiotech, Osaka Univ., Osaka, p. 419.

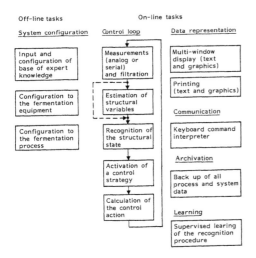

Fig. 4.   Flow-chart of the software system.

Konstantinov, K.B., and T. Yoshida. (1989).
    Physiological State Control of Fermentation
    Processes. *Biotechnol. Bioeng.*, 33, 1145–
    1156.
McLaren, R.W. (1967). A Stochastic Automation
    Model for a Class of Learning Controllers.
    In *Proc. 1967 Joint Automatic Control Conf.*,
    Univ. of Pennsylvania, Philadelphia, p. 267.
San, K.Y., and G. Stephanopoulos. (1989).
    Optimization of Fed-Batch Penicillin Fermenta-
    tion. *Biotechnol. Bioeng.*, 34, 72–78.
Saridis, G.N., and R. F. Hofstadter. (1974).
    Pattern Recognition Approach to the Classifi-
    cation of Nonlinear Systems. *IEEE Trans.
    Systems Man. Cybern.*, SMC-4, 362–371.
Shimizu, N., S. Fukuzono, K. Fujimori, N.
    Nishimura, and Y. Odawara. (1988). Fed-
    Batch Cultures of Recombinant *Escherichia
    coli* with Inhibitory Substance Concentration
    Monitoring. *J. Ferment. Technol.*, 66, 187–
    191.
Siegel, R., and D.D.Y. Ryu. (1985). Kinetic
    Study of Instability of Recombinant Plasmid
    pPLc23trpAl in *E. coli* Using Two-Stage
    Continuous Culture System. *Biotechnol.
    Bioeng.*, 27, 28–33.
Sugimoto, S., T. Seki, T. Yoshida, and H. Taguchi.
    (1986). Intentional Control of Gene Expres-
    sions by Temperature Using the Repression-
    Promoter System of Bacteriophage Lambda.
    *Chem. Eng. Commun.*, 45, 241–253.
Sworder, D.D., and R.O. Rogers. (1983). An LQ-
    Solution to a Control Problem Associated with
    a Solar Therminal Central Receiver. *IEEE
    Trans. Automat. Contr.*, AC-28, 971–978.
Tsetlin, M.L. (1973). *Automation Theory and
    Modeling of Biological Systems*. Academic
    Press, New York.
Varshavskii, V.I., and I.P. Vorontsova. (1963).
    On the Behavior of Stochastic Automata with a
    Variable Structure. *Automat. Remote Control*,
    24, 353–360.
Young, K.D., and H.G. Kwatny. (1981). Formula-
    tion and Dynamic Behavior of a Variable
    Structure Servomechanism. In *Proceedings of
    the 1981 Joint Automatic Control Conference*,
    Charlottesville, USA, 17–19 June.
Ziegler, B.P. (1972). The Base Model Concept.
    In R.R. Mohler and A. Ruberti (Ed.), *Theory
    and Applications of Variable Structure
    Systems*, Academic Press, New York and London.
    p. 69.

# REDUCTION OF BREAKDOWNS THROUGH PRODUCTIVE MAINTENANCE

## T. Miwa

*General Manager, Engineering Division, Ohkate Plant,*
*Daicel Chemical Industries Ltd., Ohtake-City, Hiroshima Prefecture, Japan*

Abstract. The activity in a chemical plant to reduce breakdowns
through total productive maintenance is explained. The causes of
breakdowns are divided into process failures and equipment failures.
Individual ameliorations are performed to process failures. For equipment failures, causes are thoroughly analyzed and recurrence was
prevented. Through these activities, breakdowns are reduced to less
than half, and total production cost was decreased.

Keywords. Productive Maintenance; Total Prodctive Maintenance;
Plant Operation; Breakdown; Failure.

### OUTLINE OF OHTAKE PLANT

Our Ohtake Plant, located next to the
western border of Hiroshima Prefecture,
has approximately ninety organic chemical
products. The principal raw materials are
naphtha, propylene, and acetaldehyde. The
plant is a member of the Iwakuni-Ohtake
petrochemical complex.

We have various technologies for synthe-
sizing organic materials: oxidation, reduction, esterification, amination, Diels-
Alder reaction, etc. Among these, technologies for producing peracetic acid and
synthetic glycerine were awarded prizes
from the Japan Invention Association,
Japan Chemical Association, and the
Society of Chemical Engineers, Japan.

Through these technologies, we are expanding our business from various high-

boiling-point solvents and general purpose
organic chemical products toward speciality chemicals such as materials for medical,
pharmaceutical, and cosmetic products. We
also have developed functional resins and
olygomers made from peracetic acid derivatives.

We have thirty-four plants as shown in
Fig. 1, with approximately 400 employees,
among which the generation around forty
years old is most dominant. The number
and types of equipment are listed in
Fig. 2. The length of piping sums up to
about 250 kilometers. Because many plants

| | |
|---|---|
| Continuous, Single Purpose | 11 |
| Continuous, Multi Purpose | 15 |
| Batch, Multi Purpose | 8 |
| TOTAL | 34 |

Fig. 1. Number of plants in Ohtake.

| | Number (under P.M.) | |
|---|---|---|
| Rotary Equipment (Pumps, Compressors) | 1,653 | (873) |
| Static Equipment (Towers, Tanks, Heat-exchangers) | 2,110 | (998) |
| Control Loops | 7,802 | (5,328) |
| Electrical Equipment | 2,343 | (1,221) |
| TOTAL | 13,908 | (8,420) (60.5%) |

Fig. 2. Number of equipment
and those under Preventive
Maintenance Program.

began operation in the 1960's, costs required for maintenance have been increasing in recent years.

## VARIOUS ACTIVITIES IN OHTAKE PLANT

Innovation of plant operation is the target of the Ohtake Plant, in order to be an "Excellent Factory" in the 21st century. As shown in Fig. 3, various activities are being performed to achieve this objective. Among these, stable operation of plants is a must.

Through analysis, about fifty-five percent of breakdowns are caused by equipment failure, and others by process failure. Whatever the cause is, things that hinder stable operation have to be eliminated. Above all, breakdowns must be avoided.

We have decided, as a main objective, to reduce "severe failures" by one half. By "severe failures," we mean the failures that cause entire or partial shutdowns of plants. We have set our target as outlined in Fig. 4.

## REDUCTION OF BREAKDOWNS CAUSED BY PROCESS FAILURES

The failures are classified by causes in Fig. 5. A large portion of process failures are caused by fouling, blockage, and polymerization. Many of these failures are caused by inappropriate design of equipment or unfavorable operating conditions.

|  | 1985 avr. | target |
|---|---|---|
| Number of severe failures | 106 -> | 50 times/half-yr |
| Loss of profit due to severe failures | 240 -> | 60 mil.¥/half-yr |

Fig. 4. Target to reduce severe failures.

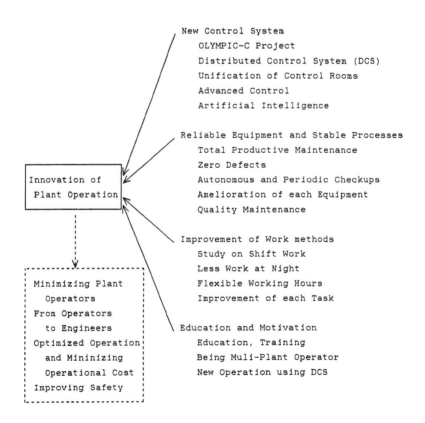

New Control System
    OLYMPIC-C Project
    Distributed Control System (DCS)
    Unification of Control Rooms
    Advanced Control
    Artificial Intelligence

Reliable Equipment and Stable Processes
    Total Productive Maintenance
    Zero Defects
    Autonomous and Periodic Checkups
    Amelioration of each Equipment
    Quality Maintenance

Innovation of Plant Operation

Improvement of Work methods
    Study on Shift Work
    Less Work at Night
    Flexible Working Hours
    Improvement of each Task

Minimizing Plant Operators
From Operators to Engineers
Optimized Operation and Mininizing Operational Cost
Improving Safety

Education and Motivation
    Education, Training
    Being Muli-Plant Operator
    New Operation using DCS

Fig. 3. Various activities for Innovation of Plant Operation.

About seventy percent of the problems of this kind were selected and listed to examine and ameloirate individually. So far, solutions have been found for more than eighty percent of the problems.

## ANALYSYS OF BREAKDOWNS CAUSED BY EQUIPMENT FAILURES

The causes of "severe failures" on equipment were analyzed using data from over the last two and a half years. The results are shown in TABLE 1. Insufficient maintenance of moving equipment, corrosion of static equipment and piping, and deterioration of electric equipment are major causes of failures.

The number of failures that occurred on the seven worst pieces of equipment were more than thirty percent of the number of overall failures. This result brought us to focus on the prevention of recurrence of failures, or that of the same kind.

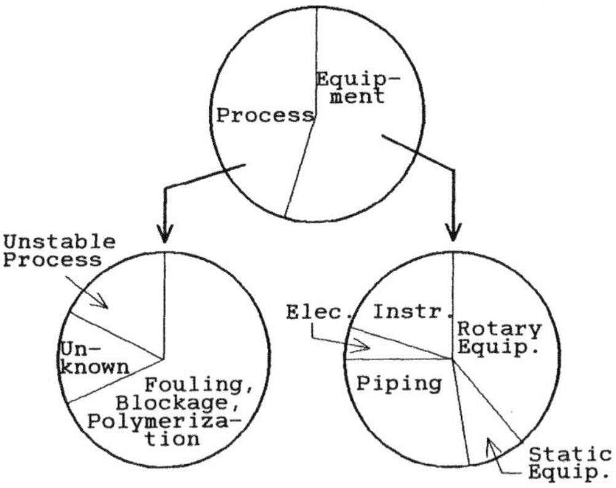

Fig.5. Analysis of Breakdowns

TABLE 1   Causes of Failures on Equipment

| Rotary Equipment | | Static Equipment | | Piping | | Elec. & Instr. | |
|---|---|---|---|---|---|---|---|
| Insufficient Inspection | 27% | Corrosion (Internal) | 65% | Corrosion (Internal) | 28% | Worn-out | 30% |
| Unmatched Seal | 13 | Corrosion (External) | 9 | Gasket | 21 | Leak, Plug | 16 |
| Corrosion | 10 | Loose Flange | 9 | Corrosion (External) | 11 | Condition changed | 10 |
| Out of spec. | 9 | Flange Distortion | 9 | Glass Lining | 11 | Environment | 8 |
| Insufficient daily checkup | 8 | Gasket Deterioration | 4 | Thermal Stress | 9 | Mal-Operation | 6 |
| Blockage | 5 | Glass Lining | 4 | Mal-Operation | 8 | Bad Design/Ill-manufacturing | 4 |
| Illmanufacturing | 5 | | | Errosion | 6 | Bad Maintenance | 4 |

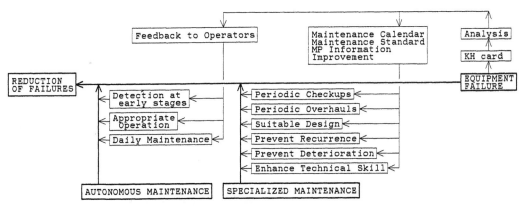

Fig. 6. Conceptual flow for reduction of failures.

SCHEDULED MAINTENANCE
ACTIVITIES FOR REDUCTION
OF EQUIPMENT FAILURES

Figure 6 illustrates the conceptual flow
of our equipment failure reduction program.
It may be said that our main goal concerning maintenance activities is to decrease
failures to zero.

The eight activities of scheduled maintenance are shown in TABLE 2. Among these,
the following points were especially
stressed: prevention of recurrence, autonomous maintenance, repair through
preventive maintenance and diagnosis of
failures.

## THOROUGH ANALYSIS OF EQUIPMENT FAILURES

In order to stop the failures, it is very
important to analyze the foregoing failures, to find out their causes and to
recognize methods to prvent and foresee
furure problems.

We have introduced the Problem-Oriented
Maintenance Record (POMR) system, for this
sake. An example of POMR sheet is shown in
Fig. 7.

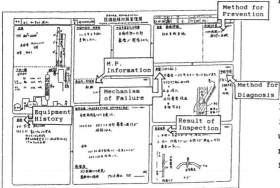

Fig. 7. An example of a Problem-Oriented
Maintenance Record

TABLE 2 Eight Activities for Scheduled Maintenance

| Activity | Objective | Key Points |
|---|---|---|
| 1. Reduction of Breakdowns by one half | Prevention of recurrence. Prevention of the same kind of failures. | Thorough analysis of failures. Promotion of Improving Maitenance. Individual Amelioration. |
| 2. Support for Autonomous Maintenance | Make Operators know about equipment. | Autonomous Maintenance Education. Publishing of "One Point Lessons." KH meeting. Distirbution of Maintenance record. |
| 3. Promotion of Preventive Maintenance | Prevention of failures. | Maintenance Calendar. Classification of Equipment and their Scheduled Maintenance. Keep Maintenance Record. |
| 4. Inspection Technology | Detection of failures in advance. Foreseeing Maintenance. | Vibration Monitoring. Non-Destructive Inspection. On Stream Inspection. Insulation Test. |
| 5. Control of Spare Parts | Cost Reduction. | Visible Control. Spare Parts Control System. |
| 6. Middle-term Maintenance Plan | Stop Deterioration. Keep Equipment healthy. | Make Maintenance Plan. Annual Re-scheduling. |
| 7. Effective Shut-down-Maintenance | Quality, Safety, and Punctuality of Maintenance work. Cut down Shut-down-Maintenance Costs. | Prepare a Shut-down-Maintenance Program at an early time. |
| 8. Specializd Maintenance Education | Training to be Specialists. | Education of DCS. Acquiring Licences. Share ones Know-how. |

## CLASSIFICATION OF EQUIPMENT AND PREVENTIVE MAINTENANCE PROGRAMS FOR EACH RANK

In view of effective promotion of the preventive maintenamce program, the equipment was classified into three ranks. The criteria are as follows:

(1) Importance of the plant
Whether a shutdown of the plant will affect other plants or not.
(2) Safety and Environment
Whether the failure will potentially cause explosion, fire, or leakage.
(3) Affect on production
The failure will end the breakdown of entire plant, reactor, or none.
(4) Accessibility to repair parts
Whether special repair parts are required or not.
(5) Skill required for repair
Whether service or direction of manufacturer is required or not.

Scores from A to C are marked for each criterion of each equipment. Overall importance of each equipment (also from A to C) is determined. If reserve equipment is redundantly installed, the importance rank becomes lower.

Autonomous and specialized maintenance programs for equipment are performed according to this importance ranking. The maintenance program for rotary equipment is illustrated in TABLE 3.

## ACHIEVEMENTS OF TOTAL PREVENTIVE MAINTENANCE ACTIVITIES

As shown in Figs. 8 and 9, the numbers of both severe and slight failures have been reduced to less than half of those in the year 1985.

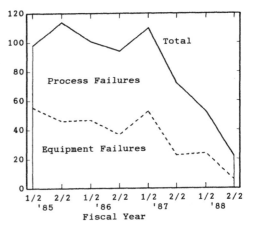

Fig. 8. Changes in the number of severe failures.

TABLE 3 Maintenance Program for Rotary Equipment

| | Rank A | | Rank B | | Rank C | |
|---|---|---|---|---|---|---|
| | by experts | autonomous | by experts | autonomous | by experts | autonomous |
| Daily Checkup | Every week, using checksheets and maintenance calendar. | Every day, using checksheets for each plant. | Every month, using checksheets and maintenance calendar. | Every week, using checksheets for each plant. | None. | Every month, using checksheets for each plant. |
| Periodic Inspection | Same as above. Cycle depends on equipment. | None. | Same as above. Cycle depends on equipment. | None. | Using checksheets. Cycle depends on equipment. | None. |
| Checking Method | Five senses, vibration monitor, tendency checked. | Five senses, sometimes vibration monitor. | Five senses, vibration monitor. | Five senses. | | Five senses. |
| Maintenance Method | Follows checking standard. | | Follows checking standard. | | Follows checking standard. | |

Figure 10 shows the number of breakdowns on rotaty equipment. As the number of equipment under preventive maintenance has increased, breakdowns have been reduced. The total number of equipment maintained under preventive maintenance once increased but is now going down.

The costs required for preventive and breakdown maintenance are shown in Fig. 11. Although the total maintenance cost has remained constant since 1985, the loss of profit by breakdowns has been reduced so that the total cost for production has decreased.

## CONCLUSIONS

Through Total Preventive Maintenance activities, we could achieve our objective of reducing breakdowns by half. Furthermore, the loss of profit caused by breakdowns reduced drastically so that the total cost of production has decreased quite a lot.

We could say that there is no special remedy for the reduction of failures. What has been changed so far is that operators have become more sensitive to abnormalities. Likewise, maintenance men have begun to look poorly upon recurrence of failures.

## ACKNOWLEDGEMENTS

The auther appreciates Daicel Chemical Industries, Ltd. for the permission of publishing this paper. He also expresses his gratitude to Japan Institute of Plant Maintenance for precious advice in performing the Total Preventive Maintenance program.

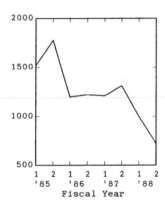

Fig. 9. Changes in the number of slight failures.

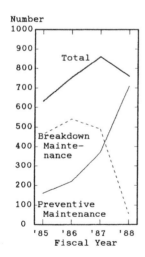

Fig. 10. Number of maintained rotary equipment.

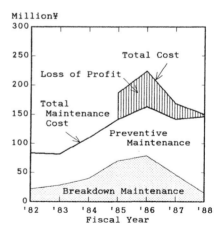

Fig. 11. Maitenance costs on rotary equipment.

# ESSENTIAL ELEMENTS OF MAINTENANCE
# IMPROVEMENT PROGRAMS

### L. H. Solomon

*Solomon Associates, Inc., Dallas, Texas, USA*

Abstract
As the petroleum refining industry restructured to cope with a collapse in
demand for products in the early 1980's most elements of operating expense
sharply declined. In contrast to these trends plant maintenance costs have
continued to increase. Today maintenance cost shows the widest variance of
any element of plant expense. But a select group of refineries in the US and
Europe have countered the industry trend. They have recorded maintenance
cost reductions each year and excel in all other aspects of plant performance
as well. Detailed consideration of each of these facilities reveals no common
physical dimensions or technological issues. The prominent common
denominators relate to organizational issues, management style and a
continuing commitment to performance improvement programs.

Key Words
Behavior Sciences; Human Factors; Management Systems; Oil Refining;
Maintenance Programs; Performance Improvement.

## INTRODUCTION

A review of the performance of the international
petroleum refining industry over the past decade
reveals many significant changes. First, the peak
operating levels of the late 1970's were followed
by a sharp decline in plant utilization resulting
from the collapse in petroleum product demand
which occurred in the early 1980's. That abrupt
collapse in demand has led to a major restructuring
of the petroleum refining industry throughout the
world. The most immediate impact was the
retirement of almost one-third of the capacity in
major refining centers. Even that effort failed to
restore satisfactory earning levels, however, and
refiners embarked on determined efforts to
restore profitability.

The profit improvement efforts initially
concentrated on the more efficient use of
resources in the operation of refineries. Early
programs concentrated on energy conservation.
Since energy represents fifty to sixty-five percent
of direct operating expense for most refiners this
was a logical beginning. And it is an effort that
has produced real dividends. Over the past nine
years refiners have reduced energy consumption
per unit of processing complexity by about twenty-
five percent.

The next major target was the level of personnel
involved in plant operation. Since 1980 the total
personnel requirements for refiners throughout the
world have declined by more than ten percent.
Surprisingly, this has not come about as a result of
increased automation, but rather from
organizational restructuring and the elimination of
low-valued tasks. When the efficiencies
associated with these changes is combined with the
sharp reduction in energy costs which occurred
over the period, the direct cost of petroleum
refining operations can be shown to have declined
by almost forty percent. Still, refining profit
margins are under pressure and refiners continue
to search for other opportunities for cost
reduction.

## TRENDS IN MAINTENANCE EXPENSE

The third largest category of refinery operating
expense is plant maintenance. In contrast to the
sharp improvements for all of the other elements
of refinery cost, the maintenance budget has
**increased** slightly over the past decade. On the
surface this might seem reasonable. After all
inflation is at work and the industry has aged
somewhat over the period. One would expect
maintenance costs to be higher for older
equipment. But maintenance is also the element of
plant expense with the greatest variation between
the best and poorest performers of the industry.
In the United States refining industry current
maintenance costs range from fifteen million to
sixty million dollars annually for facilities of
similar size and complexity. Such a wide variance
suggests that there must be opportunities for
improved performance within the industry.

### Concept of a Pacesetter
The rate of change which has occurred in the
industry further underscores this point. While the
industry average maintenance cost has increased
since 1980, a select group of refiners has

succeeded in countering the trend and registering *annual cost reductions of seven percent*. Did they sacrifice reliability or safety standards in the process? Clearly not. All of our data shows they excel in both areas. Do they have unusually skilled personnel who perform at higher productivity levels? We see no evidence of that either. We think that the key is that these *"PACESETTERS"* approach the maintenance function in an entirely different manner. They are consistently engaged in performance improvement efforts that lead them to innovations which escape the attention of most refiners.

To provide a valid basis for comparison between refineries of varying complexity and different local environments we have developed a specialized scale which combines the dimensions of unit capacity and processing complexity. This scale, EDC, basically expresses each refining facility in the size of simple topping refinery capacity which would require the same level of resources to support efficient operation.

### Pacesetter Performance
Before embarking on a description of the key elements of the *pacesetter* performance improvement efforts, perhaps it would be best to have a better understanding of the characteristics of their performance advantage. In the course of our industry studies we have identified sixteen petroleum refineries in North America and Europe which have earned our *pacesetter* designation. Each of these refineries has demonstrated excellence in every category of performance measurement over the entire period of study. Their performance levels in 1986 are summarized below.

|                                      | Industry Avg. | Avg. for Pacesetters |
|--------------------------------------|---------------|----------------------|
| Refinery EDC                         | 1800          | 1907                 |
| Average Age of Process Facilities,yrs. | 17          | 15.5                 |
| Energy Consumption, % of Goal        | 100           | 91                   |
| Personnel Level, men per 100K EDC    | 60            | 45                   |
| Maintenance Expense, $US per EDC     | 16            | 10                   |
| Mechanical Availability, %           | 93            | 94                   |
| Time Lost for On-the-Job Injuries, % | 0.37          | 0.42                 |

The *pacesetter* plants are only slightly larger than industry average facilities and they operate processing equipment which is typically a bit more recent in vintage. On average they recorded nine percent lower specific energy consumption and a dramatically lower personnel requirement in 1986. Their maintenance budget was almost forty percent lower than average for the industry at large, yet they enjoyed higher mechanical reliability and better safety records. These *pacesetters* have excelled each year since 1982, and they continue to increase their advantage over the rest of the industry. The real key is that they use less materials, expend fewer manhours and get better results. Our studies continually show that the *pacesetters* aren't smarter people who work harder. Rather, they evidence a different approach altogether to the function we call maintenance.

## PACESETTER CHARACTERISTICS

Our initial efforts to delve further into *pacesetter* characteristics was to examine the qualitative dimensions of their maintenance program. Our interaction with clients has led us to identify six features that are common to most of the *pacesetter* efforts.

### Structure
Most of the top maintenance performers are structured organizationally into geographic zones rather than oriented into craft or skill groups. The essential feature here is that the program goal is clearly defined as the promotion of plant reliability rather than to the recognition of separateness along craft lines. Individual skills are respected for what they contribute to the overall goal; however, artificial limits to craft efforts are limited. There are few rigid demarcations between skills and even fewer craft disputes.

This cooperative effort within the maintenance organization usually leads to fewer rigid constraints between operations and maintenance personnel, as well. The *pacesetters* don't require operators to perform skilled maintenance tasks. But neither do they allow arbitrary work rules to instigate barriers to cooperation throughout the entire work force.

### Orgnization
The *pacesetters* tend towards "flatter" organizations with only three or four levels of supervision separating the lowest ranking craftsman from the general manager. Less supervisory oversight means the craftsmen must be better informed of the expectations of their job. Skill levels are still established through formal apprenticeship or licensing programs. But job expectations are based upon interaction with maintenance team members and a more complete understanding of the project work plan, priority level and budget. The team concept often combines technical and craft skills to establish a dedicated group of workers that can function with a minimum of supervision.

### Priority Scheduling
The top performers have well defined priority systems. They insist on real justifications for maintenance work, and simply don't allow priority lists to become defeated by acting as if each project is an emergency or safety item. They recognize that the maintenance budget often exceeds the capital program in size and merits a comparable level of economic review.

### Information Systems
The *pacesetters* recognize the value of a maintenance information system and **use it** to plan their repair efforts. Their systems usually capture expected and actual repair costs, reasons for failure and repair frequency statistics. One important aspect of their system is the feedback of results to planners and estimators so that tools can be improved to insure that future planning is even better than the present effort.

## Failure Analysis

Failure analysis has become an established discipline for most of the *pacesetter* facilities. While many refineries offer failure analysis training to their supervisors and expect them to apply the technique, the top performers recognize that this vital effort requires a stronger emphasis. The failure analysis team examines most repetitive failure events and prepares **actionable** items for change in materials design or practices to prevent reoccurrence. Their goal is the elimination of failure and an eventual reduction in the overall maintenance effort.

## Condition-Based Repair

The top performers have graduated from time based maintenance programs and now rely extensively on examination of the current condition of operating equipment to establish maintenance requirements. This leads them to be far more selective in defining their repair needs. The end result is that their major unit turnarounds have not become institutionalized. They typically schedule the same interval between turnarounds, but expend 30% to 50% fewer manhours of repair effort and require one week shorter downtimes. As a result their turnarounds cost about one-third less than the industry average.

Does this shorter turnaround cause more unscheduled downtimes? As a matter of fact it does. But the resulting shutdowns are of much shorter duration than the average for the industry; as a result the *pacesetters* register uniformly higher mechanical availability for all process equipment. Their intentional planning of maintenance work also seems to produce a safer work environment and probably results in lower costs related to production outages.

In short, the *pacesetters* have leaner organizations, lower costs, safer work environments and report better service factors. Not only that, they also prosper with lower capital investment programs and enjoy better employee relations.

## RATE OF PERFORMANCE CHANGE

During the course of our industry studies a number of refineries have registered dramatic performance changes. Ten percent of the facilities in the United States have improved their rankings by a full quartile. Almost twenty percent of the industry has registered a full quartile decline over the same period. The net result is that the industry average performance has deteriorated slightly over the period. Over this same timeframe the *pacesetters* have managed to sustain their excellence with continued emphasis on performance improvement. This is no simple task. This group of facilities already excelled in 1982 with better reliability and costs that were only half of the industry average. But their commitment to continued improvement has allowed them to extend their advantage over their peers. With the significant advantages that accrue to the top performers it is instructive to consider the essential elements that comprise their performance improvement efforts.

## ELEMENTS OF SUCCESSFUL PERFORMANCE IMPROVEMENT PROGRAMS

There is no single recipe for dramatic improvement in a plant maintenance program. Indeed, one of the issues that frequently undermines improvement is the sense of urgency that leads to belief in a "quick fix" to a situation that is the end result of local practice and traditions accumulated over several years. But our involvement with clients has led us to identify six elements that are present in most of the improvement efforts embraced by the top performers.

## Enlightened Management

Plant management must recognize a need for change. This may result from dissatisfaction with the status-quo or be a response to corporate pressure. But it must result in a heart-felt commitment to change and a willingness to depart from past practices which have not allowed the refinery to keep pace.

Management must embrace leadership as its operating mode - not authoritarian direction. The task of performance improvement is overwhelming if assigned to a few, but can flourish if it becomes part of the work ethic of the entire organization. Management needs to objectively evaluate the resources at its command and find a comfortable way to share decision-making and goal-setting within the refinery.

## Adoption of a Plant-Wide Mission Statement

Management needs to articulate the mission for the organization - one that can be understood and embraced by all. Then the actions of management - at all levels - must be seen to be consistent with the achievement of the common goals. This calls for careful consideration of the lower supervisory levels. Often they have developed their management skills on the job after a successful performance in a craft or technical position. But they are the key to management credibility. They must have an opportunity to question, study and contribute to the mission statement for them to become committed to its success. The use of practices which are inconsistent with the mission of the plant should be quickly and firmly discouraged.

To enlist the support of the entire organization the plant mission statement should be translated into goals and objectives that are meaningful for each level of the organization. It's also helpful for the employees to receive frequent feedback on the progress towards goals for their shift, department, functional area and the refinery at large.

## Improved Competitive Awareness

It's surprising how often our clients are content with their performance without knowing how it relates to their peers. We have always found it constructive to communicate the strengths and weaknesses of overall performance throughout the organization. The effort is not only a good communications exercise, it also provides a solid basis to support the need for improvement.

## Inventory State-of-the-Art Techniques

While we find little evidence that individual productivity is a major factor which differentiates industry performance it is vital that refiners remain aware of the leading edge techniques which can lead to excellence. The keys to superior effort often involve innovation and reasoned risk-taking. Any risks associated with the implementation of new procedures can be moderated by keeping up with the latest in operating and maintenance techniques.

## Emphasis on Improved Operational Reliability

Many refiners strive to repair and return to service at the lowest cost. They view maintenance as an independent function which strives to fix equipment upon failure. But the *pacesetters* have re-defined their objective. They seek to achieve maximum operational reliability. That calls for a sharing of responsibility with operational personnel. The task of monitoring equipment condition falls to those who can meet the need regardless of functional assignment. Failure analysis addresses both maintenance and operational techniques. The primary goal becomes the elimination of failure, rather than containment of expense. The end result invariably is a more significant reduction in both expense and repair effort than could be otherwise accomplished.

The top performers are not hindered in this effort by labor productivity or bargaining agreements. We observe that their materials expenditure per manhour of maintenance effort is actually less than industry average. This attests to their increased involvement in proactive service programs to anticipate and avoid failure rather than to simply react to mechanical malfunction.

## Openness to Change

The top performers already excel, yet the investigation of new techniques has become a way of life. They do not allow maintenance or operating procedures to become institutionalized. Their thinking seems to be oriented towards emerging technology rather than the time-tested procedures. They consider and evaluate innovative and untested ways of approaching their objectives to maintain their position on the leading edge of progress. They don't evidence dissatisfaction with working systems or undue risk-taking, but they challenge the resources of all of their employees to explore the kind of improvements that will assure excellence in tomorrow's environment as well.

## CONCLUSION

The distinguishing characteristics which identify the *pacesetter* maintenance programs deal more with management style and employee attitude than with materials employed or craft skills. But the overriding consideration seems to be that the petroleum refining environment is very dynamic throughout the world. Those who can capitalize on the potential for improvement *usually already present within their organization* can approach *pacesetter performance levels*. The effort calls for substantive change in management techniques and widespread involvement of employees in the pursuit of common goals. Introduction of such an approach into many plants is not an easy task or one which can be rapidly accomplished. However, the potential benefits are exceptional. The opportunity for expense savings may be the source of initial interest, but our review of petroleum refining industry performance throughout the world suggests that the associated benefits of increased reliability, improved safety record, reduced energy consumption and improved product yields are certain to accrue as well.

# MAINTENANCE PRACTICE IN JAPANESE
# PROCESS INDUSTRIES

## E. O'Shima

*Research Laboratory of Resources Utilization, Tokyo Institute of Technology,*
*4259 Nagatsuta, Midori-ku, Yokohama 227, Japan*

Abstract. In accordance with the technical progress in mechanization or system integration, maintenance has been forced to change. When the production system is integrated, a trifle failure in a constituent element might cause a huge loss of production or even a disaster of the plant. The preventive or predictive mode of maintenance is, therefore, strongly needed, particularly in the plant of greater equipment complexity, in order to avoid such unacceptable losses. In the present paper, a new trend of maintenance activity is overviewed, where the importance of the strategic management of maintenance work is pointed out and the total productive maintenance activity is introduced as an example.

Keywords. Guidance systems; maintenance system; production control; quality control; statistics.

### INTRODUCTION

Not much effort has been directed to systematising the maintenance technologies and most of the maintenance activities are based on the experiences in the past. However, if it is attempted to improve either economical or technical performances of maintenance activities, structuring the framework of the total maintenance activities is inevitable. An attempt is made here to characterize different types of deteriorations as well as the effects of the failures, which could be the measures for selecting the maintenance policy. An introduction is made about the management activity for improving the quality of maintenance carried out in Japan. Some statistical information about the present state of application of the condition monitoring techniques in other industries is presented.

### EQUIPMENT FAILURES AND DETECTION

The Committee of Construction Material of Chemical Equipment organized in the Society of Chemical Engineers, Japan has made a survey report on failures of construction materials in chemical plants under the sponsorship of Science and Technology Agency[1]. In the first part of the report, a statistical analysis was made on the data collected from about 300 papers on material failures published during the term from 1978 to 1982.

Fig. 1 shows the cross relationships between the categories of failure mode and construction material, where it is seen that the stress corrosion cracking of stainless steel is found with a remarkably frequency.

Fig. 2 shows how or by what means the failures were detected, where the two different situations, i.e. during shutdown and during operation, are classified. It is noteworthy that about one third of the failures happened during operation and there a difficulty, as far as material failures are concerned, in detecting them without disturbing the operational conditions.

Fig. 3 indicates the correlation of the number of failures classified with respect to failure modes with the in-service time. In the second part of the survey report, the results of the questionnaire answered by 110 chemical companies on 615 failures are described. The equipment units where the failures occured are shown in Fig. 4, where the number represents the number of failures of each equipment. Though it is not easy to identify the real cause of the failure, Fig. 5 represents the causes of the failures they claimed.

The condition monitoring techniques are recently attracting the industries' attention because of the capability of quantitative monitoring of the failure trend and the possibility of its application to predicting the equipment life. Some examples but not all of the condition monitoring techniques are shown in Table 1.

The Investigation Committee on Maintenance Technologies organized by Japan Institute of Plant Maintenance made a survey on the present state of application of the condition monitoring techniques by the method of sending questionnaire[2]. Table 2 shows how the various methods of condition monitoring are practically applied in different industrial sectors. According to the analysis, the failure modes which are subject to the condition monitoring techniques are distributed as shown in Fig. 6.

### APPROACHES TO MAINTENANCE PROBLEMS

Oftentimes maintenance is simply defined as the activity to recover the deterioration of a facility to maintain the required functions of the facility. There is, however, a wide range of variety in the maintenance policy depending on the characteristics of deterioration mechanism or the importance of the equipment in the total system.

The patterns of deterioration of an equipment can be generally categorized as shown by the following scheme:

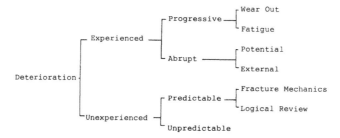

As to the experienced deteriorations, there seems
to be, as far as objectively observing, two different types of revelation of deterioration, or
progressive mode and abrupt mode.

In the progressive mode, the deterioration which
has appeared prior to degradation of the function
of the equipment is observable quantitatively in
the course of time. The deterioration of progressive type is further classified into two
types; wear out and fatigue. Wearing is here
defined as the type where the progressive deterioration can be predicted where to occur such as
degradation of bearing or gasket. On the other
hand, fatigue is not predictable where to occur,
like a pinhole or crack, but once having discovered, the progress of the deterioration can be
observed.

It is generally true that the progressive type of
deterioration is the easiest to carry out maintenance, since the life expectancy of the equipment
or the time of failure are predictable by
extrapolating the process of the progressing
deterioration. The condition monitoring techniques are the typical example of the maintenance
strategies which can be effectively applied to
the progressive type of deterioration.

Similarly, the deterioration of abrupt type is
divided into two cases. The type meant by potential is the case where the cause of the failure
has already been built-in within the equipment
and is triggered by some chance. The deterioration of the external type is such that is
attributed to some external condition exerting on
the equipment.

The abrupt type of deterioration is characterized
by occurrence without a sign. However if it
occurs periodically, the approach of probabilistic prediction is expected to be effective. On
the other hand, for such type of deterioration
that the frequency function of occurrence is
broad, there seems no effective method to avoid
the failure to occur.

The classification mentioned above is merely
based on the observed patterns and may have
nothing to do with the real mechanism of the
deterioration. In the other words, if the
observation is made at any deterioration
sufficiently precisely, all the process of deteriorations might be observed as the progressive
type. It is this fact that gives the basis for
the argument that the system of maintenance
technology should be based on the fracture
physics rather than the statistical approach.

Another factor which should be taken into account
in making decision of the maintenance policy is
the effect of deterioration. The seriousness of
the effect can be graded, for example, into five
levels of the extent;

(1) to give the fatal damage to the system,
(2) to give an influence on the system
    performance,
(3) to bring a degradation in the function of an
    equipment,
(4) to give a fear of causing a trouble in the
    future, if it is left unrepaired, and
(5) to give no positive fear of causing a
    trouble, but preferably to be repaired.

The purpose of evaluating the effect of the
deterioration is to rationalize the allocation of
the maintenance efforts by taking into consideration the sensitivity of the deterioration to the
system performance. In other industries, the
economical measure is often applied for evaluating the effect of the deterioration, such as in
terms of losses of damages, shutdown etc., which
cannot be applicable to the case of nuclear power
plant due to the extremely strict requirement of
safety.

The next problem is how to determine the maintenance policy based on the characterization of the
failure mechanism and the evaluation of the effect of the deterioration. It is generally
understood that there are two different principles for maintenance policy, i.e. time-based
maintenance and condition-based maintenance.

        Time-based Maintenance
            Periodical Inspection
            Periodical Repair
            Overhaul
            Preventive Parts Replacement

        Condition-based Maintenance
            Corrective Maintenance
            Failure Trend Control
            Condition Monitoring
            On-Condition Maintenance

The policy of time-based maintenance is, as seen
from the examples shown, such system where every
procedure of the maintenance activities is
scheduled in accordance with the calendar time.
This principle is basically effective for such
type of deterioration that the time constant of
progress is far long compared with the interval
of the maintenance. From the viewpoint of the
effect of the failure, the types of deterioration
categorized above as (4) and (5) are considered
to be subject to the time-based maintenance.

The principle of time basis is of little effect
on the random type deterioration which takes
place with a random interval having nothing to do
with the next occurrence. In practice, however,
it is expected in the process of time based
inspection that such type of failures that cannot
be predicted to happen at the time of designing
could be found out preferably at a preliminary
stage.

As to overhaul, a typical time-based maintenance, where every constituent equipment is dissembled regardless of the condition of the equipment and retrieved or replaced if necessary, it is not necessarily true that the more frequent the better, not only from economical reasons but even from technical reasons. It is often pointed out from the experience of maintenance of the chemical plant or aircraft that the policy of overhaul is not always the best method of maintenance because the overall failure rate as well as the failure rate of each equipment increases just due to the maintenance activity itself.

It is noteworthy that an attempt was made in the field of airplanes to construct a new concept for guiding maintenance strategies, that is Reliability Centered Maintenance. MSG-3, which is the maintenance guidelines specifically compiled for B767, is a typical example of application of RCM. RCM has been, in the meantime, gradually accepted by the other sectors of industry.

RCM recognizes that more often than not, maintenance requirements have to be determined in the absence of meaningful historical information. So, RCM enables meaningful discussion to be made now, rather than wait for information which is often too little, too late, and sometimes of no use at all by the time it becomes available.

It is claimed that RCM enables to determined the maintenance requirements of each item of equipment in its operating context with an unprecedented dgree of confidence and precision. It also leads to substantial reductions in maintenance workloads. RCM works at three levels:
-it forces a structured evaluation of the context of equipment in a way which integrates decisions about safety, operating economics and maintenance costs
-it recognises that all of the six possible approaches to maintenance (condition-based maintenance, routine overhauls, scheduled discard, failure finding tasks, so-called "breakdown maintenance" and redesign) have a part to play in modern maintenance systems, and it provides a robust set of decision-support tools for deciding which is most appropriate in any situation.
-it combines both activities into a single decision-making process of quite extraordinary power.

Although some attempts are being made to implement RCM to the practical problems, no systematic approach, at least in the computerized form, has been established. As generally seen, the maintenance activities are based, to a large extent, on the experience which are quite specific to the plant. This situation tends to make the maintenance people reluctant to accept the generallized guidelines.

### DEREGULATING ATTEMPT FOR HIGH PRESSURE GAS PLANTS

According to the law on safety of high pressure gas production plant, it is obligated to assure the safety condition by carrying out a regular inspection of the plant, which has been customary understood as to carry out the annual shutdown inspection of the whole plant. Ministry of International Trade and Industry, which is in charge of that particular law, has changed its policy to deregulate the annual shutdown maintenance for the plants which are qualified by their committee investigation. This attempt would be translated as the qualification for autonomous safety management.

The rationales for deregulation are as follows;
(1) The annual shutdown maintenance is not necessarily regarded as the best way, particularly from the viewpoint of cost, to assure the satisfactory safety condition of the plant.
(2) There are many examples in foreign countries to show the evidence that exactly the same type of plant can be satisfactorily safely maintained without carrying out shutdown maintenance so frequent as once every year.
(3) From the technical point of view, it is quite confident as to the qualified plant that the same or even higher level of safety can be achieved by introducing newly developed inspection techniques instead of carrying out the annual shutdown inspection.

In the process of qualification, the investigation is carried out with respect to application of the on-stream inspection techniques, extent of accomplishment of the plant data utilization, and establishment of the scheduling of maintenance strategies, as well as attitude philosophy toward the safety management of the company.

Now that the system has been operated for about two years, almost all the refinery plants and some of the petrochemical plants have been qualified to carry out the autonomous safety management, and they are achieving a remarkably good record in managing the plants.

### PRODUCTIVE MAINTENANCE MOVEMENT

At the early stage of production activities, maintenance was not completely isolated from operation and maintenance was considered to have a simple role of fixing the failure. It, generally speaking, was after the World War II that the Japanese industry rationalized the organizational structure, along the suggestion of the United States, to have maintenance division separately from the operation division to be specialized in the maintenance activities. This concept is based on the observation that the specialist is more reliable in making a judgement and performing the work due to his abundant specialized knowledge and experience. The efficiency of the total organization was expected to increase by forming the structure comprising of subdivided specialized task forces.

After 30 years since then, there is now an argument as an evil effect of the rationalization about the inefficiency due to the distribution of responsibilities and the depressive atmosphere for productivity. The problems lying at the border between maintenance and production divisions tend to be left unsolved and sometimes cause some big troubles. The concept of productive maintenance was proposed with the purpose to make the maintenance activities contributory to improving the productivity by overcoming the said dilemma.

The productive maintenance is proposed to be realized as the form of companywise movement where any of the maintenance activities is understood to contribute for supporting the productivity rather than to be uneconomical or unproductive. In other words, the main concern of the movement is how to prove the saying that maintenance is good buisiness. The productive maintenance movement is promoted under the initiative of the cell teams of operators of the plant, similarly to the Total Quality Control movement.

The slogans of the productive maintenance movement can be summarized as the following five items:

(1) Increasing of Productivity;   Unplanned stop or trouble of operation is one of the biggest factors to reduce the productivity of the plant particularly of continuous operation. Many examples in fabricating plants, assembly plants and even in chemical plants have shown that the productive maintenance movement is quite effective in reducing the unplanned stop of operation.

(2) Upgrading of Product Quality;   With increase of extent of automation, the causes of off graded or deviated quality of product are more attributed to failures of the facility or equipment rather than the variation of skills of the human operators. The defect of equipment to the cause of degrading the product quality can best be found by the daily inspection conducted by the operators.

(3) Evaluation of Cost;   Cost control is essential concern of the productive maintenance movement.   The issue is how much money would be the most appropriate amount to be spent for maintenance cost, and how the cost effectiveness of maintenance can be increased. Life Cycle Costing is regarded one of the interesting approaches to the cost evaluation.

(4) Voluntary Maintenance;   In contrast with the rationalization by structuring the organization of specialized divisions, the productive maintenance movement expects the operators to take care of daily maintenance work like oiling, greasing or daily inspections and sometimes even easy repairs, this is considered to be very effective for preventing the failure to grow at a very early stage.

(5) Capability of Improvement;   It happens quite often that the improper design is discovered or recognized during the practical operation. The modification proposal system, where any kind of proposal for modifying the equipment made by any personnel is taken up for investigation and the number of proposals is filed individually to be taken into consideration in the process of evaluation, is an effective way of encouraging operators to perform maintenance jobs and motivating them to get interested in the plant they are dealing with.

The key for success of the productive maintenance movement is the cell team activity of the operators for performing voluntary maintenance, which is promoted stepwise taking a fairly long period of time.

(1) Cleaning Plant;   A thoroughgoing cleaning of the plant is required.   Cleaning oil dirt, repainting, neatly arrangement of tools, etc.

(2) Modification of Equipment;   When the clean plant is realized, the sources of dirt, dust, debris or oil would become explicit.   The modification of the equipment for eliminating the sources would realize, at the same time, better performance of the equipment.

(3) Standardizing Procedures;   After settling the rearrangement of cleaning and maintenance procedures, the operators are required to compile the standard of the voluntary maintenance procedures by themselves, because it is they that should follow the standard.

(4) Maintenance Activity;   The effect of the cell team activity is usually quite remarkable when it is started.   In order to keep the activity ongoing, an overall investigation and evaluation of the activity should be made once in a while.

## CONCLUDING REMARKS

The maintenance policies are manifold in different sectors of industry and should, of course, depend on the characteristics of the equipment to be maintained.   At the same time, a fairly large part of the differences seem to be attributable to the lack of communication or technology transfer among the sectors.

The bullet train, Shinkansen express, is highly evaluated for the fact that no fatal accident has ever happened since the start of its commercial operation.   Partly bacause the scale of the system is too large or because too much diversified techniques are involved, the organizational structure for the maintenance activities is clearly subdivided having little interaction among them.   The good record of safe operation and high performance is, however, explained as the result of the policy of early replacement of system components, which is in contrast with the policy for the aircraft maintenance.

It is noteworthy to point out that the new developments encompass changes in the way we maintain equipment, we organize and manage maintenance labor, and we manage spares and materials.   These changes encompass a whole new generation of thought, which was originated in the field of maintenance of commercial aircrafts with the name of Reliability Centered Maintenance being becoming known as the Third Generation of Maintenance Management.

In Reliability Centered Maintenance, it is attempted to integrate classical administrative system with modern thinking on downtime, safety, labor and the management of plant modifications. It incorporates the results of extensive research into the failure patterns of complex equipment, which have changed the basis of preventive maintenance.   Finally, it includes new methods for assessing criticality and for deciding the condition monitoring technique.

There is no doubt about the fact that the maintenance system for aircraft is the most advanced technically and systematically.   The maintenance procedures of every parts of the aircraft are precisely specified by the plane manufacturer with quantitative instructions on such as inspection period, which is continuingly updated based on the field data collected from the users of all over the world.

## REFERENCES

Committee of Construction Material of Chemical Equipment (1984).  Failure analysis and life evaluation of chemical plant materials. Kagaku Kogaku, 48, 158.
Nakajima, Seiichi (1988).  Introduction to TPM., Productivity Press.
Survey Report on Maintenance Technologies of Production Plants (1985).  Japan Institute of Plant Maintenance.

| | Iron Steel | Low Alloy Steel | Stainless Steel | Nickel Alloy | Aluminium Alloy | HK 40 | Titanium | Refractory | Others |
|---|---|---|---|---|---|---|---|---|---|
| Abrasion, Melt | | | 1 | 1 | | | | 20 | |
| Mechanical Fracture | 3 | 3 | 9 | | | 3 | 1 | | |
| Other Errosion | 3 | | 1 | 1 | | | | | |
| Carbonization, Nitrogenation | 2 | | 1 | 1 | | 5 | | | |
| Oxidation, Molten Salt Errosion | 1 | 1 | 6 | 2 | | | | | |
| Creep Cracking | 1 | | 2 | 1 | | 4 | | | |
| Erosion Corrosion | 11 | | 6 | 6 | 3 | | 1 | | 1 |
| Hydrogen Attack | 6 | 11 | | | | | | | |
| Hydrogen Embrittlement, Blister | 11 | 3 | | | | | 1 | | 1 |
| Stress Corosion Cracking | 11 | 2 | 62 | 1 | 5 | 7 | | | |
| Localized Errosion | 26 | 1 | 19 | | 4 | | 6 | | |
| Genral Corrosion | 17 | 1 | 6 | 1 | 1 | | | | |

Fig. 1  Cross relationships between failure mode and construction material

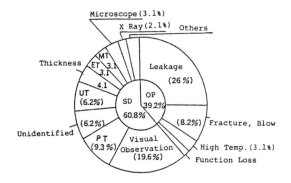

Fig. 2  Means of failure detection

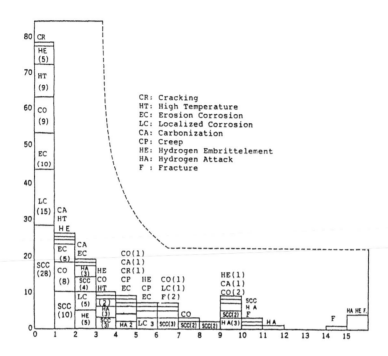

CR: Cracking
HT: High Temperature
EC: Erosion Corrosion
LC: Localized Corrosion
CA: Carbonization
CP: Creep
HE: Hydrogen Embrittelement
HA: Hydrogen Attack
F : Fracture

Fig. 3  Failure rates versus In-service time

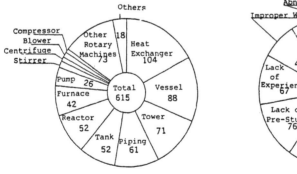

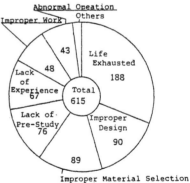

Fig. 4  Number of failures of equipment units        Fig. 5  Causes of failures

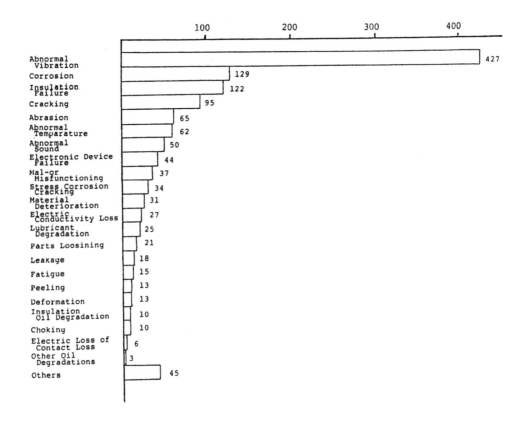

Fig. 6  Failure modes subject to condition monitoring maintenance
applied in industries

TABLE 1  Examples of Condition Monitoring Techniques

| Principles | Detecting Methods | Detecting Items |
|---|---|---|
| Radiation | Radiation Penetration<br>Fluorescence X ray Analysis<br>X ray Defraction<br>radiotherapy | Tchickness<br>Thickness, Element<br>Crystal Structure<br>Residual Stress<br>Deffect |
| Sound,<br>Vibration | Ultrasonic<br>Critical Angle, Decrement<br>Horography<br>Acoustic Emission<br>Sound<br><br>Vibration | Thickness, Deffect,<br>Crystal Structure<br>Deffect<br>Cracking<br>Deterioration,<br>Abrasion<br>Deterioration,<br>Abrasion |
| Magnetic,<br>Electric | Magnetic Particle<br><br>Magnetic Characteristics<br>Eddy Current<br>Magnetic Flux Leakage<br>Electric Resistance | Surface Deffect<br>Material<br>Surface Deffect<br>Surface Deffect<br>Corrosion, Cracking<br>Strain, Residual<br>Stress |
| Penetration | Color Penetration<br>Gas Penetration | Surface Deffect<br>Cracking, Pinhole |
| Chemical<br>Analysis | Gas Analysis<br>Liquid Analysis, Corrosion<br>Monitoring<br>Solid Analysis,<br>Oil Analysis | Cracking, Pinhole<br>Corrosion, Deffect<br><br>Cracking,<br>Deterioration |
| Light | Visual Observation<br><br>Optical<br><br>Optical Elasticity<br><br>Moire Effect | Surface Deffect,<br>Shape<br>Surface Deffect,<br>Shape<br>Internal Stress,<br>Strain<br>Surface Shape,<br>Deffect |
| Heat<br><br><br><br>Pressure | Temperature<br><br>Liquid Crystal<br><br>Thermography<br><br>Pressure | Structure Deffect,<br>Work Failure<br>Structure Deffect,<br>Work Failure<br>Furnace Inspection<br><br>Deffect,<br>Deterioration |

TABLE 2 Application of Condition Monitoring Techniques in Industries

| INDUSTRIES | Vibration | Elec. Insulation | NDI | Other Electric | Temperature | Sound | Pressure | Oil Analysis | Optical | Others | No Answer | Total |
|---|---|---|---|---|---|---|---|---|---|---|---|---|
| Foods | 12 / 29.3 | 6 / 14.6 | 5 / 12.2 | 5 / 12.2 | 2 / 4.9 | 2 / 4.9 | 0 / 0.0 | 0 / 0.0 | 0 / 4.9 | 0 / 0.0 | 7 / 17.7 | 41 / 100.0 |
| Pulp and Paper | 38 / 70.4 | 4 / 7.4 | 6 / 11.1 | 0 / 0.0 | 3 / 5.6 | 0 / 0.0 | 1 / 1.9 | 1 / 1.9 | 0 / 0.0 | 1 / 1.9 | 0 / 0.0 | 54 / 100.0 |
| Chemical, General Pharmaseutical | 109 / 33.7 | 21 / 6.5 | 88 / 27.2 | 18 / 5.6 | 17 / 5.3 | 9 / 2.8 | 1 / 0.3 | 6 / 1.9 | 20 / 6.2 | 16 / 5.0 | 18 / 5.6 | 323 / 100.0 |
| Chemical Fats, Paints, Ink | 14 / 43.8 | 2 / 6.8 | 6 / 18.8 | 1 / 3.1 | 0 / 0.0 | 0 / 0.0 | 2 / 6.3 | 0 / 0.0 | 4 / 12.5 | 0 / 0.0 | 3 / 9.4 | 32 / 100.0 |
| Fiber, Plastics | 34 / 43.0 | 4 / 5.1 | 18 / 22.8 | 6 / 7.6 | 0 / 0.0 | 3 / 3.8 | 0 / 0.0 | 0 / 0.0 | 9 / 11.4 | 5 / 6.3 | 0 / 0.0 | 79 / 100.0 |
| Petroleum, Coal | 19 / 21.1 | 7 / 7.8 | 23 / 25.6 | 10 / 11.1 | 5 / 5.6 | 6 / 6.7 | 5 / 5.6 | 1 / 1.1 | 4 / 4.4 | 1 / 1.1 | 9 / 10.0 | 90 / 100.0 |
| Glass, Rubber, Ceramics | 31 / 50.8 | 5 / 8.2 | 11 / 18.0 | 2 / 3.3 | 3 / 4.9 | 0 / 0.0 | 0 / 0.0 | 3 / 4.9 | 2 / 3.3 | 4 / 6.6 | 0 / 0.0 | 61 / 100.0 |
| Iron and Steel | 110 / 36.3 | 48 / 15.8 | 40 / 13.2 | 25 / 8.3 | 14 / 4.6 | 7 / 2.3 | 8 / 2.6 | 8 / 2.6 | 9 / 3.0 | 24 / 7.9 | 10 / 3.3 | 303 / 100.0 |
| General Machinery | 11 / 25.0 | 7 / 15.9 | 8 / 18.2 | 8 / 18.2 | 1 / 2.3 | 1 / 2.3 | 0 / 0.0 | 2 / 4.5 | 1 / 2.3 | 2 / 4.5 | 3 / 6.8 | 44 / 100.0 |
| Transportation | 34 / 35.8 | 18 / 18.9 | 6 / 6.3 | 8 / 8.4 | 4 / 4.2 | 1 / 1.1 | 5 / 5.3 | 2 / 2.1 | 2 / 2.1 | 11 / 11.6 | 4 / 4.2 | 95 / 100.0 |
| Others | 64 / 28.7 | 23 / 10.3 | 46 / 20.6 | 26 / 11.7 | 18 / 8.1 | 8 / 3.6 | 3 / 1.3 | 9 / 4.0 | 6 / 2.7 | 13 / 5.8 | 7 / 3.1 | 223 / 100.0 |
| Total | 476 / 35.4 | 145 / 10.8 | 257 / 19.1 | 109 / 8.1 | 67 / 5.0 | 37 / 2.8 | 25 / 1.9 | 32 / 2.4 | 59 / 4.4 | 77 / 5.7 | 61 / 4.5 | 1345 / 100.0 |

Copyright © IFAC Production Control in the
Process Industry, Osaka, Japan 1989

# AN EXPERIMENTAL EVALUATION OF TWO REALTIME FAULT DIAGNOSIS SYSTEMS IN A REFINERY PLANT DISTRIBUTED CONTROL SYSTEM

## J. Shiozaki,* J. Nishimura,* A. Kikutani* and T. Moriya**

*Yamatake-Honeywell Co. Ltd., Tokyo, Japan
**Idemitsu Kosan Co. Ltd., Chiba, Japan

Abstract. Two real-time fault diagnosis systems were built into the distributed control system of a refinery plant to evaluate and compare two diagnostic techniques, i.e., the signed directed graph based diagnosis technique and rule-based system technique. Field tests were done to examine the diagnostic accuracy and the performance of these systems. From the results, we recognized that the two diagnosis systems have different characteristics of diagnostic accuracy but both are capable of diagnosing a system failure in two minutes in the distributed control system environment.

Keywords. Diagnosis; Graph theory; Artificial Intelligence; Computer Control; Expert System.

## INTRODUCTION

Recently, the importance of operator-assistance systems has increased because of the decreasing number of well trained operators and the increasing scpe of work which a single operator should handle

One of the most important parts of an operator-assistance system is the real-time fault diagnosis system.

Many IF-THEN rule based diagnosis systems (Ex. Andow(1985) ) have been built, and some have had great success. But, in the area of large chemical plants, there seem to be only a few such systems. We think that one reason for this is the difficulty to make large rule-base. Another reason is that the common sense knowledge is not used in such systems.

## SDG BASED DIAGNOSTIC SYSTEM

The name of the signed directed graph based diagnosis system was LOOSUS ( LOgical Operation SUpport System ). There are several types of SDG based diagnosis algorithms. FADUS used the algorithm in Shiozaki(1985).

LOOSUS reads all sensor values every 60 second, as does LIOREX. These values are compared with thresholds set by operators. If at least one sensor violates the threshold, the diagnosis algorithm check the consistency between the pattern and the SDG and shows the operators possible cause es graphically. The computation time for diagnosis was within one minute.

The SDG of the process had 754 nodes and 1380 branches. The base number of possible causes which were included the SDG data was 892.

Fig. 5 shows an example of the SDG subgraphs. The nodes represent the state variables. The branches represent the cause-effect relationships between two variables. The solid (dotted) line branches correspond to plus (minus) branches.

A three step approach was used to create an SDG for the plant.

Step 1. Two knowledge engineers of YH created equipment SDGs corresponding to all types of equipment. All equipment graphs were checked by IDEMITSU process engineers.

Step 2. Several engineers of YH and IDEMITSU created the entire SDG, combining the equipment graphs and making pipeline graphs.

Step 3. YH knowledge engineers checked the SDG, and modified it to increase diagnostic accuracy and speed.

## FIELD TEST RESULTS

The object plant was the desulfurization process at Idemitsu's Chiba Refinery. The process has one reactor, two distillation towers, two absorption columns and many heat exchangers for heat integration (see Fig.1).

We intentionally produced two kinds of failures for each systems to evaluate and compare the two

diagnostic systems. The followings are the failures we produced.

A. An absorption liquid inlet valve of an absorption column V8 closed.

B. An absorption liquid outlet pump of the absorption column V8 stopped.

These failures were generated by the operators carefully and safely.

### LIOREX Results

The diagnostic results for LIOREX are shown in Table 1. The two failures are represented by failure A (VALVE CLOSED) and failure B (PUMP STOPPED). The sets of cause candidates in Table 1 were obtained successively from the diagnostic system, as process value data were collected every 60 seconds by LIOREX.

The top five cases, from No.1 to No.5, correspond to failure A. In these five cases, the candidate set in No.3 includes the cause of the failure, but the candidate sets in the other four cases do not include the cause of the failure.

The bottom five cases, from No.6 to No.10, correspond to failure B. The candidate sets from No.6 to No.8 include the cause of the failure, but the candidate sets in the other two cases did not include the cause of the failure.

The numbers of candidates in the candidate sets were from 1 to 3.

In Table 1, the 60 % sets of the cause candidates did not include the cause of the failure. This does not necessarily mean that we cannot use this system because of low accuracy. At least one set of cause candidates for each failure (A,B) included the cause of the failure; therefore, it can be said that if the operator checked the union of all candidate sets, the cause of the failure could be found. Nevertheless, the operators could not know when the union of all candidates sets included the cause of the failure.

Table 1. Diagnostic Results for LIOREX

| Case | Time | Failure | No. of Candidates | Includes Cause ? |
|------|------|---------|-------------------|------------------|
| 1 | 11:12 | A | 1 | NO |
| 2 | 11:13 | A | 1 | NO |
| 3 | 11:14 | A | 1 | YES |
| 4 | 11:15 | A | 1 | NO |
| 5 | 11:16 | A | 1 | NO |
| 6 | 13:36 | B | 3 | YES |
| 7 | 13:37 | B | 3 | YES |
| 8 | 13:39 | B | 2 | YES |
| 9 | 13:41 | B | 2 | NO |
| 10 | 13:41 | B | 2 | NO |

Failure A: Valve Closed
Failure B: Pump Stopped

LOOSUS ( LOgical Operation SUpport System ) on the distributed control system of the plant.

LIOREX is an IF-THEN rule based diagnostic system. LIOREX used EXPERT SHELL, which was an expert system tool developed by YH.

LOOSUS diagnoses failures using a qualitative model of the process, i.e. a Signed Directed Graph. LOOSUS used FADUS ( Fast Diagnosis Using Signed directed graph ), which was an SDG-based diagnostic program developed by Kyushu Univ. and YH.

### IF-THEN RULE BASED DIAGNOSITIC SYSTEM

LIOREX reads all sensor values every 60 seconds. The sensor output values are compared with the thresholds set by operators. If at least one sensor violates the threshold, the rule tree ( Ex. Fig.4 ) corresponding to the sensor will be checked to locate the cause of the failure. The operator can see the diagnostic results on USs graphically. The computation time for diagnosis was within one minute.

The 3206 number of rules were created in tree structure format by operators and process engineers using their own words, as in Fig.4. After that, the rules were rewritten in EXPERT SHELL rule format by YH application engineers. The base number of possible causes which were included in the rules was 362.

There exist two trees for every sensor. One tree corresponds to a case in which the sensor value is higher than the upper threshold. The other tree corresponds to a case in which the sensor value is lower than the lower threshold. These thresholds can be set by the operators.

Fig. 4 shows an example of the rule trees. A part of the rule tree can be read as follows.
"If NF_20 is larger than the normal value, check NH_P_6. If NH_P_6 is increasing, check NF_19. Otherwise the NF_20 sensor failure can be considered."

On the other hand, there are other techniques using common sense knowledge of the object system. Iri (1981) and Shiozaki (1985,1989) presented a diagnosis algorithm for real-time diagnosis of chemical plants. The algorithm used Signed Directed Graphs as the qualitative model of the object plant.

In this paper, we evaluate and compare two techniques , i.e. the IF-THEN rule-based technique and the SDG-based technique after applying two prototype systems to a distributed control system and doing field tests in Idemitsu Kosan Co., Ltd., Chiba Refinery.

### IDEMITSU REFINERY PLANT

Idemitsu Chiba Refinery is the largest refinery in Japan, built in 1963. It processes 310,000 bbl/day of oil.

The object process was the desulfurization process in the refinery, built in 1967. The processing capacity of the desulfurization process is 25,000 bbl/day. The input material of the desulfurization process is the naphtha from the topper and the output is the desulfurized naphtha, which is transformed to gasoline in later processes.

Fig. 1 shows the desulfurization process. The naphtha, charged in the surge drum V30, is heated and sent to the reactor V1. In the reactor, S in the naphtha changes to $H_2S$. The naphtha output from the reactor goes to the stripper V4 and is separated into light and heavy naphtha. In the rectification tower V6, LPG is separated from the light naphtha. The gas from the condenser V7 includes LPG, therefore the two absorbers (V8, V9) are used to collect the LPG.

## DIAGNOSIS USING
## SIGNED DIRECTED GRAPHS

A Signed Directed Graph(SDG) is a qualitative causal model of the system. An SDG consists of nodes and branches. A node represents the state variable. A branch represents the direct influence between two state variables, and its branch is assigned "+" if it represents a positive influence (reinforcement) or "-" if it represents a negative influence (suppression).

The value of a state variable being normal, higher or lower than the normal value is represented as "0", "+" or "-", respectively. The combination of the signs assigned to the nodes of the SDG is defined as "pattern" and represents the state of the system. Patterns on all the sensors are generated by the diagnosis system which compares the process values with the thresholds. The diagnosis algorithm (Iri(1981), Shiozaki(1985,1989)) checks the consistency between the SDG and a pattern and outputs the possible causes of the failure.

Example.

Fig.2 (a) is a tank system. F0, F1 and F2 represent the flow rates. L1 and L2 represent the liquid levels in the tanks. The SDG of this system is shown in Fig.2 (b). The arrow in solid line represents the branch with "+" whereas the arrow in broken line indicates the branch with "-". The branch with "+" from node F0 to node L1 indicates that when F0 is increased (decreased), L1 is also increased (decreased). The branch with"-" from node F1 to node L1 indicates that when F1 is increased (decreased), L1 is decreased(increased). For instance if blockage occurs in the pipeline between Tank 1 and Tank 2, it may generate the pattern that is shown in Fig.2 (b).

## DIAGNOSIS SYSTEM STRUCTURE

The distributed control system used in the plant was Yamatake-Honeywell TDC3000. The diagnosis system was built into the distributed control system.

Fig. 3 shows the diagnosis system structure. In Fig. 3, diagnosis programs worked on a mini computer ( CM60: Computer Module 60 ). The engineering work, such as making rules and SDGs, were done on the CM60. The operator interface system was built into USs( Universal Station ), so that the cause of the failure could be seen graphically as the output of the diagnosis system. The USs were also used for system startup and threshold setting to distinguish whether a sensor was showing an abnormal state or not. An HG( Highway Gateway) was used for data gathering.

## TWO DIAGNOSIS SYSTEMS

We created two diagnostic systems, The LIOREX ( Little Owl as Revolutionary Expert System ) and

The main reasons why the 60 % sets of the cause candidates in Table 1 did not include the cause of the failure are as follows:

- We did not consider the controller effects when we made rules, because it was difficult to predict all cases which might occur with the effect of the controllers;

- It was difficult to create rules that the diagnostic system could output correct cause candidate sets at any time, because, when we make a rule, we usually consider a period immediately following the failure, but it does not include all periods.

### LOOSUS Results

The diagnostic results for LOOSUS are shown in Table 2. The two failures are represented by Failure A (VALVE CLOSED) and Failure B (PUMP STOPPED), as in Table 1. The sets of cause candidates were obtained successively from the diagnosis system, since process value data were collected every 60 seconds, just as with LIOREX.

The top five cases, from No.1 to No.5, correspond to failure A and the bottom five cases, from No.6 to No.10, correspond to failure B. All candidate sets included the cause of the failure. The number of candidate sets were from 21 to 75.

The number of candidates for LOOSUS were many, compared with that of LIOREX. The reasons are considered below.

- The base number of causes which LOOSUS can distinguish was larger than for LIOREX.

- LOOSUS was not using operator experiences which were used in LIOLEX rules.

### Accuracy of the Two Systems

The two results, shown in Table 1 and Table 2, had very different characteristics. To understand these characteristics, we defined the following two types of accuracies, because we felt that the usual accuracy concepts were not enough.

First kind of diagnostic accuracy:
The first kind of diagnostic accuracy of system A is higher than that of system B, because probability that the cause candidate set would include the cause of the failure is larger for system A than for system B.

Second kind of diagnostic accuracy:
The second kind of diagnostic accuracy of system A is higher than that of system B, because the number of cause candidates is smaller for system A than for system B.

The first kind of diagnostic accuracy of LOOSUS was larger than that of LIOREX. The sets of LOOSUS candidate causes always included the cause of the failure, but, the 60 % sets of LIOREX candidate causes did not include the cause of the failure.

The second kind of diagnostic accuracy of LIOREX was larger than that of LOOSUS, because the number of LIOREX candidates was smaller than that of LOOSUS candidates.

Table 2.  Diagnostic Results for LOOSUS

| Case | Time | Failure | No. of Candidates | Includes Cause ? |
|------|------|---------|-------------------|------------------|
| 1 | 12:04 | A | 73 | YES |
| 2 | 12:07 | A | 75 | YES |
| 3 | 12:10 | A | 68 | YES |
| 4 | 12:12 | A | 71 | YES |
| 5 | 12:15 | A | 21 | YES |
| 6 | 14:40 | B | 32 | YES |
| 7 | 14:42 | B | 32 | YES |
| 8 | 14:43 | B | 27 | YES |
| 9 | 14:45 | B | 39 | YES |
| 10 | 14:46 | B | 28 | YES |

Failure A: Valve Closed
Failure B: Pump Stopped

## CONCLUSION

The two prototype diagnostic systems, which were using the rule- based system technique and the signed directed graph technique, successfully diagnosed two failures within two minutes, which were intentionally generated in a refinery process. The two systems had different characteristics in diagnostic accuracy.

It would be effective if we could develop a system which combines the advantages of these two systems.

## REFERENCES

Andow, P. K. (1985). Fault Diagnosis Using Intelligent Knowledge Based Systems. Chem. Eng. Res. Des., 64, 368-372.

Iri, M., K. Aoki, E. O'Shima and H. Matsuyama (1981). An Algorithm for Diagnosis of System Failures in the Chemical Processes. Computers and Chem.Eng., 3, 485-493.

Shiozaki, J., H. Matsuyama, K. Tano, and E. O'Shima. (1985). Fault diagnosis of chemical processes by the use of signed directed graphs. Int. Chem. Eng., 25-4, 651-659.

Shiozaki, J. , B. Shibata, H. Matsuyama and E. O'Shima ( 1989). Fault Diagnosis of Chemical Processes Utilizing Signed Directed Graphs. IEEE Trans. on Ind. Elec. To be published.

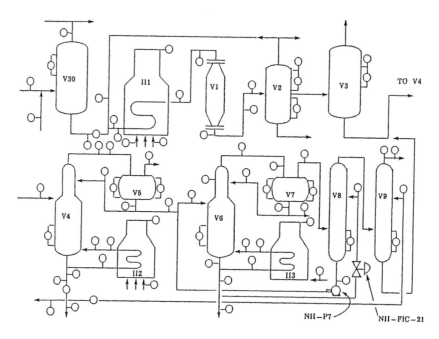

Fig. 1 Desulfurization Process of Idemitsu Chiba Refinery

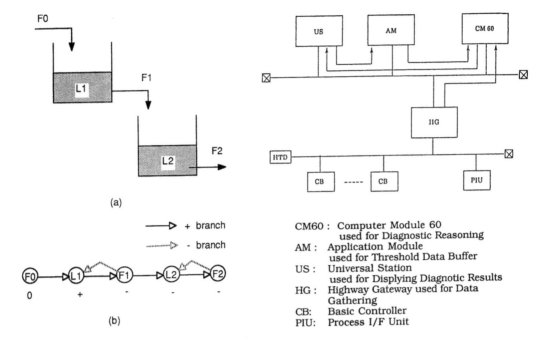

(a)

⟶▷ + branch

┈┈┈▷ − branch

(b)

CM60 : Computer Module 60
used for Diagnostic Reasoning
AM : Application Module
used for Threshold Data Buffer
US : Universal Station
used for Displying Diagnotic Results
HG : Highway Gateway used for Data
Gathering
CB: Basic Controller
PIU: Process I/F Unit

Fig. 2 Tank System (a) and Its Signed Directed
Graph (b)

Fig. 3 Diagnostic System Structure

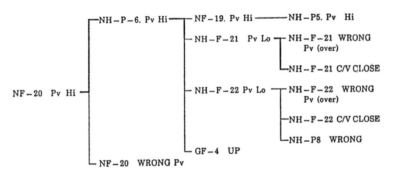

Fig. 4 Rule Tree for "NF_20 Pv Hi"

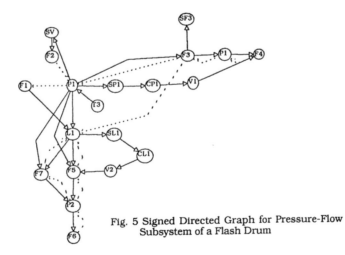

Fig. 5 Signed Directed Graph for Pressure-Flow
Subsystem of a Flash Drum

# DISCUSSION

## Session 4: Maintenance Strategy

*Chairmen*

**J. König**

*Bayer, FRG*

**H. Matsuyama**

*Kyusyu University, Japan*

In Session 4, the discussion was focused on
(1) Improvement of the maintenance organization
(2) Maintenance manpower
(3) Prediction of failures
(4) Relation between design and maintenance

## 1. Improvement of the maintenance organization

There are two sides in improving the maintenance organization; keeping the plant in the better state and performing better maintenance tasks. The former is much more important than the latter. Though a lot of people focus their effort to improve their skills for detecting and repairing failures, it is fundamental to make detecting and repairing unnecessary.

## 2. Maintenance manpower

Proper maintenance requires to gather more data from the objective plant. This task tends to be time consuming and requires manpower. Successful examples were shown as PACESETTERS in the refinery industry in USA and TPM movement in Japan where this task are made not by specialists of maintenance but by operators in the field. The problem to be solved in near future is who is responsible to this task when automation or unmanned operation goes on.

It is one of solutions for the problem that a group of highly trained engineers manages operation, maintenance and all tasks concerning production in the highly automated plant. Computer aided maintenance systems will play an important role in future, which are constructed on the basis of Reliability Centered Maintenance.

## 3. Prediction of failures

The prediction of the state of the objective equipment will play an important role in maintenance nowadays and the computer aided maintenance system in failure. The prediction of the state of the equipment will be able to be made by evaluating the degree of the individual deterioration occurring in the equipment. Deteriorations are classified into experienced and unexperienced ones into progressive and abrupt. The degree of the progressive deterioration can be easily evaluated if there exist suitable means to observe parameters representing it. The abrupt deterioration, however, cannot be predicted by condition monitoring. The probabilistic approach will be effective if it occurs periodically. Unexprienced deteriorations and abrupt ones without periodical characteristics should be evaluated by analyzing the physicochemical mechanism of the deterioration.

Looking at the tragic accidents happened in the past, it seems to be difficult to avoid the accidents due to unexperienced failures by the maintenance activity. Progress of technology of predicting unexperienced failures is indispensable for maintenance to contribute in assuring safety of the objective plant.

## 4. Relation between design and maintenance

The appropriate level of the dependency on the initial design varies according to the objective. Televisions or refrigerators, for instance, will 100% depends on the initial design. In the process industry, however, the appropriate level of the dependency on the initial design has not yet been determined.

Chemical plants with more than thirty years life should not 100% depend on the initial design because the plants may be used in the environments which could not be imagined by designers 30 years ago. Operation alone cannot cope with these changes of environments. Design review should be always repeated and modifications of plants should be performed going back to the design phase.

# APPLICATION OF J.I.T. FOR AUTOMOTIVE
# PARTS PRODUCTION

## T. Torii

*Assistant General Manager, Production Control Department, Production Group,*
*Aisin Seiki Co. Ltd., Japan*

**Abstract**. Many types of cars are produced today to satisfy the various tastes of
car owners. Automotive parts manufacturers must automate their production processes
and implement information networks to supply parts to automakers in a "just in time"
(J.I.T.) manner. AISIN SEIKI's Shiroyama Factory (located in Nishio, Aichi Prefecture), where truck transmissions are produced, established a so-called "specific
plant for specific product" system. This system allows desired types of products
to be made efficiently at any time. Each production line in the plant was also
improved. As a result, lead time was reduced by one-half.

**Keywords**. Assembling; conveyors; grinding; machining; integrated production;
order delivery report.

## INTRODUCTION

Automakers have been taking measures to reduce
the lead time from receiving an order to delivering the car, as well as to increase the types
of cars produced, in an attempt to increase their
market share.

To meet various requirements from these automakers, AISIN SEIKI implemented a new production system called the "specific plant for
specific product" system, which allows desired
types of products to be produced efficiently
at any time. This system is a result of improvements or modifications of existed facilities and production systems.

## BASIC CONCEPT OF PRODUCTION

The basic concept of automotive parts production
at AISIN is to efficiently use labor, materials,
equipment and information through the effective
application of the Toyota Production System,
and thus to increase total productivity. In
other words, the concept is to promote integrated production through the Toyota Production
System and to increase productivity through a
JIDOUKA, or self-diagnosable system.

### 1. What Is the Toyota Production System?

The Toyota Production System consists of two
elements: J.I.T. and JIDOUKA.

J.I.T. is a system and concept to produce or
handle a required amount of required materials
or goods at the required time.

JIDOUKA is the concept of a system where any
abnormal condition in equipment or product
quality is automatically detected, the equipment or production line is stopped and the
abnormal condition is automatically reported,
thus producing 100% complete products.

### 2. Production Efficiency Enhancement Activity Based on the Toyota Production System

The history of the production efficiency enhancement activities at AISIN consists of the following three stages. During the first stage (1973 ∿
1978), production efficiency was increased by
improving small-lot production processes and
material handling procedures. In the second
stage (1979 ∿ 1985), production efficiency was
innovatively increased by applying AISIN FMS
wherein equipment and information networks were
automated by applying well-engineered production
systems. AISIN is in the third stage at present,
in which the work improvement of the first stage
and the equipment improvement of the second stage
are effectively combined for further enhancement
of productivity from the viewpoint of a total
production system. A typical example is the
"specific plant for specific product" system in
the Shiroyama Factory.

## EXAMPLE OF PRODUCTIVITY ENHANCEMENT ACTIVITIES

### 1. Outline and Features of the Shiroyama Factory

The Shiroyama Factory is one of the leading
transmission manufacturing plants in the world,
where various types of transmissions are produced for compact cars and industrial and construction vehicles, as well as for trucks.

From the viewpoint of products, this factory
manufactures various types but small lots of
products. A transmission consists of 200 or
more precision parts to offer required function
and performance. From the viewpoint of production systems, on the other hand, this factory has
two main features.

One is its two-shift operation, which covers
operations from machining to assembly, based on
the Toyota Production System. The other is its
24 deliveries a day at intervals of 40 minutes,
according to the vehicle assembling sequence at
the customer's plant. (See Fig. 1.)

| Area | 112,000 m$^2$ |
| Building floor | 66,000 m$^2$ |
| Employees | 890 |
| Equipment | 3,200 units |
| Annual sales | ¥63 billion/year |

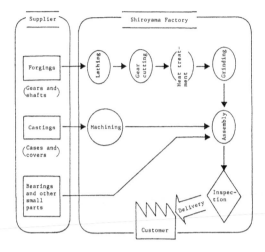

Fig. 1. Outline of Shiroyama Factory and
its production process

## 2. Role of Partsmakers in Car Production

(1) Car production line. Automakers assemble various types of cars on the same line in order to
efficiently meet customers' demands in a timely
manner. The automotive parts (such as engines
and transmissions) are received as a unit for
a car. Among these parts, such large parts as
transmissions, tires and seats must be delivered
to respective assembly lines in a J.I.T. manner
in response to the "Order Delivery Report," made
according to the progress of car assembly on the
lines.

(2) Flow of information on production and products. Figure 2 shows how the order delivery
reports are issued according to the progress of
car assembly on a production line.

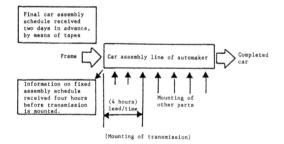

Fig. 2. Issuance of order delivery report
and lead time

As illustrated in the above Fig., the final car
assembly schedule is supplied from the customer
to the Shiroyama Factory two days in advance by
Vehicle Linkage Tapes (=V.L.T.). The factory
also receives information on final car assembly
sequence, when a body frame is charged to the car
assembly line. Based on this information, the
factory delivers the parts to the assembly line.

The relation between the order delivery report
and product flow is described below, taking a
transmission as an example.

As shown in Fig. 2, the transmission is mounted
four hours after a corresponding frame is charged
onto the assembly line. This "4 hours" of lead
time is the key to the J.I.T. production at the
factory. If it is possible to make the transmission and deliver it to the customer's production line within four hours, AISIN's transmission production line and the customer's car
assembly line can synchronize precisely. Thus
the J.I.T. system between the supplier and user
can be established. (The Shiroyama Factory is
located within about a one hour truck drive to
the customer's site.) Based on this concept,
each production system and unit of equipment was
improved or modified at the Shiroyama Factory.
As a result, a successful J.I.T. system for
transmission production and delivery was established.

## 3. Basic Concept of the "Specific Plant for Specific Product" System

(1) Ideal configuration of a production line.
The ideal configuration of a production line
manufactures the products in a J.I.T. manner by
synchronizing the production and information
flows. There are three key factors to realize
this configuration. The first is to minimize
material flow in terms of time and quantity to
minimize lead time. The second is to harmonize
manpower and equipment to realize well-balanced
automation. The third is to efficiently process
information to establish a real-time processing
system. The Shiroyama Factory has put these
factors into its practical production lines by
establishing a "specific plant for specific
product" system. In this system, each of the
specific plants is configured based on a specific product unit (such as medium size truck
transmission or four-wheel drive vehicle transmission) and its assembly line is organically
integrated with the related process lines.
Various types of transmissions are processed
and assembled on this line according to customer
car assembly schedules, and delivered to the
customer in a J.I.T. manner.

(2) Plant management. To allow a plant for a
specific product to function optimally, the
product and information flow systems must be
automated under the conditions of "Zero Defects"
and "Zero Failures." To put these conditions
into practice, productive aimed design, product
quality line, failure-free activity, work-area
environment improvement and other basic activities must be performed, as shown in Fig. 3.

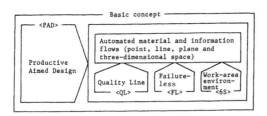

Fig. 3.  Four basic activities

## 4.  Description of Each Activity

(1) Productive Aimed Design (PAD) activity. PAD activity is promoted under the cooperation of the new product development, manufacturing and sales divisions. This activity can be divided into two subactivities: simultaneous multi-dimensional planning and high quality/productivity assurance design. The former provides a new type of transmission with required functions, taking into account the expansion of application of the transmission to a certain series under a definite range. The latter maintains close relations between the engineering and manufacturing divisions from the very beginning of a new product's planning, so that the subject transmission design can fulfill both quality and productivity requirements. As the result of these activities, machines could be integrated, assembly lines could be automated, their length could be shortened, and most of the assembly lines could be organically combined. Thus the concept of a "specific plant for specific product" system could be realized. Production costs were also reduced. Further automation of the lines is also expected.

(2) Quality line. The "Zero defect" quality level is required to maintain highly efficient, orderly production at a specific plant for a specific product. If a defective product is found during final product inspection process or after products have been delivered, product flow is disturbed, causing eventual suspension of plant operation.

To avoid such an unfavorable condition, the Shiroyama Factory has been promoting zero defect activity called the "QL-Promotion." Results show that the defective product rate has been decreased to one-seventh its previous level.

(3) Failure-less. A specific plant for a specific product loses its ability to function if any of its equipment fails. Generally speaking, maintenance is mainly focused on heavily-loaded machines and machines with high failure rates. At the Shiroyama Factory, however, we have been promoting "F.L. (Failure-less) line" activity to keep all of the machines in the plant free from failure, considering every machine important. Results show that machine failure rate has been reduced to 1/15.

(4) 6S (Initials of Japanese terms meaning good order, proper arrangement, cleanliness, cleaning, instruction and completeness). The 6S activity at its initial stage was to keep the work areas clean by preventing machine oil from spilling and by sweeping out chips. 6S promotion teams in the factory made chip covers and oil drop pans.

Their cover/pan making technique (or 6S technique) was conducted for years. This developed technique was helpful in mechanization of workpiece loading/unloading and workpiece transferring. When a specific plant for a specific product was built, this 6S technique was well-linked with state-of-the-art automation technology. As a result, highly efficient, clean production lines were installed.

## 5.  Specific Plant for Specific Product

The "specific plant for specific product" activity, which aims at the realization of a J.I.T. production system, is divided into two subactivities: "automation of product flow" (product line construction) and "automation of information flow" (production system establishment).

(1) Automation of product flow (product line construction). Even if a part of the equipment and production lines operates efficiently, a plant for a specific product is slightly influenced because it consists of much equipment and many production lines. Partial efficiency increase does not always increase total productivity. Based on this fact, we have been working to increase total efficiency by expanding automation from equipment (point) to production lines (line) and plants (plane). The basic concept of production line construction as well as an example of such construction is described below.

For the equipment, appropriate types of machines are selected for a specific process, depending upon the production rate and the quantity of works in process, as shown in Fig. 4. Special attention has been paid to automate the machines and set-up.

| Production Aim / Process | Fewer Types/ Large Volume | Medium Types/ Medium Volume | Diversified Types/Small Volume |
|---|---|---|---|
| | Compact/High Speed | Integration/ No Set-up | Flexible |
| Assembly | · Automatic assembly machine | · Semiautomatic assembly machine | · Manual assembly |
| machining | · Single purpose machine · Transfer machine | · NC machine · Multipurpose machine | · Machining center · General purpose machine |
| Gear cutting | · Single purpose high speed machine | · Single spindle NC machine · Multipurpose machine | · General purpose machine · Multispindle NC machine |
| Grinding | · Single purpose high speed machine | · Dual-spindle NC machine | · General purpose machine · Dual-spindle NC machine |

Fig. 4.  Process, production and machine

The basic concept for production line construction is to minimize and streamline the path of worker movement, to make product flow straight and to automate the equipment and process. As shown in Fig. 5, production lines are grouped into three types, according to types of parts and production. We have been trying to increase productivity, paying special attention to ① and ②.

| Produc-<br>tion<br>Line | Fewer Types/<br>Large Volume | Medium Types/<br>Medium Volume | Diversified<br>Types/Small<br>Volume |
|---|---|---|---|
| Machining | · Automatic set-<br>up<br>· Full automatic/<br>unmanned<br>· System control | · Selected auto-<br>matic set-up<br>· Semi-automatic,<br>automatic re-<br>moval and con-<br>veyance<br>· Semi-control | · Single step<br>set-up<br>· Manual (except<br>machining)<br>· ————— |
| Assembly | · Automatic set-<br>up<br>· Semi-automatic<br>assembly<br>· Automatic pro-<br>duction control | · Selected pre-<br>paration<br>· Semi-automatic<br>assembly<br>· Automatic re-<br>port on pro-<br>duction | · Single step<br>set-up<br>· Manual assem-<br>bly<br>· Production<br>based on<br>Kanban card |

Fig. 5.   Production line

The goal in plant construction is to minimize
the paths of material flow by organically linking the equipment (point) and the production
lines (line), as well as to automate the equipment and lines, thus increasing total productivity.

Based upon the previously described considerations, we have built nine specific plants for
specific products.  Figs. 6 and 7 show the
plant layout for FJ type transmission production, before and after the improvements.  Before
the improvements, plant layout had been based
on process.  All materials were received at one

station and all of the products were shipped
from another.  Materials inside the plant flowed
in one direction.  All products in a process
were gathered to the same station for the same
processing.

As the quantity and types of products increased,
however, it was found that the above plant layout could not satisfactorily solve problems
concerning half-finished products stocked between processes, labor forces required for
material movement and product rate variation.
Improvement from the existing plant layout to
a new plant for each product was required.
The new plant needed to be systematized so as
to fully function by making most of product
features.  To meet this requirement, we started
to build high efficiency plants, putting special
emphasis on rationalization of material flow
between processes, automatization of manual
work and automatic information processing by
use of computers.  As a result, we successfully
built new production lines in which product
machining and assembly are integrated into one
block.  The effective use of computers links
the transmission assembly lines and parts machining lines at our factory with customers'
car assembly lines, as if they are connected
by a pipe.

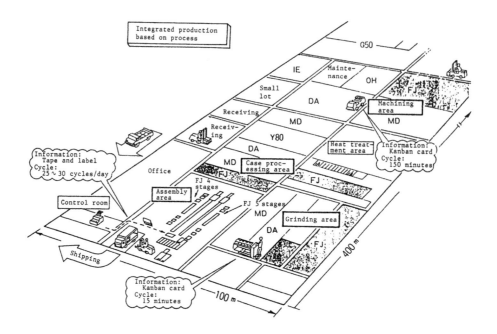

Fig. 6.   Plant layout before improvements

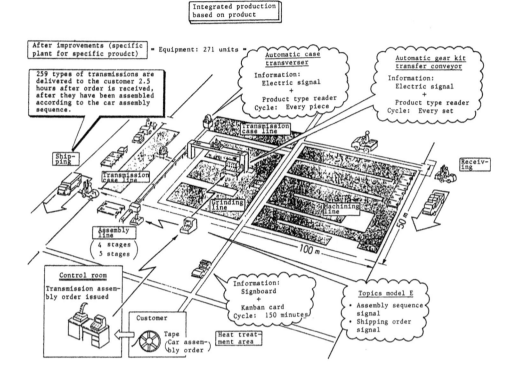

Integrated production
based on product

After improvements (specific
plant for specific proudct) = Equipment: 271 units =

259 types of transmissions are
delivered to the customer 2.5
hours after order is received,
after they have been assembled
according to the car assembly
sequence.

Automatic case
transverser
Information:
Electric signal
+
Product type reader
Cycle: Every piece

Automatic gear kit
transfer conveyor
Information:
Electric signal
+
Product type reader
Cycle: Every set

Transmission
case line

Ship-
ping

Transmission
case line

Receiv-
ing

Grinding
line

Machining
line

Assembly
line
4 stages
5 stages

50 m

100 m

Control room
Transmission assem-
bly order issued

Customer

Tape
(Car assem-
bly order)

Heat treat-
ment area

Information:
Signboard
+
Kanban card
Cycle: 150 minutes

Topics model E
• Assembly sequence
  signal
• Shipping order
  signal

Fig. 7.  Plant layout after improvements

(2) Automation of information flow (production
system establishment). In an ideal production
system, the product and information flows in
various plants for specific products are handled efficiently as a network. To efficiently
manage the 9 plants at the Shiroyama Factory,
a great volume of information must be processed
in a short time. For this purpose, we introduced a computer system to establish an information management system that can control all
of the plants simultaneously by linking necessary information to office equipment. This
information management system consists of two
subsystems: the production report subsystem

and the parts report subsystem. As shown in
Fig. 8, the production report subsystem simultaneously manages two or more plants integrally.
When two or more orders are received from two
or more customers, this subsystem prepares a
product assembly schedule for each product and
transmits it to each plant. This subsystem
also receives information on product inspection,
shipping and stock on an online basis. The information is processed for accurate P/L control.
This subsystem permits complete synchronization
of customer's assembly lines with our transmission assembly lines.

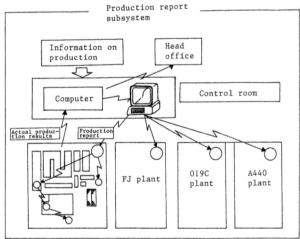

Production report
subsystem

Information on
production

Head
office

Computer

Control room

Actual produc-
tion results

Production
report

FJ plant

019C
plant

A440
plant

Fig. 8.  System for controlling product
         information

The parts report subsystem consists of three
elements and 17 lines, as shown in Fig. 9.

| Control element | Number of Line |
|---|---|
| ① Parts machin-<br>ing report | 1 |
| ② Parts trans-<br>fer report | 3 |
| ③ Parts unload-<br>ing report | 13 |

Fig. 9.

This subsystem reports machining, transferring
or unloading of parts according to the sequential
production order, and supplies necessary parts
one at a time as needed. This subsystem eliminates manual parts selection, thus minimizing
erroneous assembly work. As a result, product
quality is remarkably improved.

6. Result

As a result of building the specific plants for
specific products, product quality and productivity have been remarkably improved, leading to
increased market power of AISIN SEIKI. Regarding productivity, the workforce was reduced by
410, lead time was reduced by one half as was
total stock. From the viewpoint of quality,
defect rate was reduced by 80% by replacing
manual work with equipment as well as by promoting "Q-Compo-Control" activities. Q-Compo
refers to equipment portion or component that
greatly affects qualtiy.

CONCLUSION

More diversified types of products will have to
be made in the future. To further increase our
production efficiency under such industrial conditions, it is most important to shorten lead
time. We will continue further studies on
hardware (how to improve production process)
and software (how to transmit information
efficiently).

# DISCUSSION

## Session Plant Tour 1: Aisin Seiki Co. Ltd.

*Chairman*
**J. Clemmons**
*ChemShare, USA*

### 1 Introductory Presentation

The presentations composed of a brief profile of
Aisin Seiki Co., Ltd. and the principles of production management were given by Mr. S. Aiki,
President and Mr. T. Torii, Assistant General
Manager, respectively. According to the speakers, the business lines of the company include
production of automotive components such as
transmissions, clutches, brakes etc. and household equipment. The production system is of a
typical assembly manufacturing. The most fundamental basis is to keep high quality of products.
So, it is necessary to try to achieve higher
quality by pursuing technical development to make
the creation of better products. The company
always sets up a five-year vision with respect to
ideal operations every five years. In addition,
in order to cooperate constituency, it implements
TQC, JIT or Toyota production system, and TPM.
Finally, though the strategies to advance JIT and
Jidouka were introduced, the details are involved
in the paper.

### 2 Visit to Transmission Assembly Plant at
Shiroyama

The transmission assembly plants at Shiroyama in
Aichi were established in 1973 and have been
developed based on JIT and Jidouka since 1975.
The plants have been improved in terms of efficiencies of work, equipment and total system.
The features of these plants are to produce a
wide variety of transmissions with small lot
production. For example, the plant for four-
wheel drive vehicles can produce 259 types of
transmissions and 600 units per day.

### 3 Discussion

Mr. Clemmons (ChemShare, USA) asked whether one
production line is used or not in order to produce many kinds of transmissions and in addition,
asked about inventory of finished transmissions.
Mr. Torii answered that every line is specified
for specific products such as transmissions of
four-wheel drive vehicles and in the case of M150
line involved in this plant tour, it produces 120
types of transmissions. The number of the inventory is equivalent to the amount of one track,
that is, about 24 to 40 units. Professor
McGreavy (Leeds University, UK) also asked about
the stocks that the production has to hold so
that the plant can be operated. The answer was
that transmission cases, bearings, bolts and so
on were purchased from outside suppliers but
almost parts are produced in house. The purchased parts are delivered once a day. But, big
parts are delivered two or three times a day. If

all the parts are used up, a blank kanban (order
form) is sent to the suppliers and required parts
are received next day.

Dr. White (E.I. Du Pont, USA) asked about mechanization of manufacturing, especially, whether
the company will develop robots internally or
not. Taking into account keeping good balance
between manpower and machines in terms of economy, Aisin makes decision. The running robots in
M150 plant were mostly purchased from outsides,
but otherwise were made in house. Dr. König
(Bayer, FRG) stated that some activities to standardize something like a language for definition
of orders, bills and so on in Europe at moment
and then asked about such standardization. The
answer was that there is no such activities except for each product has a different number.

Mr. Hop (Shell, NL) asked what sort of maintenance policy Aisin has in order to hold the
availability of machinery as high as possible.
The answer was that two activities for maintenance have been carried out: preventive maintenance to establish failure-free operation and
training and education to detect a failure as
quickly as possible. For the former activity,
the professional staffs make a scheduling for
conducting the preventive maintenance and perform
regular inspection on certain check-up areas
where there are priority facilities. For the
latter, operators have to have training and education so as to identify a failure area. On the
other hand, Aisin has two aspects for runability:
one is how many products a line can actually
produce in responding to the orders, and the
other is how long a line has a running period.
These can be modified by changing preparatory
operations and equipment. A specific plant for
specific product has both of advantage and disad-
vantage on profitability. For example depending
on the quantity of orders, some plants have to be
operated with high runability and others do not.
As the production demands often have fractuation,
such system may not be perfect. In the future,
the plants should have flexibility so as to meet
with the varying demands.

Professor O'Shima (Tokyo Inst. of Tech., JPN)
gave comments that as assembly manufacturing
would be automatized and mechanized more and
more, the form of operations would be getting
closer to process industry. But, the reason why
the assembly manufacturing is able to handle many
product is that it has definite processes. For
process industry, there is batch production system, where a different equipment has to be selected depending on a product. That is, process
industries have more difficult hurdles to overcome than assembly and machining industries.

# REFINERY MANAGEMENT AND OPERATION
# BY ADVANCED TECHNOLOGY

## T. Matsueda

*Manager of Refining Coordination & Engineering Section, Idemitsu Kosan Co. Ltd.,*
*Aichi Refinery, 11, Minamihama-cho, Chita-city, Aichi-prefecture, Japan*

## 1. Idemitsu Refineries Produce Excellent Products

Idemitsu's refineries are distributed well throughout the country so that our clients may not be inconvenienced by supply shortages of our products.

In 20-odd years since Idemitsu first constructed a refinery at Tokuyama in 1957, it has completed four refineries, each in Chiba, Hyogo, Hokkaido and Aichi prefectures.

The total capacity of the five refineries reaches 640,000 barrels/day(about 100,000 kiloliters/day).

| Refinery | Capacity (barrels/day) |
|----------|------------------------|
| Tokuyama | 100,000 |
| Chiba | 210,000 |
| Hyogo | 110,000 |
| Hokkaido | 90,000 |
| Aichi | 130,000 |
| Total | 640,000 |

## 2. Aichi Refinery with the Latest Equipment

### 2.1 Supplying Energy to Major Industrial Cities

Aichi refinery is located on the seacoast of Ise Bay at Chita City, Aichi Prefecture.

This central part of Japan between Tokyo and Osaka is called the Chubu district and is comprised of an enormous concentration of industries and a large population.

Favorably located in the Chubu distrct, the refinery has easy access to one of the most industrialized centers of Japan.

It lies only 30 kilometers south of Nagoya, the capital city of Aichi Prefecture. The role of Aichi Refinery is to maintain a stable supply of energy to this economic zone and population center.

Aichi refinery's production capacity is 130,000 barrels(about 20,000 kiloliters) per day, which meets some 15% of the overall demands of the Chubu district(Aichi, Shizuoka, Gifu and Mie prefectures).

### 2.2 The Latest Equipment in the World

From the standpoint of efficiency, energy-saveing, compactness and safety, Aichi Refinery employs the most advanced equipment in the world today.

Equipment in operation includes the residual crude hydrodesulfurization unit which reduces the sulfur content in residual crude to 0.3% or less and the residue fluid catalytic cracking unit.

Aichi Refinery is also one of the largest importers of LPG, with accompanying receiving and shipping systems, storage capacity, etc..

### 2.3 Concern for Environmental Conditions and Safety

Idemitu has given much consideration to environmental protection and public and employee safety measures in accordance with its basic policy that the company should prosper together with the local community.

Contrary to ordinary patterns, Aichi Refinery was built adjacent to the water and a greenbelt containing a virtual forest of trees was planted between the facility and the nearest residential areas.

With regard to safety and disaster prevention, the goal is to keep the refinery completely accident-free by employing the latest techniques and equipment and providing stringent training in their use.

### 2.4 Capacity of Refining Unit

| | |
|---|---|
| atomosperic distillation | 130,000 bbl/day |
| naphtha hydrodesulfurization | 26,000 bbl/day |
| naphtha reformer | 18,000 bbl/day |
| kero & gas oil hydrodesulfurization | 48,000 bbl/day |
| residual crude hydrodesulfurization | 55,000 bbl/day |
| residue fluid catalytic cracking | 30,000 bbl/day |
| alkylation | 10,000 bbl/day |
| LPG recovery | 550 t/day |
| sulfur recovery | 400 t/day |

## 3. Efficient Refinery with New High Technology

Aichi Refinery has been making efforts to establish the system for production optimization.

In order to improve production facility and reduce production cost, RFCC(residue fliud catlytic cracking) and a coal boiler are employed.

Moreover, for survival from the competition we are developing "Efficient Refinery With New High Technology" to optimize refinery production.

### 3.1 Efficient Refinery

To build up efficient refinery, we have to pay special attention to the following.

(1)Production plan

To optimize production facilities, we make and execute best, flexible and consistant production plans from oil receiving to product delivery.

(2)Operation control

We continue efficient and safe operation to optimize the facility.

(3)Equipment management

We maintain the highest performance of facility and safe and reliable condition.

(4)Safety control

We respond immediately and precisely in case of abnormal operation and prevent any further unpreferable happenig.

On the other hand, technological innovation is making rapid progress and we are living in the information society.

In Aichi Refinery, we have been introducing new technology such as advanced micro-electronics computer.

For example,
  · digital instrument control system and CRT operation
  · optimising control by computers
  · more efficient monitor/control system in the plant by using new sensors

Recently, optics technology, communication, information technology and AI are making rapid progress.

These high technology will play major roles in the efficient refinery.

Production, operation, facility and safety in our refinery are important to us.

In coordinating these for the best results, we are utilizing high technology for more efficiency. Only then, with multiplied effects due to them, low cost and minimum operator are possible.
(fig.1)

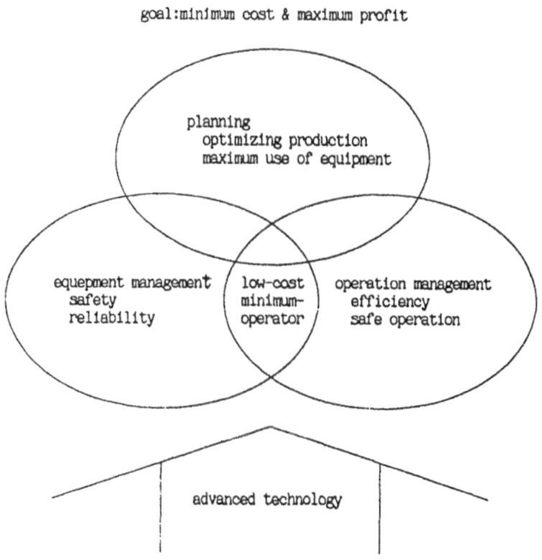

3.2 "Efficient Refinery " PLAN
  · IRIS(Idemitsu Refining Information System)
  · Computer Optimizing Control system
  · Automated and Integrated Operation
  · Preventive Maintenance System
  · Expert System

The structure and the functions are shown in fig.2 and 3. and they are inseparable. Let's look at more specific details.

goal:minimum cost & maximum profit

fig.1 Concept of Efficient Refinery

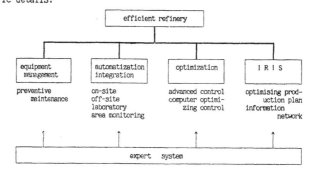

fig.2 Structure of Efficient Refinery

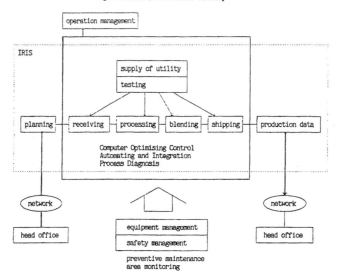

fig.3 Function Flow

### 3.2.1 IRIS

In considering restrictions regarding production, IRIS makes a flexible, efficient and optimized plan and execute timely. (fig.4)

(1)Making Reliable Data for Efficient Planning

We need highly reliable data for production planning.

In oder to acquire such data, we use accurate simulators which connect all refineries.

Thus, we all get the same reliable data under different operation condition more smoothly.

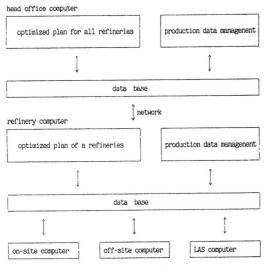

fig.4 Design of IRIS

(2)Support for Optimizing Planning

Planning tools are provided to optimize production plans at a refinery.

We utilize Linear Programming, Process Model and Blend Simmulator for optimizing production planning, operation conditions and for choosing blending oil.

(3)Efficient Management of Production Data

We accumulate various data such as operation, crude oil, product quality, blend, shipping and recieving,and stock oil.

They are data-based and readily available for analysis.

In addition, the data which is reported to outside company should be controled by same method.

(4)Utilizing Production Information

Anybody is able to utilize production information at a refinery whenever it is necessary.

We establish fast information network by connecting head office to refineries, and refinery to refineries by terminal units. (fig.5)

### 3.2.2  Computer Optimizing Control System

According to optimized production plans, using process computers and control system with digital instrument, we operate most efficiently.

Various restrictions such as quantity, quality and facility conditions are considered.

Then we finds the optimum operation condition and control by different purpose.

Thus, "Increase of High value Added", "Energy Saving", "Stable and Flexible Operation", "Minimizing Operator" can be achieved smoothly. (fig.6)

Earlier,Computer Optimizing Control depended on technology from abroad, but recently we are developing our own technology in this ares.

Also, we employed Dynamic Simulator, through which we will be able to simulate the process dynamicaly, analyze the dynamism and develop the precice control system.

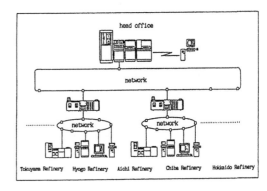

fig.5 Information Network System

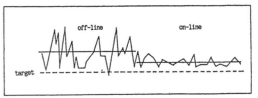

Control is improved and stable at the target.

fig.6 Example of Computer Oputimizing Control

### 3.2.3 Automated and Integrated Operation

Present automated operations are included as followings.

· Optimized operation based on production plans
· Highly efficient operations according to computer optimizing control
· Operation monitor and control through advanced control system with digital instrymentation
· Safety check during opration

It is always necessary to make appropriate judgements and take actions accoringly.

Thus, whether in normal operation or in emergency, these automatization and instrumentation not only help to improve efficiency but also help to operate more safely. (fig.7)

Also, for more efficient quality control,we introduced LAS(Laboratory Automation System).

It automated testing and centralized the management of test data which are essential to optimum quality control and quality design. (fig.8)

Furthermore, when machines and equipments go beyond mulfunction and there is a possibility of disaster, Area Monitor System will detect such danger and prevent any further happening.

### 3.2.4 Preventive Maintenance System

In order to continue stable production, it is necessary to maximize the performance of equipment, To do so we need to improve the equipment management,guarantee reliability and safety,prevent troubles, and decrease maintenance cost.

Thus, we have to change periodical maintenance for each machine.

Instead, we check the conditions of facilities and estimate the rest of life span to do the necessary maintenance at the appropriate time.

This is called Preventive Maintenance and we are working on this system. (fig.9)

### 3.2.5 Expert System

More anb more works at a refinery are being automated. Also, Expert System with AI is said to play the major role in the future.

In Expert System, the knowledge and experience of experts in the area is arranged and inputed in the computer so that it judges the condition and solves the problems.

Although there are many areas in which Expert System can play major roles, we start with more important support systems for production planning, operation and equipment management.
Some refineries in Idemitsu started working on the research and devedopment of the system.
RFCC Operation Support System (process diagnosis) at Aichi Refinery had started its operation in in April, 1987 and has been highly evaluated in the field. (fig.10)
Currently, we are gathering all accumulated technology at each refinery to examine Expert System at different levels. (fig.11)
We are also participating in "AI Research Room" in Petrorium Energe Center where 14 petrorium companies are jointed in the reseach.

### 3.3 Our goals

Thus,with the use of high technology, we will improve the current approaches and reduce the cost dramaticlly at a refinery.
They are our goals and we are doing our best toward our goals.

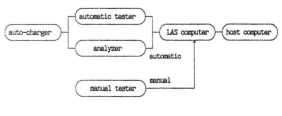

fig.8 LAS Function

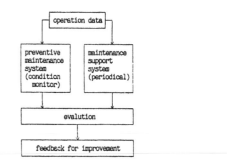

fig.9 Preventive Maintenance

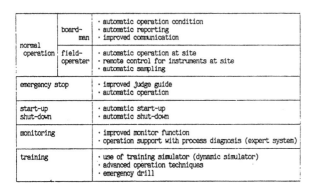

fig.7 Automation and Integration

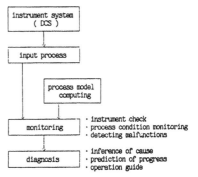

fig.10 RFCC Operation Support System Function
(Process Diagnosis)

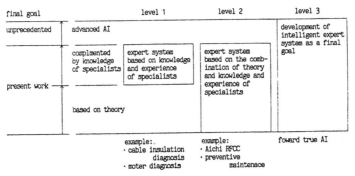

fig.11 Expert System, Level of Development

# DISCUSSION

## Session Plant Tour 2: Idemitsu Kosan Co. Ltd.

*Chairman*
### N. Jensen
*Technical University of Denmark*

The session Plant Tour 2 at the Workshop focussed
on production control in a continuous process
industry - a refinery. The session consisted of
three parts:

1. An introductory presentation about Idemitsu
   Kosan by Mr. Matsueda.
2. A visit to the Aichi Refinery of Idemitsu
   Kosan.
3. A discussion period involving two representa-
   tives from the Aichi Refinery.

Each of these parts are reviewed in the following
sections followed by some general comments on the
session.

## 1 Introductory Presentation

This presentation was given by Mr. Matsueda,
manager of the Refining Coordination and Engineering Section at the Aichi Refinery of Idemitsu
Kosan. An overview of Idemitsu Kosan was given
together with the concepts behind refinery management and operation by advanced technology as
described in Mr. Matsueda's paper. The business
philosophy of the company is, that business means
seeking human potential. In line with this philosophy the company do not have specialist
groups, such as an optimizing control group, but
work on advanced technology tools and applications is done in project teams by line staff.
Two key items in the advanced technology programme are the integrated refinery information
system (IRIS) and the use of expert systems in
refinery operations (e.g. process diagnostics,
exemplified by PDIAS).

## 2 Visit to Aichi Refinery

At the Aichi Refinery a presentation on the
structure, functionality and operation of the
process diagnostic and alarming expert system
PDIAS was given by the developer of the system.
PDIAS is a rule based expert system for alarming
abnormal conditions in a residue fluid catalytic
cracker (RFCC). It synthesizes audible alarms
based on process conditions input automatically
via distributed control system (DCS).

The tour of the refinery included a visit to the
RFCC control room, where the user interface to
PDIAS could be observed as well as the traditional DCS. PDIAS unfortunately was not working
due to a communications problem after a software
upgrade.

## 3 Discussion

Mr. Hop (Shell, The Netherlands) asked what

training the line staff received in order to
participate in the development of systems such as
IRIS. Mr. Matsueda answered that development is
usually occur in a project team with participation from the head office. Team members concerned with production control study for example
at MIT, while others are trained at equipment
vendors and computer manufactures. However,
there is no in house training system or department for this kind of training.

Professor McGreavy (Leeds University, England)
asked how the audible alarms in PDIAS was synthesized. The PDIAS developer answered, that the
basic unit in the computer data base are words.
However, the word pattern for a given fault is
predetermined and also stored in the computer.
The selection of a given pattern is based on
analysis of plant conditions as input via the
DCS. Mr. Narita (Asahi Chemical, Japan) wondered
how PDIAS helped prevent the recurrence of accidents. This was done by simply modifying the
appropriate threshold limits after an abnormal
event. Mr. van Rijn (Shell, The Netherlands)
asked what the experience is with respect to the
reliability of PDIAS (e.g. false alarms). The
reliability of PDIAS was secured by working on
each step in a continuous improvement, using
various process and equipment models and using
standard threshold setups for different process
conditions. Dr. König (Bayer, West Germany)
wondered how many avoidable shut downs of the
RFCC had been experienced. The answer is none.
However, the operators have difficulty identi-
fying problems with for example the level meters
on the RFCC. PDIAS can do this, but the system
is not perfect, and the final evaluation and
decision on what action to take always rest with
the operator. Professor Umeda (University of
Tsukuba, Japan) asked if the information on time
series symptom - consequence propagation is prepared in advance and how the data base is kept
current. The PDIAS developer answered, that the
symptom - consequence rules are deterministic and
not adaptive. The data base is continuously
being maintained by line staff. Professor Shioya
(Osaka University, Japan) asked if a dynamical
model was used in PDIAS and how the parameters
was tuned to real plant. PDIAS uses theoretical
flow relationships, but parameters and thresholds
are tuned to the process.

Finally Professor Jensen (Technical University of
Denmark, Denmark) asked how many of the Aichi
Refinerys 370 employees were engineers and how
the engineers were distributed among the different engineering disciplines. Mr. Matsueda repli-
ed, that there are 20 female clerical employees,
50 maintenance employees, 35 staff employees, 60
employees associated with the RFCC control room
and 60 with the other control room at the refinery.

## 4  Reviewers Comments

The production control systems in a continuous operation are in general not as visible as in descrete parts manufacturing. One cannot show the LP's used in production planning and in maintenance planning. The session nonetheless, especially the introductory presentation on IRIS, showed the concepts behind advanced production control in a continuous process and the type of tools being developed for production control at Idemitsu Kosan.

The visit and the discussion focussed on the PDIAS expert system for process alarming. PDIAS, however, is a process control tool - not a production control tool. Unfortunately this ment that question surrounding the IRIS system was not addressed, e.g. data base integrity and data base security.

# CURRENT STATUS OF CHEMICAL PLANT OPERATIONS IN JAPAN

## E. Nakanishi

*Department of Chemical Engineering, Kobe University, Nada-ku, Kobe 657, Japan*

Abstract. The current status of plant operations in Japan was investigated by the joint research committee of academy and industry in the Society of Chemical Engineers, Japan. A survey on the plant operations classified into five categories was put into practice by a questionnaire method and the answers were collected from 65 leading chemical companies and/or factories. This paper presents the analysis centered on the plant automation among the 50 items of the questionnaire.

Keywords. Chemical plant automation; factory automation; CRT operation, operator training for emergencies; future plant operation

## Introduction

Today one of the most imortant topics on plant operations in chemical industries is associated with the highly automated systems termed as FA and CIM by which plant operation is of efficient operability to improve the plant performances simultaneously in both safety and productivity.

The joint research committee of academy and industry on Plant Operations Engineering in the Society of Chemical Engineers, Japan, investigated the current situation on chemical plant operations in 1987-1988 by a quetionnaire method. The answers to the questions containing 50 items in total, 10 items in each of the following 5 categories, were collected from the 65 leading chemical industries and/or factories whose classification is shown in Table 1

1. Plant management in the CRT operation
2. Plant operations in the highly automated control system
3. SOP and its application to the plant operations
4. Education and training of operators
5. Activation of operator's spirit

### Table 1 Classificatons of factories

| type of factory | number of answers |
|---|---|
| A: chemical(cont.) | 20 |
| B: chemical(batch) | 18 |
| C: refinery | 15 |
| D: town gas | 7 |
| E: pharmacy & food | 5 |

### Survey of current situations of chemical plant automation

Figure 1 shows that the labor saving is considered to be the direct merit of plant automation and then is of primary concern on introducing the automation system. The original purpose of employing the automation system is firstly to achieve the stable operation without human error by which the stability in operation is reinforced to protect the stable product quality and secondly to accomplish the sophisticated operation by which higher productivity of plant is attainable. In fact, the items 2-4 next the item 1 score high points in Fig.1

The needs for the plant automation is shown in Fig.2. It is noted that plant automation is keenly desired especially in the batch processes. Two reasons are pointed out for strong needs of plant automation in batch processes where the handling operations by man-power are still widely found. The plant automation, firstly, contributes directly to the labor saving owing to the decrease of operators and , secondly, reduces considerably the chance for operator to participate in the plant operation by which the accident due to human error is avoided as much as possible. In group C to which refinery and town gas industries belong, the further needs for plant automation is not so strong as compared to the others since plant automation is considered to be already attained in high level in these industries.

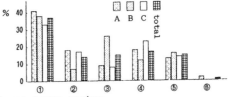

1. man-power saving
2. energy saving
3. improvement of productivity
4. improvement of safety
5. stability of product quality
6. others

Fig.1 Objectives of plant automation

Figure 2 indicates on the whole that the needs for plant automation are strong in every process industry. The further cost reduction of hardwares including computer which is the main element composing of automation system will increasingly promote the use of plant automation in the wide range of process industries in the near future.

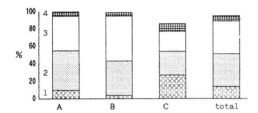

1.unnecessary
2.preferentially necessary
3.abosolutely necessary
4.others

Fig.2   Needs for plant automation

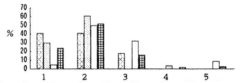

1.steady state operation
2.instrumentaion management
3.proper emergency counteraction
4.alarm management
5.supplement to the instrumentaions
6.others

Fig.4   Main use of CRT

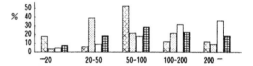

Fig.5 Number of CRTs managed by an operator

## CRT operation of chemical plant

With the promotion of introducing DCS in the automation system in chemical industries, the CRT operation has been popularized rapidly in recent years in the wide range of chemical processes. In fact, the number of CRT operations employed is drastically increasing especially in these 5 years as shown in Fig.3 which demonstrates the trend of the CRT operations introduced in these ten years. It is noted that the rapid number of employment of the CRT operations are found not only in the large scale continuous plants but also in the small scale batch processes. It is reported in a leading chemical company that the number of CRT operations introduced recently in the batch processes is by overwhelming majority.

Fig.6 Number of control loops managed by an operator

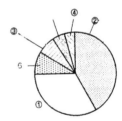

1.fully displayed
2.partially displayed
3.not displayed
4.others
5.none
6.unidentified

Fig.7 CRT display in emergencies

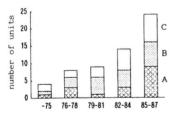

Fig.3   Trend of CRT operations employed

The several statistical results on the CRT operation are summarized in Figs.4 to 6. The most important task imposed on the CRT operation today is to establish the effective conterplan/action to secure the safety of plant in an emergency. It is said that the operator's patter recognition to the overall situation in the plant is rather difficult in the CRT operation than in the panel operation. To overcome this shortcoming, the improvements of CRT dispay (Fig.7) and SOP(Fig.8) are attempted while the operator training for emergencies is essential (Fig.9).

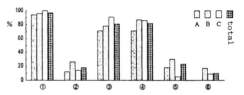

1.guideline for emergencies
2.daily inspection of plant
3.counteraction for trouble of CRT
4.cultivation of form and style
5.others

Fig.8 Improvement of SOP for emergencies

A standard guideline in introducing the CRT operation is proposed by the Kanagawa Prefecture in the following items.

1. number of CRTs required/CRT display fitted for emergencies
2. prevention of miss-operation
3. provision against earthquakes
4. environment of console room
5. backup system for accident and trouble in DCS system
6. DCS maitenance and management
7. education and training for operators

This guideline may be helpful in introducing or renewing DCS/CRT in chemical plant operations.


### Plant automation in the future

Thanks to the recent progress of LSI technology, the use of low cost and high performance computer is widely realized in the process control from large scale continuous plants to small scale batch processes. Being different from the analog process control system, the most important function of digital control system is to be able to practice the high level control strategy such as modern control theory and AI. The AI applications in the CRT operation is shown in Fig.10. The utilization and application of high level control strategies have just started in chemical process control and much yet remains to be done until practical results are confirmed.

As explained in Fig.1, one of the main objectives to employ a automation system into plant operation is the labor saving. In fact, most of chemical plants equipped with DCS are operated by rather small numbers of operators. Then is ultimately operatorless plant operation possible or realized? The result of quetionnaire for this matter is seen in Fig.11 which shows the minor opinion for the possibility of operatorless plant operation. On the other hand, the majority considers that operatorless plant will not be realized in the future in views of safety of plant and economy of operation.

In the near future, most of chemical plants will be operated by a limited number of operators with the help of highly automated computer control system. The accidents will seldom occur in such plants and there are very few chances that the operators have effective OJT for emergencies. In such operation environment, it is said to be rather difficult how to activate the operator's spirit. Various trials to do this are attempted in each of industries and/or factories as seen in Fig.12.

As mentioned above, the plant operation will be highly automated in the future chemical plants, but the realization of operatorless plant may be negative excluding exceptional cases. Thus the future chemical plants may be operated by extremely small number of operators who are required to have high level of knowledge and understanding for complicated and sophisticated plant operation systems. Figure 13 indicates that the future operator should have the high capability not only for technique of plant operation but also for knowledge and understanding of the operation systems of highly automated plant. In this context, it is noted in Fig.13 that the operation engineers who gradute from universities or colleges should play the leading role in the future chemical plant operation.

### REFERENCES

Research Committee on Plant Operation Engineering(1988). A survey of chemical plant operation in Japan, Chem.Eng.Symp. Ser.19, The Society of Chemical Engineering, Japan

Research Committee on Safety and Management of Plant(1988). A guideline of CRT operation system, Technical Report, Kanagawa Prefecture

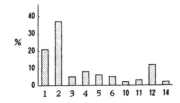

frequencies/year

Fig.9a Training frequencies for emergencies

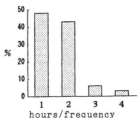

hours/frequency

Fig.9b Training time for emergencies

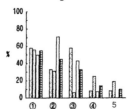

1. sophisticated operation
2. safety operation
3. labor-saving operation
4. others
5. no responses

Fig.10a  Objectives of employing AI

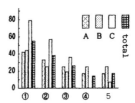

1. Fault diagnosis
2. operation gide
3. sophisticated control
4. others
5. no responses

Fig.10b  Items of AI applications

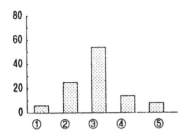

1. probable
2. conditionally probable
3. improbable in view of safety
4. economically improbable
5. others

Fig.11  Possibility of operatorless factories

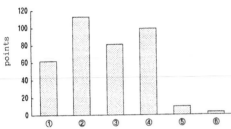

1. enhancement of self-activity
2. level-up of skill
3. level-up of knowledge
4. level-up of plant management
5. encouragement of group information
   exchange
6. others

Fig.12  Activation of operator's spirit

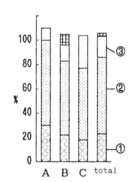

1. skill free operator
2. operation engineer having high level of
   knowledge graduated from universities
3. others
4. no responses

Fig.13  Operators in the future

# THE DEVELOPMENT OF COMPUTER INTEGRATED MANUFACTURING AT A CHEMICAL COMPANY

## E. Wada

*Production Process Development Department,*
*Kawasaki Works, Asahi Chemical Industry Co. Ltd., 1-3-2, Yakoh, Kawasaki-ku,*
*Kawasaki-City, Kanagawa, Japan*

Abstract. A view, rough sketch and outlines for the development of CIM in a typical chemical complex are presented. Drastic improvement on the productivity of the complex is foreseen by this development in its existing facilities. Key objectives are increased productivity of the physical plants through provision of information and tools to the people who manage and operate the plants by means of a centralized plant control system and the consistent production of quality products in an efficient manner that is responsive to the external environment through the integrating and automating of the information processing activities. At the initial stage there should be depicted the future image and models of the restructured complex with CIM implementation. The stepwise development toward final goal is followed. Although the development of CIM in the existing facilities is a greater challenge than in new facilities, necessitating more sophisticated technologies within existing constraints, efforts should be payed for a future goal of the CIM world.

Keywords. Chemical industry; distributed digital control; automated materials handling; integrated plant control; centralized plant control; simplification of the shift operator task; automated information processing.

## INTRODUCTION

Top management of chemical companies is generally keen on the implementation of CIM in addition to the other important business activities in order to keep the competitiveness of its existing operations. The development of CIM in the chemical complex, however, has been found to be difficult for some reasons discussed later.

This article presents the outlines of the CIM development feature in exisisting facilities of a typical chemical complex.

Final goal would be to up-grade all activities in the complex by means of the automated information processing activities and the automated production systems.

## REASONS FOR CIM APPLICATION

Some of the important reasons for CIM application include the following:

Automation of manufacturing operations holds the promise of increasing the productivity of Labor.

Labor shortages stimulate the development of automation as a substitute for labor. Current trend of labor toward the service sector accelerates the labor shortages.

By automating the operation and transferring the operator from handling jobs to the control room tasks, work is made safer.

Improvement of product quality is attained by means of producing the products with greater consistency and conformity to specifications.

CIM allows the company to reduce the time between customer order and product delivery. This gives the company a competitive advantage in promoting good customer service.

Improved capital asset utilization is attained by the inventory reductions and the more effective maintenance of existing plant and equipment.

The entire system contributes to lower overhead costs and higher staff productivity by better information flow and better tools to manage the information.

## FUTURE COMPUTER INTEGRATED MANUFACTURING SYSTEM

On the basis of the above-mentioned advantages, the company would visualize the future CIM system architecture which integrates the number of operations to a consistent structure. Fig. 1, shows the simplified CIM architecture diagram.

It is foreseen that this innovated system will lead to the significant reduction in the manpower requirement, the optimized production, the curtailment of inventories and the better customer services.

## PROBLEMS TO BE SOLVED FOR CIM IMPLEMENTATION

There are difficulties in realizing the highly advanced operations by the employment of the automated production systems and information-processing systems within existing constraints of equipment, buildings, spaces and operating personnel while continuing the production.

### Summary of the Typical Problems

The safety preservation is one of the important issues to be studied carefully and thoroughly.

How to improve the safety assurance with the reduced number of operators is a difficult question to be solved.

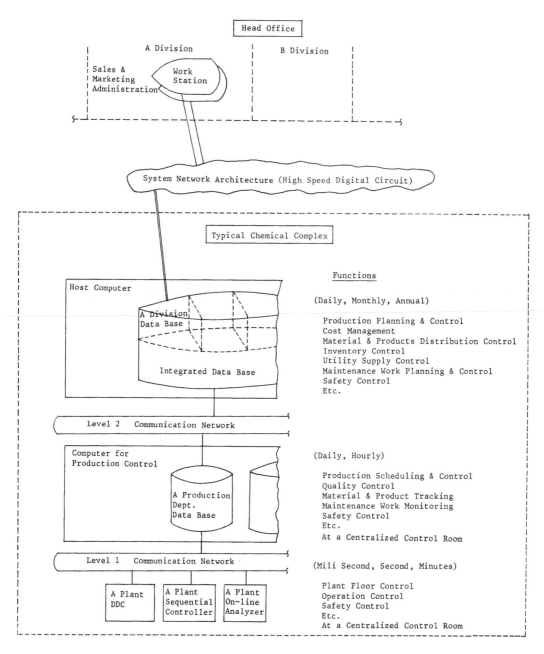

Fig. 1.  Typical CIM architecture model.

Improvement in the equipment and machine reliability, and the more stable operation of the plants.

The automating of the material handling jobs and the other routine monotonus tasks.

The development of the censors including the automatic analysis and inspection devices.

The development of the expert systems to assist the human judgement and diagnosis.

The supply of engineers for the analysis of process characteristics and the preparation of process models.

The sufficient supply of engineers for the design and development of the computer software.

The training of the aged workers for the tasks requiring new skills.

Restructuring of the organization.

Economic justification of the large incremental investment.

### INAUGURATION OF THE ACTIVITIES
### FOR CIM APPLICATION

In view of the complicated nature of this development it is necessary to proceed toward the future CIM goal step by step taking sufficient time. It is important to organize the efforts of each segment to be directed toward the future unification of multi-systems.

Following guide lines are desirable, along which long term and sequential development would be continued for the accomplishment of centralized computer integrated manufacturing.

### Long Term Sequential Development Strategies

Replacement of analogue controllers with distributed digital controllers.

Partial integration of control rooms.

Automating as far as possible of the material handling and other routine tasks of shift operators. Switching of the remaining routine tasks to the daytime operator jobs. Then shift operator jobs will be simplified and up-graded to such an extent as the watching of CRT consoles, safety related task and so on.

Whole plants should be controlled at one central station.

The organization of the entire complex will be restructured.

Completion of the automatic information-processing network integrating the sales, production and product delivery activities.

### CONCLUSION

The development of computer integrated manufacturing at the existing chemical complex is a tough project. The applied technologies must be incorporated for the existing constraints of equipment, buildings, spaces and personnel.

The important point is to set and depict the future image or models of the CIM architecture and organize the efforts of each segment headed for the final goal.

Though it may take long period of time to realize a complete computer integrated manufacturing complex, continual efforts should be payed for the final goal which will display a drastic improvement in the productivity, a significant improvement in the capital asset utilization, and better customer services.

### REFERENCES

Groover, M.P. (1987). Automation, Production Systems, and Computer Integrated Manufacturing, Prentice-Hall.
White, D.C. (1986). Manage better with a computer, Part 1. Hydrocarbon Processing April 1986, 125-130.
White, D.C. (1986). Manage better with a computer, Part 2. Hydrocarbon Processing May 1986, 73-82.

# DISCUSSION

## Session 5: Final Discussion

*Chairmen*
### C. F. H. van Rijn
*Koninklijke/Shell Laboratorium, The Netherlands*
### E. O'Shima
*Tokyo Institute of Technology, Japan*


In the final discussion session, there was a very
lively discussion on integrated systems approach
to the problems of production control.


Rationale for Computer Integration

Although the concept of computer integrated manufacturing system or computerized production / system seems quite attractive, there are various
problems in the phase of practical application
such as human resources to maintain the system,
rationale for return on investment, justification
of man-hour saving, etc.

Computer integrated manufacturing must have business purpose, where it can only have the significance of reality. Management people are not
convinced, in general, of the fact that they have
to work business-wise in a integrated fashion,
which may make the computer integrated manufacturing system difficult to be realized very soon.


Methodology of System Integration

The purpose of this Workshop that is to provide
the opportunity of interaction among the difference approaches of systems engineering toward the
common objectives of production control seems to
have been well appreciated by the participants.
As far as the personal observation, however, very
few papers have touched upon the methodology of
how to coordinate the different approaches.

For example, the following problems are relevant:
(1) How to structure the integrated computer
system from the viewpoint of production control.
(2) How to incorporate the data of different
kinds to be obtained from different sections.
(3) How to achieve a high performance of the
overall process rather than the efficiency of the
individual units. (4) What kind of mathematical
tools to be effectively applied to the complicated problems. (5) How to redesign the process
itself to meet the conditions required for computer integration.


Role of University Education

At least the university level, there seems to be
no such a concept as integration, they are mostly
indoctrinated in one specialized discipline.
There are some group of people who are aware of
the importance of methodologies of integration
and actually trying to establish such a core of
methodology like Process Systems Engineering.

Teaching the methodology of integration is not
the direct purpose of education, but the student
would find how to integrate the knowledge he has

been taught when he faces the problem in his
future carrier.

The most necessary thing in the real business
activity is teamwork rather than each of the specialists working in his specialized field.
Therefore, teaching integration or multidisciplinary education at university is absolute necessary. The real purpose of education is not to
educate the people how to overcome the specific
problems but to educate the young people how to
solve the general problem of a long term.


Comment from Academia

It is believed that the question should be seen
in more general terms. University education
should not concentrate on instructing students in
specific skills. This is more appropriately done
on training courses for which companies have a
responsibility. One reason for this is that such
special training, while important for immediate
use, can be relatively quickly rendered out of
date by technological developments. The responsibility of educators is to ensure that graduates
have a sufficiently broad understanding of fundamental principles that they can enter a wide
range of industries and quickly acquire the specialist knowledge to contribute effectively.
This includes the ability to make well-informed
decisions and judgements, although this will
obviously develop with experience.

It is not feasible or even realistically possible
to believe that an educational programme can
produce fully experienced engineers in a particular field at the time students complete their
studies. The success of an educational programme
should be judged on how effectively graduates can
apply their basic knowledge to solving problems
in a specialized domain. Particular methodologies or techniques for attacking problems should
only be seen as applications of more general
principles. Then as the detailed nature of the
problems changes, because of, for example, a
change in technology, new procedures can be developed and there is a continuous evolution of
skills.

In summary, education should focus on teaching
general principles which can be applied to any
situation. However, this should not preclude
taking as examples problems of immediate concern.
Experience will then enable the basic skills to
develop and evolve.

# DISCUSSION

## Session 2: Quality Control and Quality Assurance

*Chairmen*
**O. A. Asbjornsen**
*University of Maryland, USA*
**I. Hashimoto**
*Kobe University, Japan*

The importance of quality control is now widely recognized in every facet of chemical industry. Ever since Shewhart proposed a statistical approach to the improvement of product quality, various methods of so-called statistical quality control have been developed and widely accepted in many segments of industries. Further, this approach has been expanded and grown to what is now called total quality control (TQC) and is playing a crucial role in maintaining a high standard of product quality.

Japan, in particular, is well known for company-wide quality control (CWQC) covering a broad range of issues that emerge from managerial and engineering domains. One of the essential elements in TQC/CWQC is quality assurance (QA). The activities of QA can be characterized as follows:

| Step | 1 | Market research (marketing) |
|------|---|------------------------------|
|      | 2 | Product planning |
|      | 3 | Research & development and design |
|      | 4 | Preparation for mass production |
|      | 5 | Manufacturing |
|      | 6 | Sales and after-sale services |
|      | 7 | Order and purchase control |

The concept of "quality" in this QA is based on the so-called consumerism as described by Feigenbaum as follows:

"Quality is a customer determination, not an engineer's determination, not a marketing determination or a general management determination. It is based upon the customer's actual experience with the product service, measured against his or her requirements--stated or unstated, conscious or merely sensed, technically operational or entirely subjective--and always representing a moving target in a competitive market."[1]

In Session II, discussions centered around the issues related to the definition of quality.

When an emphasis is placed on customer satisfaction and when quality is defined as such, it becomes essential that through an iterative process the supplier and the customer cooperate in making the definition of quality clearer in terms of quantitative measurements.

In Japan, quality function deployment (QFD) is currently used to include the needs of customers into specification of quality planning, and to ensure that product quality meet the customer's needs. This way, the results can be evaluated on the basis of customer satisfaction.

In any event, it may be concluded that the general definition of quality has not been completely clarified despite the fervent dispute on this issue in the past years and probably in the coming years. Even though the definition of quality is not completely clear, each company has to set its own goal with regard to product quality and endeavor to develop new methodology to enhance the quality of each product.

On a more technical plane, many problems related to quality arise. One such problem is how to measure product quality. Clearly, there is a need for some suitable measuring devices or sensors. Unfortunately, however, we often encounter those cases in which there is not a suitable testing and measuring system available for ensuring that product quality meet the customer requirements. Even when such a device is available, it is often the case that only destructive or off-line measurements are possible and that the product's properties can be measured and analyzed only through lengthy, time-consuming processes. Needless to say, development of non-destructive and on-line measuring devices or sensors is indispensable. However, it usually requires considerable time to invent and develop such devices or sensors for actual industrial use. Thus, we need to pay more attention to the development of computer-aided estimation systems, which enable us to measure those qualities that cannot be measured directly, by processing data on many other measurable variables.

MacGregor published the paper titled "On-Line Statistical Process Control" in the October, 1989 issue of *Chemical Engineering Progress*. In the paper, he discussed the aspect of on-line quality control in the area of statistical process control (SPC), which covers all the statistical techniques for improving process productivity and product quality. He clarified the interface between process control and on-line quality control, and stressed that process control can and must play more important roles in the field of SPC.

"In short, the groups appear to be nearly incompatible with each other, having almost no common base of knowledge on which they can build a relationship. It is the purpose of this article to try to explore the interface between these areas of knowledge and to present a common base. Not unexpectedly, stochastic control theory provides a means of achieving this."[2]

In the session, discussions also centered on the topics related to the application of advanced process control techniques to the improvement of product quality. While the statistical approach is likely to continue to play an important role in quality control, other approaches based on operation research and process control will and must be developed and proven to be effective in further improving product quality. Much closer cooperation between statisticians and control engineers will be needed for future innovation in quality assurance techniques.

1) A.V.Feigenbaum, Total Quality Control, P.7, McGraw-Hill, New York, 1986.

2) J.F. MacGregor, Chemical Engineering Progress, P.22, October, 1988.

# AUTHOR INDEX

# KEYWORD INDEX

## SYMPOSIA VOLUMES

ADALI & TUNALI: Microcomputer Application in Process Control

AKASHI: Control Science and Technology for the Progress of Society, 7 Volumes

ALBERTOS & DE LA PUENTE: Components, Instruments and Techniques for Low Cost Automation and Applications

ALONSO-CONCHEIRO: Real Time Digital Control Applications

AMOUROUX & EL JAI: Control of Distributed Parameter Systems (1989)

ATHERTON: Multivariable Technological Systems

BABARY & LE LETTY: Control of Distributed Parameter Systems (1982)

BALCHEN: Automation and Data Processing in Aquaculture

BANKS & PRITCHARD: Control of Distributed Parameter Systems (1977)

BAOSHENG HU: Analysis, Design and Evaluation of Man–Machine Systems (1989)

BARKER & YOUNG: Identification and System Parameter Estimation (1985)

BASANEZ, FERRATE & SARIDIS: Robot Control "SYROCO '85"

BASAR & PAU: Dynamic Modelling and Control of National Economies (1983)

BAYLIS: Safety of Computer Control Systems (1983)

BEKEY & SARIDIS: Identification and System Parameter Estimation (1982)

BINDER & PERRET: Components and Instruments for Distributed Computer Control Systems

CALVAER: Power Systems, Modelling and Control Applications

Van CAUWENBERGHE: Instrumentation and Automation in the Paper, Rubber, Plastics and Polymerisation Industries (1980) (1983)

CHEN HAN-FU: Identification and System Parameter Estimation (1988)

CHEN ZHEN-YU: Computer Aided Design in Control Systems (1988)

CHRETIEN: Automatic Control in Space (1985)

CHRISTODULAKIS: Dynamic Modelling and Control of National Economies (1989)

COBELLI & MARIANI: Modelling and Control in Biomedical Systems

CUENOD: Computer Aided Design of Control Systems†

DA CUNHA: Planning and Operation of Electric Energy Systems

DE CARLI: Low Cost Automation

De GIORGIO & ROVEDA: Criteria for Selecting Appropriate Technologies under Different Cultural, Technical and Social Conditions

DUBUISSON: Information and Systems

EHRENBERGER: Safety of Computer Control Systems (SAFECOMP '88)

ELLIS: Control Problems and Devices in Manufacturing Technology (1980)

FERRATE & PUENTE: Software for Computer Control (1982)

FLEISSNER: Systems Approach to Appropriate Technology Transfer

FLORIAN & HAASE: Software for Computer Control (1986)

GEERING & MANSOUR: Large Scale Systems: Theory and Applications (1986)

GENSER, ETSCHMAIER, HASEGAWA & STROBEL: Control in Transportation Systems (1986)

GERTLER & KEVICZKY: A Bridge Between Control Science and Technology, 6 Volumes

GHONAIMY: Systems Approach for Development (1977)

HAIMES & KINDLER: Water and Related Land Resource Systems

HARDT: Information Control Problems in Manufacturing Technology (1982)

HERBST: Automatic Control in Power Generation Distribution and Protection

HRUZ & CICEL: Automatic Measurement and Control in Woodworking Industry — Lignoautomatica '86

HUSSON: Advanced Information Processing in Automatic Control

ISERMANN: Automatic Control, 10 Volumes

ISERMANN: Identification and System Parameter Estimation (1979)

ISERMANN & KALTENECKER: Digital Computer Applications to Process Control

ISIDORI: Nonlinear Control Systems Design

JANSSEN, PAU & STRASZAK: Dynamic Modelling and Control of National Economies (1980)

JELLALI: Systems Analysis Applied to Management of Water Resources

JOHANNSEN & RIJNSDORP: Analysis, Design, and Evaluation of Man–Machine Systems

JOHNSON: Adaptive Systems in Control and Signal Processing

JOHNSON: Modelling and Control of Biotechnological Processes

KAYA & WILLIAMS: Instrumentation and Automation in the Paper, Rubber, Plastics and Polymerization Industries (1986)

KLAMT & LAUBER: Control in Transportation Systems (1984)

KOPACEK *et al.*: Skill Based Automated Production

KOPACEK, TROCH & DESOYER: Theory of Robots

KOPPEL: Automation in Mining, Mineral and Metal Processing (1989)

KUMMEL: Adaptive Control of Chemical Processes (ADCHEM '88)

LARSEN & HANSEN: Computer Aided Design in Control and Engineering Systems

LEININGER: Computer Aided Design of Multivariable Technological Systems

LEONHARD: Control in Power Electronics and Electrical Drives (1977)

LESKIEWICZ & ZAREMBA: Pneumatic and Hydraulic Components and Instruments in Automatic Control†

LINKENS & ATHERTON: Trends in Control and Measurement Education

MACLEOD & HEHER: Software for Computer Control (SOCOCO '88)

MAHALANABIS: Theory and Application of Digital Control

MANCINI, JOHANNSEN & MARTENSSON: Analysis, Design and Evaluation of Man–Machine Systems (1985)

MARTOS, PAU, ZIERMANN: Dynamic Modelling and Control of National Economies (1986)

McGREAVY: Dynamics and Control of Chemical Reactors and Distillation Columns

MLADENOV: Distributed Intelligence Systems: Methods and Applications

MUNDAY: Automatic Control in Space (1979)

NAJIM & ABDEL-FATTAH: System Approach for Development (1980)

NIEMI: A Link Between Science and Applications of Automatic Control, 4 Volumes

NISHIKAWA & KAYA: Energy Systems, Management and Economics

NISHIMURA: Automatic Control in Aerospace

NORRIE & TURNER: Automation for Mineral Resource Development

NOVAK: Software for Computer Control (1979)

O'SHEA & POLIS: Automation in Mining, Mineral and Metal Processing (1980)

OSHIMA: Information Control Problems in Manufacturing Technology (1977)

PAUL: Digital Computer Applications to Process Control (1985)

PERRIN: Control, Computers, Communications in Transportation

PONOMARYOV: Artificial Intelligence

PUENTE & NEMES: Information Control Problems in Manufacturing Technology (1989)

RAMAMOORTY: Automation and Instrumentation for Power Plants

RANTA: Analysis, Design and Evaluation of Man–Machine Systems (1988)

RAUCH: Applications of Nonlinear Programming to Optimization and Control†

RAUCH: Control of Distributed Parameter Systems (1986)

REINISCH & THOMA: Large Scale Systems: Theory and Applications (1989)

REMBOLD: Robot Control (SYROCO '88)

RIJNSDORP: Case Studies in Automation Related to Humanization of Work

RIJNSDORP *et al.*: Dynamics and Control of Chemical Reactors (DYCORD '89)

RIJNSDORP, PLOMP & MÖLLER: Training for Tomorrow— Educational Aspects of Computerized Automation

ROOS: Economics and Artificial Intelligence

SANCHEZ: Fuzzy Information, Knowledge Representation and Decision Analysis

SAWARAGI & AKASHI: Environmental Systems Planning, Design and Control

SINHA & TELKSNYS: Stochastic Control

SMEDEMA: Real Time Programming (1977)†

STRASZAK: Large Scale Systems: Theory and Applications (1983)

SUBRAMANYAM: Computer Applications in Large Scale Power Systems

TAL': Information Control Problems in Manufacturing Technology (1986)

TITLI & SINGH: Large Scale Systems: Theory and Applications (1980)

TROCH, KOPACEK & BREITENECKER: Simulation of Control Systems

UHI AHN: Power Systems and Power Plant Control (1989)

VALADARES TAVARES & EVARISTO DA SILVA: Systems Analysis Applied to Water and Related Land Resources

van WOERKOM: Automatic Control in Space (1982)

WANG PINGYANG: Power Systems and Power Plant Control

WESTERLUND: Automation in Mining, Mineral and Metal Processing (1983)
YANG JIACHI: Control Science and Technology for Development

YOSHITANI: Automation in Mining, Mineral and Metal Processing (1986)
ZWICKY: Control in Power Electronics and Electrical Drives (1983)

## WORKSHOP VOLUMES

ASTROM & WITTENMARK: Adaptive Systems in Control and Signal Processing
BOULLART et al.: Industrial Process Control Systems
BRODNER: Skill Based Automated Manufacturing
BULL: Real Time Programming (1983)
BULL & WILLIAMS: Real Time Programming (1985)
CAMPBELL: Control Aspects of Prosthetics and Orthotics
CHESTNUT: Contributions of Technology to International Conflict Resolution (SWIIS)
CHESTNUT et al.: International Conflict Resolution using Systems Engineering (SWIIS)
CHESTNUT, GENSER, KOPACEK & WIERZBICKI: Supplemental Ways for Improving International Stability
CICHOCKI & STRASZAK: Systems Analysis Applications to Complex Programs
CRESPO & DE LA PUENTE: Real Time Programming (1988)
CRONHJORT: Real Time Programming (1978)
DI PILLO: Control Applications of Nonlinear Programming and Optimization
ELZER: Experience with the Management of Software Projects
GELLIE & TAVAST: Distributed Computer Control Systems (1982)
GENSER et al.: Safety of Computer Control Systems (SAFECOMP '89)
GOODWIN: Robust Adaptive Control
HAASE: Real Time Programming (1980)
HALME: Modelling and Control of Biotechnical Processes
HARRISON: Distributed Computer Control Systems (1979)
HASEGAWA: Real Time Programming (1981)†
HASEGAWA & INOUE: Urban, Regional and National Planning—Environmental Aspects
JANSEN & BOULLART: Reliability of Instrumentation Systems for Safeguarding and Control
KOTOB: Automatic Control in Petroleum, Petrochemical and Desalination Industries
LANDAU, TOMIZUKA & AUSLANDER: Adaptive Systems in Control and Signal Processing
LAUBER: Safety of Computer Control Systems (1979)

LOTOTSKY: Evaluation of Adaptive Control Strategies in Industrial Applications
MAFFEZZONI: Modelling and Control of Electric Power Plants (1984)
MARTIN: Design of Work in Automated Manufacturing Systems
McAVOY: Model Based Process Control
MEYER: Real Time Programming (1989)
MILLER: Distributed Computer Control Systems (1981)
MILOVANOVIC & ELZER: Experience with the Management of Software Projects (1988)
MOWLE: Experience with the Management of Software Projects
NARITA & MOTUS: Distributed Computer Control Systems (1989)
OLLUS: Digital Image Processing in Industrial Applications—Vision Control
QUIRK: Safety of Computer Control Systems (1985) (1986)
RAUCH: Control Applications of Nonlinear Programming
REMBOLD: Information Control Problems in Manufacturing Technolo (1979)
RODD: Artificial Intelligence in Real Time Control (1989)
RODD: Distributed Computer Control Systems (1983)
RODD: Distributed Databases in Real Time Control
RODD & LALIVE D'EPINAY: Distributed Computer Control Systems (1988)
RODD & MULLER: Distributed Computer Control Systems (1986)
RODD & SUSKI: Artificial Intelligence in Real Time Control
SIGUERDIDJANE & BERNHARD: Control Applications of Nonlinear Programming and Optimization
SINGH & TITLI: Control and Management of Integrated Industrial Complexes
SKELTON & OWENS: Model Error Concepts and Compensation
SOMMER: Applied Measurements in Mineral and Metallurgical Process
SUSKI: Distributed Computer Control Systems (1985)
SZLANKO: Real Time Programming (1986)
TAKAMATSU & O'SHIMA: Production Control in Process Industry
UNBEHAUEN: Adaptive Control of Chemical Processes
VILLA & MURARI: Decisional Structures in Automated Manufacturing

†Out of stock—microfiche copies available. Details of prices sent on request from the IFAC Publisher.

*IFAC Related Titles*

BROADBENT & MASUBUCHI: Multilingual Glossary of Automatic Control Technology
EYKHOFF: Trends and Progress in System Identification
NALECZ: Control Aspects of Biomedical Engineering

Printed and bound by CPI Group (UK) Ltd, Croydon, CR0 4YY

03/10/2024

01040323-0015